Vegetation of inland waters

Handbook of vegetation science

FOUNDED BY R. TÜXEN
H. LIETH, EDITOR IN CHIEF

Volume 15/1

Vegetation of inland waters

Edited by
J. J. SYMOENS

Kluwer Academic Publishers
DORDRECHT — BOSTON — LONDON

Library of Congress Cataloging in Publication Data

```
Vegetation of inland waters / J.J. Symoens, editor.
      p.    cm. -- (Handbook of vegetation science ; pt. 15/1)
   Includes index.
   ISBN 9061931967
   1. Freshwater flora.    I. Symoens, J. J. (Jean-Jacques)
II. Series: Handbook of vegetation science ; 15/1.
QK105.V44 1988
581.5'2632--dc19
```

87-33796
CIP

ISBN 90—6193—196—7

Kluwer Academic Publishers incorporates the publishing programmes of
Dr W. Junk Publishers, MTP Press, Martinus Nijhoff Publishers, and D. Reidel
Publishing Company.

Distributors
_____

for the United States and Canada: Kluwer Academic Publishers, 101 Philip Drive,
Norwell, MA 02061, USA
for all other countries: Kluwer Academic Publishers Group, P.O. Box 322, 3300 AH
Dordrecht, The Netherlands

Copyright
_____

PRINTED IN THE NETHERLANDS

CONTENTS

SERIES EDITOR'S PREFACE

By 1988 the Handbook of Vegetation Science is well on its way to completion. With 7 volumes in circulation, 3 volumes in the press, and most of the remaining volumes in preparation it appears that the total task can be completed in the early 'nineties. I am especially thankful to Professor Symoens for accepting the task of editing the volume on aquatic vegetation. The main emphasis of work in phytosociology is devoted to land plants, yet the landscape analysis remains incomplete without the consideration of rivers and lakes. A volume on inland aquatic vegetation must therefore be most helpful to the land vegetation analyst and not only to the specialist on aquatic vegetation.

Professor Symoens succeeded in drafting the most competent team for his task. I am sure that all colleagues working in vegetation analysis will be grateful to them that they have taken the time and energy to complete their chapters. Handbook articles are not easy to write and certainly not easy to edit.

The major aquatic components in the landscape are treated. The vegetation analysts will welcome the fact that certain physiological and ecological processes of water plants are covered for which otherwise they would have to consult the limnological literature. This volume, together with the forthcoming volume on wetlands, should completely cover the inland aquatic vegetation problematic.

Again I thank all contributors and the editor for this volume. Mr Wil Peters and the staff of Dr W. Junk Publishers deserve our special gratitude for their continuous co-operation. It is very gratifying to see and feel the service Dr W. Junk grants our profession. This was recently expressed during the editorial board meeting of the journal *Vegetatio* held in Berlin on the occasion of the 14th Botanical Congress. On behalf of the editors of the handbook I can certainly express the same gratitude.

I expect for this volume the same success as the previous volumes have had. I expect it to be of help to the professionals in the field as well as to the research student.

Osnabrück, October 1987   H. LIETH

FOREWORD

Water is life!

It is generally accepted that life on our planet originated in water, even though it is not possible to state without doubt whether it arose in sea, estuarine or inland waters. When some living organisms left the aquatic environment, they retained an aquatic medium within their cells; the average water content of biomass, terrestrial organisms included, is about 70% and many species, such as man, can only tolerate a very limited decrease in their hydration level.

However, the aquatic environment often escapes man's full perception and consequently man's understanding and research effort. Whilst during the neolithic revolution, man replaced his gathering activities by cultivating the soil and his hunting activities by raising domesticated animals, he mostly continued to exploit the aquatic resources by capturing natural stocks; fishing is but a form of gathering. In spite of the huge extent of water ecosystems — more than two thirds of our planet — aquaculture, i.e. the culture or farming of aquatic plants and animals, has lagged far behind agriculture and, even now, is but a relatively limited source of human food on a worldwide scale; the application of science and technology to aquaculture is very recent. Several reasons have been advanced to explain this backward state of the aquatic sciences: it has been suggested that man, as a terrestrial animal, has achieved a better understanding of the life strategies and a greater mastery of the production mechanisms of terrestrial plants and animals than of the organisms which dwell in an aquatic environment.

This part of the Handbook of Vegetation Science should contribute to a better understanding of the environmental conditions offered to plant life by inland waters, of the characteristics and adaptations of the plants which colonize the latter, and of the structural and functional features of their plant communities.

Although the general principles of ecology and biocenotics are common to both aquatic and terrestrial communities, the water environment is less accessible to the observer's eye: it requires specialized methods and equipment of investigation as soon as the depth goes beyond a few decimeters.

Although the biochemical pathways of photosynthesis are fundamentally the same in water and land plants, the assimilation of carbon dioxide from several forms of dissolved inorganic carbon is a feature peculiar to aquatic plant life.

The most important forms of life and growth of aquatic plants were identified long ago by the analysts of vegetation structure, but a more detailed analysis of the morphological responses of the plants to the totality of environmental factors shows the adaptative richness of water vegetation. It is noteworthy that water, due to its high density, is a very favourable medium for sustaining plants; submersed plants have no need to develop a large mechanical structure to provide support and can thus concentrate their activity in the development of productive chlorophyll-containing cells. This explains the rapid growth of many aquatic plants. It also explains why aquatic plants can so easily become troublesome weeds.

The phytosociology of aquatic plant communities — macrophytic or microphytic — is faced with problems which do not arise in terrestrial communities. In the aquatic habitat where many macrophytes and most microphytes are not bound to the soil, the plants may be transported by water currents and the assemblage present in any given water sample may consist of populations derived from different habitats and adapted to respond to substantially different sets of conditions. Several communities may occur in the same place as a mixture and should be treated with a synusial approach.

Two environmental factors peculiar to the aquatic medium play a decisive role in two types of water habitats: the highly variable hydrological regime in the flooded plains and the water current in the fluviatile habitats.

Some species of water plants are able to colonize extreme aquatic environments: e.g. autotrophic prokaryotes, the cyanobacteria (most of them commonly known as blue-green algae), live in waters with temperatures up to 73 °C!

All this justifies devoting separate volumes of the Handbook of Vegetation Science to Aquatic Vegetation; this volume, 15/1 to the Vegetation of Inland Waters and volume 15/2 will be devoted to the Wetlands of the World.

We thank the authors of the individual chapters in this volume for their excellent treatment. There was an unforeseen delay in putting together all chapters due to the fact that they were written by over-committed specialists who nevertheless agreed to give us the fruits of their knowledge, their experience and their reflections on perspectives for future research. The diversity of their approaches contributes to the richness of this volume, even if they do not hold identical views on some points. As the editor of this volume, I wish to express once more my sincere thanks to all of them for their contributions.

J. J. SYMOENS

LIST OF CONTRIBUTORS

P. J. Ashton, National Institute for Water Research, C.S.I.R., P.O. Box 395, Pretoria 0001 (South Africa).

B. Beltman, Department of Plant Ecology, University of Utrecht, Lange Nieuwstraat 106, 3512 PN Utrecht (The Netherlands).

E. P. H. Best, Centre for Agrobiological Research, P.O. Box 14, 6700 AA Wageningen (The Netherlands).

C. M. Breen, Institute for Natural Resources, University of Natal, P.O. Box 375, Pietermaritzburg 3200 (South Africa).

R. S. Clymo, School of Biological Sciences, Queen Mary College, University of London, London E1 4NS (England).

F. H. Dawson, River Laboratory, Freshwater Biological Association, East Stoke, Wareham, Dorset BH20 6BB (England).

C. den Hartog, Laboratory of Aquatic Ecology, Catholic University, Toernooiveld, 6525 ED Nijmegen (The Netherlands).

J. P. Descy, Unité d'Ecologie des Eaux douces, Facultés universitaires N.-D. de la Paix, rue de Bruxelles 61, B-5000 Namur (Belgium).

H. L. Golterman, Station biologique de la Tour du Valat, Le Sambuc, F-13200 Arles (France).

W. Koerselman, Department of Plant Ecology, University of Utrecht, Lange Nieuwstraat 106, 3512 PN Utrecht (The Netherlands).

E. Kusel-Fetzmann, Institut für Pflanzenphysiologie, Universität Wien, Althanstrasse 14, A-1091 Wien (Austria).

J. Lauga, Centre d'Ecologie des Ressources renouvelables, Centre national de la Recherche scientifique, 29, rue Jeanne Marvig, F-31000 Toulouse (France).

J. M. Melack, Department of Biological Sciences and Marine Science Institute, University of California, Santa Barbara, California 93106 (U.S.A.).

K. H. Rogers, Botany Department, University of the Witwatersrand, 1 Jan Smuts Avenue, Johannesburg 2001 (South Africa).

J. J. Symoens, Laboratorium voor Algemene Plantkunde en Natuurbeheer, Vrije Universiteit Brussel, Pleinlaan 2, B-1050 Brussel (Belgium).

M. Søndergaard, Roskilde University, Institute of Life Sciences and Chemistry, P.O. Box 260, DK-4000 Roskilde (Denmark).

G. van der Velde, Laboratory of Aquatic Ecology, Catholic University, Toernooiveld, 6525 ED Nijmegen (The Netherlands).

J. T. A. Verhoeven, Department of Plant Ecology, University of Utrecht, Lange Nieuwstraat 106, 3512 PN Utrecht (The Netherlands).

R. G. Wetzel, Department of Biology, The University of Michigan, Ann Arbor, Michigan 48109 (U.S.A.).

G. Wiegleb, Fachbereich 7 Biologie, University of Oldenburg, P.O.B. 2503, D-2900 Oldenburg (West Germany).

WATER AS AN ENVIRONMENT FOR PLANT LIFE

ROBERT G. WETZEL

INTRODUCTION

The immigration of terrestrial plants into the freshwater environment presented numerous physiological barriers. Relatively few angiosperms (<1 percent) and pteridophytes (<2 percent) have successfully adapted to total submersion in water. Although certain environmental parameters (e.g. temperature) in water are more constant than in the terrestrial habitat, most conditions below water and in water-saturated sediments are acutely hostile to normal growth and reproductive characteristics of macrophytes. Certain plants, such as the emergent angiosperms that grow in wetlands in continuously water-saturated sediments, have adapted very well and are among the most productive plants in the biosphere. Macrophytes and sessile algae within the water, however, exhibit greatly reduced productivity because of the inhibitory effects of water itself to gaseous transport, reduced light availability, and other factors. Phytoplankton algae are the least productive of all aquatic plants, even under the most optimal conditions, as a result of the restrictive nature of physical, chemical, and in some instances biotic properties of the environment.

The ensuing discussion presents a brief summary of the freshwater environment for aquatic plants. Emphasis is on the sessile forms, either emergent, rooted but surface floating, or submersed, and major environmental factors that influence their growth. Summary reviews of the biology and diversity of the macrophytes and attached algae are given in Schulthorpe (1967), Hutchinson (1975), and Wetzel (1983a, 1983b), and the ecological ramifications of littoral productivity by Westlake *et al.* (1980) and Wetzel (1983a).

LIGHT AVAILABILITY AND QUALITY

Aquatic macrophytes utilize between 4 and 9 quanta of light energy per

1

J. J. Symoens (ed.), Vegetation of inland waters. ISBN 90—6193—196—7.
© 1988, Kluwer Academic Publishers, Dordrecht. Printed in the Netherlands.

molecule of carbon dioxide reduced. The absorption peaks of the receptor pigments are 640 nm and 405 nm for chlorophyll *a* and 620 nm and 440 nm for chlorophyll *b*. Numerous other accessory pigments are found among certain algae that enhance the efficiency of photosynthesis at the longer wavelengths (red light) of lower frequencies.

The amount and spectral composition of solar energy impinging on emergent or floating plant tissue on the water surface is influenced by many dynamic factors (Gates 1980, Wetzel 1983a). The angle of light incidence is more perpendicular in equatorial regions, during summer months, and at midday than in regions of higher latitude, winter months, or at time of the day closer to dawn or dusk. Light must pass through greater distances of atmosphere as the angle of incidence increases from the perpendicular. Selective absorption by oxygen, ozone, carbon dioxide, and water vapor occurs as light passes through the atmosphere, and shorter wavelengths of higher frequencies are rapidly removed. Indirect solar radiation from the sky results from atmospheric molecular scattering. Short wavelength radiation of higher frequency is scattered more than longer wavelength light.

A portion of light reaching water is reflected away from the surface. The greater the departure of the angle of the sun from the perpendicular, the greater the extent of reflection (Wetzel 1983a). When the surface of the water is disturbed by wave turbulence, reflection increases (10 to 20 percent). Cloudy ice (containing air bubbles and irregular crystals) and snow greatly increase the amount of light that is reflected from the lake surface; clear, colorless ice, however, absorbs little more light than does water.

Of the total light energy entering the water, a portion is scattered in the water, and the remainder is absorbed by the molecules of water itself, by dissolved inorganic and organic compounds, and by suspended particulate matter. Scattering of light results from deflection of light energy by molecules of water, its solutes, and suspended inorganic and organic particulate materials. As a result, the extent of this internal reflection varies greatly with depth, season, and the dynamics of inorganic and organic loading in relation to water stratification, productivity and distribution of organisms, and other factors.

Light absorption, i.e. the diminution of light energy with depth by transformation to heat, is selective as a result of the molecular structure of water. Percentile absorption of light by pure water alone is very large in the infrared and red, lowest in the blue, and increases again somewhat in the violet and ultraviolet portions of the visible spectrum. Dissolved organic compounds, especially humic compounds of plant origin, markedly increase the absorption of light. Selective absorption is greatest in the infrared and red wavelengths; coupled to the high infrared absorption of water itself, the combined effect is rapid absorption and heating of surface water. Very low concentrations of dissolved organic compounds absorb energy greatly in the UV, blue, and green wavelengths. In stained waters, therefore, the dominant energy available to submersed plants at significant depths is in the yellow and red portions of the spectrum. Light absorption by particulate suspensoids, when not in extremely high concentrations, is relatively unselective at different wavelengths.

Light attenuation with increasing water depth is the total diminution of radiant energy by both scattering and by absorption mechanisms. Vertical attenuation of light is exponential. The depth at which extinction of irradiance occurs varies considerably over a year in a given body of water in relation to changes in internal (autochthonous) and external (allochthonous) loadings. The annual mean value, however, is reasonably constant from year to year and is representative of the water transparency of a particular water body.

Emergent macrophytes, exemplified by the ubiquitous reed *Phragmites* and cattail *Typha*, are morphologically ideally suited to maximize inception of available light. Leaf morphology is largely linear with dominating vertical orientation that minimizes self-shading. Growth, population development seasonally, and high densities commonly result in maximizing individual species productivity with high competition by restricting light availability for competing species, particularly in the understory (e.g. Rodewald-Rudescu 1974, Grace & Wetzel 1981, 1982a, 1982b, Dickerman & Wetzel 1985).

Floating-leaved macrophytes are attached to the substratum and possess leaves that float on the surface, exemplified by the ubiquitous water lilies. Floating leaves are well developed dorsoventrally, with differentiation of the mesophyll into an upper photosynthetic palisade tissue and an extensive lower lacunate tissue. Leaves lying parallel to the water surface create a vigorous competition for space to expose maximum leaf area to incident light without leaf overlap. In water lilies between water depths of 10 cm and 4 m, a complete proportionality occurs between water depth and petiole length. Petioles are about 20 cm longer than water depth, which permits leaves to remain on the surface among undulating waves. Petiole growth is arrested at the surface when the leaves begin to lose ethylene to the atmosphere and hormonal activity is reduced (reviewed in Wetzel 1983a). Under conditions of high population densities and crowding, petiole elongation frequently continues until leaves extend somewhat above the water surface.

Extreme plasticity and adaptability is found among submersed macrophytes to the highly variable underwater light conditions. Because of the rapid attenuation of light in water, numerous morphological and physiological mechanisms have developed to maximize light acquisition.

Leaves tend to be much more divided and reticulated than those of terrestrial or other aquatic plants. The cuticle is extremely thin if present, leaves are usually only a few cells in thickness, and chloroplast density is high in epidermal areas. Shade-adapted leaves are finely divided, whereas leaves on the same plant growing at higher light intensities of shallower depths can be larger and more lobate. Photosynthetic rates can be significantly higher near the surface in cases where overlying leaves of dense populations shade the lower leaves. Shade-adapted plants have been shown to be sensitive to ultraviolet light, which is rapidly absorbed in the first meter of water, whereas photosynthetic rates of plants adapted to high light intensities were insensitive to ultraviolet light found in surface strata of water.

Although net photosynthetic capacity per unit area commonly increases per unit area of shade-adapted leaves (e.g. Spence 1982), the capacities and compensation points of submersed plants growing at low light intensities are highly variable. Light compensation points are commonly at 1 to 3 percent of full sunlight. Values as low as $1.5\ \mu E$ $m^{-2}s^{-1}$ have been found (Wetzel *et al.* 1984). Although respiration rates are commonly low among shade-adapted species, these types of submersed plants also exhibit adaptations to lower temperatures often found at lower depths (e.g. Sand-Jensen 1978, Titus & Adams 1979, Barko & Smart 1981).

TEMPERATURE

Because of the very high specific heat of liquid water, thermal conditions within standing waters change more slowly, are more stable, and permit a longer growing season than is commonly found in terrestrial aerial habitats. The heat-absorbing and heat-retaining properties of bodies of water also modify the climate of immediately adjacent land masses and surrounding wetland habitat of emergent macrophytes.

Growth and life history patterns of emergent and surface-floating macrophytes are clearly influenced by air temperatures much as among terrestrial plants. Temperatures within water-saturated sediments of wetlands fluctuate less rapidly and to a lesser extent than in terrestrial soils.

Large differences in photosynthetic rates have been found among submersed macrophytes in relation to temperature. Although certain species may have similar optimal temperatures (25 to 35 °C) for photosynthesis, they often differ greatly in photosynthetic capacities at lower temperatures (10 to 15 °C) (e.g. Titus & Adams 1979, Barko & Smart 1981). Despite temperature interactions with light availability and marked species differences in photosynthetic capacities to high and low light adaptation, temperature is often a controlling factor in competitive success. Some submersed species are capable of maintenance of basal metabolism at water temperatures below 10 °C and overwinter under ice cover, but no appreciable growth occurs, regardless of adequate light availability. In small lakes of sufficient depth to thermally stratify, growth of submersed angiosperms is limited by colder (< 15 °C) water of the metalimnion even though irradiance is adequate for net photosynthesis (Moeller & Wetzel, in preparation).

HYDROSTATIC PRESSURE

Hydrostatic pressure likely interacts with low light intensities to limit the distribution of submersed freshwater angiosperms to depths less than 10 meters. Normal growth can be inhibited when certain submersed angiosperms are exposed to moderate increases in hydrostatic pressure of between 0.5 to 1 atm above surface atmospheric pressure (approximately 5 to 10 m depth equivalency) (reviewed in Wetzel 1983a). The physiological ramifications of increases in hydrostatic pressure remain obscure, but it is probable that hydrostatic pressure interacts with other factors, particularly light and temperature, to commonly restrict the distribution of freshwater angiosperms to depths of less than ca. 1 atm. hydrostatic pressure. For example inhibitory pressure effects on the growth of *Hippuris* could be overcome by warm temperatures (> 15 °C) and high light intensities (> 100 μE m^{-2}s^{-1}) (Botkin *et al.* 1980). Shoots of *Myriophyllum spicatum* were unaffected by increased hydrostatic pressure (Dale 1981, Payne 1982). Payne (1982) demonstrated that a lacunar arch system in the stems of *Myriophyllum* expands the gas lacunae against external pressure by increasing cellular turgor pressure. Older stems and roots were weaker and were limited by low hydrostatic pressure increases. Macroalgae (*Chara, Nitella*), lacking intercellular gas lacunal systems, are unaffected by pressure differences (Golubić 1963).

CURRENTS AND WATER TURBULENCE

The surface of the water of lakes is a habitat subject to severe mechanical stresses from wind and water movements. Within streams the greatest water velocity within the channel is central and somewhat below the water surface (cf. Hynes 1970); velocities of water currents decrease markedly toward the sediments and the banks. The hydraulic resistance to water movement offered by rooted submersed or floating-leaved plants depends on their morphology and size.

Adaptations to water movement stresses by floating-leaved macrophytes include a tendency towards peltate leaves that are strong, leathery, and circular in shape with an entire margin. The leaves often have hydrophobic surfaces, are buoyant, and have long, pliable petioles. A similar suite of adaptations is found among floating-leaved macrophytes from many taxonomically unrelated groups. In spite of these adaptations, severe winds and water movements restrict these macrophytes to relatively sheltered habitats in which water movements are small.

Leaves of submersed macrophytes occur in three main types: entire, fenestrated (rare), and dissected (cf. Sculthorpe 1967, Hutchinson 1975). The entire leaf form is the most common throughout all taxonomic groups and habitats. Entire leaves are often elongated to ribbonlike and filiform (threadlike) morphology in which length greatly exceeds width, even among the more lanceolate-like leaves. Elongated, pliable leaves resist shearing in moving water, maximize utilization of the reduced available light, and increase the ratio of surface area to volume. Plants of this leaf morphology are most common in habitat areas of streams of appreciable current velocity (cf. Haslam 1978).

Dissected leaves are common among submersed dicotyledonous angiosperms. The most common form is extreme dissection with leaf segments in whorls radiating from the petiole. The dissected leaf forms greatly increase the surface-area-to-volume ratios for light and nutrient absorption, but simultaneously increase hydraulic resistance to water movements. Hence plants of this morphology are restricted to relatively quiescent areas (deeper, along margins of streams, or sheltered shallow habitats). Plants which are most tolerant of water turbulence usually have deep underground root and/or rhizome systems and are rooted in stable hard substrata. Tolerance to water currents is increased somewhat with increased vegetation cover, with carpet and clump-forming populations being less vulnerable than single plants.

SEDIMENTS: THE ROOTING MEDIUM

Sediments of the littoral zone of lakes and of running waters are sorted by water movements into particle size gradients. Substratum particle size increases with increasing water velocities, and of course is influenced by the geomorphology of the drainage basin. In most habitats conducive to the development of significant aquatic vegetation, the sediments contain an appreciable content of organic matter in combination with inorganic mineral matter. Lake sediments are the main sites of microbial degradation of detrital organic matter and biogeochemical recycling of nutrients. As discussed below, rooted plants depend to a large extent upon the sediments for nutrients, even though the sediments constitute an anaerobic environment that is hostile to normal plant metabolism.

Sediments are composed of organic matter in various stages of decomposition, particulate mineral matter, and an inorganic component of recent biogenic origin (e.g. photosynthetically and bacterially induced $CaCO_3$ precipitation, diatom frustules). Particulate organic matter formed by phytoplankton in the pelagial zone of lakes is decomposed more readily and completely during sedimentation than is the more refractory structural tissue of littoral and wetland macrophytes (cf. review of Wetzel 1983a). As senescing littoral vegetation is decomposed and fragmented by the metabolic activities of successive communities of fungi and bacteria, redistribution of the particulate detritus often occurs by water turbulence. In this manner, the partly degraded detrital material is displaced among both littoral and pelagial sediments.

Bacterial populations and metabolic activities are several orders of magnitude larger in the superficial layer of sediments than are those in the overlying water column. Bacterial population densities and their metabolic activity decrease precipitously in the transition from the sediment-water interface to lower depths (10 to 20 cm) within the sediments. Bacterial numbers and metabolic activities of sediments are often positively correlated with lake productivity and the quantity of particulate organic matter settling onto the sediments. The respiratory rates of sediment-dwelling bacteria are commonly inversely related to the surface area of particles (particle size) of sediments, and directly related to the organic nitrogen content of the sedimenting organic matter.

Most sediments of lakes and many areas in streams contain significant amounts of organic matter and experience limited intrusions of dissolved oxygen from the overlying water. The intense bacterial metabolism rapidly produces anoxic, reducing conditions. Under the usually pre-vailing anaerobic conditions, rates of mineralization are slower than under aerobic conditions. In addition under anoxic conditions, various substances and intermediate metabolic endproducts function as alternate electron acceptors and are reduced instead of molecular oxygen. Often the same compound can function as a hydrogen acceptor or donor, depending on the environmental conditions of reducing conditions and organic substrate availability (cf. Wetzel 1983a). Large amounts of organic matter are anaerobically decomposed by methane fermentation. Organic matter is converted to methane and CO_2 in two stages (Figure 1). Facultative and obligate anaerobic bacteria ('acid formers') convert proteins, carbohydrates, and fats primarily to fatty acids, particularly acetate, by hydrolysis and fermentation. Obligately anaerobic metha-nogenic bacteria then convert the organic acids to CO_2 and CH_4;

8

Fig. 1. Stages of metabolism of complex organic compounds in anaerobic sediments. (From Wetzel, 1983a after several sources.)

alternatively, they can reduce CO_2 to CH_4 by enzymatic addition of hydrogen derived from the organic acids.

In addition to the organic acids and carbon gaseous products formed in prevailing anaerobic sediments, the rooting tissue of aquatic plants also is often exposed to high concentrations of hydrogen sulfide. Sulfate is the primary dissolved form of sulfur in oxic waters. Hydrogen sulfide accumulates in anoxic zones of intensive decomposition in productive lakes or in reducing sediments where the redox potential is lowered below about 100 mv. Although most of the sulfur in lake water is stored as dissolved sulfate and in anoxic strongly reducing strata as hydrogen sulfide, in sediments the primary constituents are sulfur-containing proteins, sulfate esters, and dissolved sulfides. Organic sulfate esters are mineralized by an active sulfhydrolase system in the sediments.

The internal gas lacunal system is highly variable among freshwater angiosperm species, but is commonly extensive and constitutes a major portion (often exceeding 70 percent) of total plant volume. The possibility of the lacunal system becoming filled with water and non-functional if the tissue were damaged at any point is protected against by a number of types of lateral plates and watertight diaphragms, permeable to gases, that interrupt the lacunal passages at intervals (cf. Arber 1920, Sculthorpe 1967). Some of the oxygen produced during photosynthesis in submersed plants (and from the atmosphere in emergent plants) is retained in the lacunal system and diffuses from the leaves through the petioles and stems to underground root and rhizome systems, where respiratory demands are high.

Gases within the intercellular lacunae of submersed angiosperms diffuse along gas-partial-pressure gradients. In floating-leaved plants (e.g. *Nuphar*), however, the internal gas spaces function as a pressurized flow-through system (Dacey 1981). Ambient air enters the youngest emergent leaves against a small gas-pressure gradient as a result of physical processes driven by gradients in temperature (thermal transpiration) and water vapor (hygrometric pressure) between the lacunae and the atmosphere. Lacunal gas spaces are continuous through young emergent leaves, petioles, rhizomes, and the petioles of older emergent leaves. The older leaves vent the elevated pressure generated by the younger leaves. The resulting flow-through ventilation system accelerates both the rate of oxygen supply from the atmosphere to the rooting tissue and the rate of CO_2 and methane transport from the roots to the atmosphere.

Movement of oxygen from the atmosphere in emergent and floating-leaved plants to the root tissue is essential to prevent accumulation of toxic respiratory endproducts (e.g. ethanol produced during glycolysis). Wetland emergent plants and submersed angiosperms often increase the volume of the lacunal gas system when sediments are more reducing along a littoral gradient (e.g. Katayama 1961, Armstrong 1978, Penhale & Wetzel 1982). Some aquatic plants have developed metabolic adaptations to cope with anaerobic rooting conditions: diverse nontoxic end products (malate, shikimic acid) can be produced during glycolysis (Crawford 1978, Penhale & Watzel 1982, Sale & Wetzel 1982), or ethanol can be released to the environment (Bertani *et al.* 1980).

Bacterial endproducts of fermentation (volatile fatty acids, hydrogen sulfide) in anaerobic, reducing sediments are toxic to many plants. Some oxygen diffuses from the roots and forms an oxidized microzone in the rhizosphere sediments (cf. Armstrong 1978). This oxidized microzone in the rhizosphere reduces the toxicity of fermentation products, but simultaneously reduces the availability of certain nutrients, such as iron and manganese (Tessenow & Baynes 1978). Certain sulfur bacteria (*Beggiatoa*) of the plant rhizosphere can oxidize hydrogen sulfide to sulfur granules and lower H_2S concentrations in the root rhizosphere (Joshi & Hollis 1977).

NUTRIENT AVAILABILITY AND UTILIZATION

The topic of the distribution, biogeochemical cycling, and availability of nutrients for utilization by aquatic plants in a spectrum of aquatic habitats (wetlands, littoral regions, and open water of lakes, streams, bogs, and other habitats) is a subject much too involved to treat adequately in a brief summary. Comprehensive treatises have devoted much of their discussion to nutrient biogeochemical cycling (cf. particularly Hutchinson 1957, 1967, 1975, Hynes 1970, Wetzel 1983a). Several summary points, however, are possible in order to cast an overview of physiological problems that aquatic plants face in the aqueous environment and general adaptive mechanisms that have evolved to cope with a relatively hostile environment.

In general the nutrient concentrations and biogeochemical regeneration rates of nutrients in the free water of natural lakes and streams are low to extremely low in comparison to concentrations of available nutrients in most soil systems of terrestrial habitats. Even in the most eutrophic standing water bodies or running waters, the concentrations of nutrients in the water are usually several orders of magnitude less than in the interstitial water of water-saturated or submersed sediments. It is therefore germane to discuss whether plants utilize nutrients from the water, the sediments, or both.

Secondly, it is important to stress that most water-saturated sediments are anaerobic, and that these anoxic conditions are for all practical purposes continuous (cf. review of Wetzel 1983a). Most freshwater sediments of wetlands and standing waters contain significant amounts (10 to 90% or more) of organic matter. Microbial degradation rates in sediments are several orders of magnitude greater than in the much more organic substrate-dilute overlying water. Utilization rates of electron acceptors are very high in sediments; intrusions of oxygen from physical or biotic sources of turbulence are rapidly (minutes to hours) consumed. Even in streams with oxygen saturated water flowing over sediments, organic matter decomposition within sediments is often adequate to maintain anoxia, particularly along stream edges and in backwater areas where flow is reduced or absent.

Under reducing conditions in sediments, nutrient solubility is increased markedly in comparison to oxic conditions (cf. Wetzel 1983a). As discussed above, however, endproducts, such as H_2S and volatile fatty acids, of fermentative metabolism can be toxic to plants and compensatory mechanisms must be available to counter these compounds. Moreover, oxygen must reach rooting tissue from above-sediment tissues in order to meet normal respiratory demands.

Whether rooting systems of aquatic macrophytes function merely as organs of attachment or in absorption of nutrients from the substratum has been a long-standing question in limnology (reviewed in Gessner 1959, Wetzel 1964, 1983a, Sculthorpe 1967, Bristow 1975, Hutchinson 1975). Among emergent and rooted floating-leaved angiosperms with active transpiration-mediated root-pressure systems, nutrient absorption and translocation from the roots to the foliage are clearly operational. Practically no nutrients are absorbed from the water by the submersed portions of these plants. In a floating-leaved water lily (*Nuphar luteum*), phosphorus absorption rates were much higher by submersed leaves; floating leaves absorbed little (Twilley *et al.* 1977). About four times more phosphorus moved from roots to leaves than from the leaves to the roots during the summer period. Relative to terrestrial soils, large quantities of nutrients in available forms are contained in the highly reducing anaerobic sediments, although the plants require a well-developed internal aeration system to the rooting tissue from aerial organs. Macro-algae, liverworts, mosses, ferns, and other macrophytes, including angiosperms, that float on the surface or in the water or are variously rudimentarily attached or resting on the substratum obtain nutrients by direct foliar absorption rather than by rhizoidal structures.

An exception is the rhizoid-bearing macroalga *Chara*, which absorbs phosphorus equally efficiently by all parts; phosphorus absorbed is translocated to other parts of the plant (Littlefield & Forsberg 1965). Freely-floating angiosperms absorb nutrients both by roots suspended in the water and to varying degrees through the underside of leaves floating on the surface of the water.

Despite total immersion, rooted submersed angiosperms maintain a positive absorptive capacity. Guttation always occurs in small quantities, and absorption increases from the root tips to the basal portions of the plant and increases from the apical parts to the base (cf. review of Stocking 1956). Water exits via hydathodes on the upper portions of many submergents, and nutrients can enter leaves apically through numerous porous specialized cell aggregates termed hydropoten. The absorption of phosphorus by submersed rooted angiosperms both from the water and from the sediments by roots is well-known (cf. reviews of Gessner 1959, Schwoerbel & Tillmanns 1964, Bristow 1975, Wetzel 1983a). Phosphate absorption rates by foliage are proportional to and dependent upon the concentrations in the water. Very large quantities of several mg 1^{-1}, found only in heavily polluted aquatic ecosystems, are rapidly assimilated in excess of requirements until concentrations in the water (batch cultures) were reduced to about 10 µg 1^{-1}. This type of luxury consumption is common for phosphorus but has also been observed for nitrogen.

Recent experimental analyses have demonstrated that most rooted submersed angiosperms obtain most of their required phosphorus from sediment interstitial water by active absorption and translocation to the foliage. Even when exogenous nutrient concentrations in the water are high, uptake can be limited by the rate of diffusive influx across the gradients in the 'unstirred' microlayer at the leaf-water interface (10 to 100 µm in thickness). For nutrients such as phosphorus that are commonly in chronically limiting concentrations in many natural waters, the nutrients absorbed from the surrounding water can be effectively utilized and recycled within the epiphytic community of algae and bacteria. The epiphytic community can compete with the supporting macrophyte tissue for exogenous nutrients in the water (cf. reviews of Wetzel & Grace 1983, Wetzel 1983c). High concentrations of nutrients, particularly phosphorus, in the water can encourage prolific development of epiphytic algae to the detriment of the supporting submersed macrophytes. This detriment can occur not only by the competition for nutrients but also for light (Wetzel 1983c, Losee & Wetzel 1983).

Nitrate assimilation rates by the foliage of submersed macrophytes was found to be considerably less than rates of ammonia assimilation, particularly at high pH values. However, ammonium ions are in very high concentrations in many anaerobic sediments and are readily absorbed by roots and translocated to foliar tissue. In addition combined nitrogen from N_2-fixing bacteria of the sediments and plant rhizosphere can serve as a major nitrogen source for both emergent and submersed plants.

Large variations have been found in the inorganic chemical composition of aquatic plants (cf. review of Hutchinson 1975), even among closely related species and from site to site within the same species. Metals and alkalies are absorbed both from the water by leaves and to a greater extent from the interstitial water of sediments by roots, and many elements are concentrated in the macrophytes in great excess of metabolic requirements.

In summary, it is apparent that ion absorption in submersed macrophytes occurs both from the water by foliage and from the interstitial water of sediments by root and rhizoid systems. Translocation occurs in both directions. In most cases, however, roots serve as the primary site of nutrient absorption from the sediments. The rooted plants can function as a nutrient 'pump' from the sediments. These nutrients can be released to the water, if not used by the epiphytic microfloral community, during active growth and particularly during senescence and decomposition of the macrophytes. It should be noted that most of the analyses of nutrient absorption and exchange reaction rates are based on relatively simple direct uptake analyses. Relatively few kinetic assimilation analyses have been determined to give insights on nutrient assimilation capacities and compensation points for different species under controlled environmental conditions. Therefore, at this time it is difficult to know the competitive advantages among species resulting from differing nutrient assimilation capacities, even though such differences certainly exist and are important in community dynamics.

INORGANIC CARBON AVAILABILITY AND UTILIZATION

Rates of diffusion of gases in water are several orders of magnitude slower in water than in air. The availability of CO_2 from aqueous solutions can rapidly become limiting to submersed plants under other environmental conditions conducive to high photosynthetic rates.

CO_2 is very soluble in water and obeys normal solubility laws. When water is in equilibrium with air, dissolved free CO_2 is approximately 10 µmol. Dissolved CO_2 hydrates to carbonic acid (H_2CO_3) at a low concentration (1/40th of free CO_2). H_2CO_3 rapidly dissociates to HCO_3^- ions (Wetzel 1983a, Wetzel & Grace 1983). Final concentrations in the water are controlled by the partial pressure of CO_2, temperature, pH, and ionic strength of the water solution. Concentrations of all forms of inorganic carbon can be predicted by solubility and dissociation constants for any concentration of atmospheric CO_2 by assuming equilibrium between the atmosphere and water. Increasing atmospheric CO_2 will result in a linear increase in dissolved CO_2. This increase will result in an increase in HCO_3^- concentration, but the increase is not as great as, nor is it linearly proportional to, the CO_2 increase because of a simultaneous reduction in pH (Wetzel and Grace 1983).

Both marine and most fresh waters are near to equilibrium with atmospheric CO_2 much of the time. In marine waters, the inorganic carbon pool contains about 2.5 mM C, largely as HCO_3^-. In freshwater ecosystems, the total inorganic carbon is much more variable (50 µM to 10 mM); in addition, pH is more variable because of differing alkalinity and high photosynthetic carbon demands. Free CO_2 (dissolved) is predominant in water at pH 5 and below, while above pH 9.5, CO_3^- is quantitatively significant. Between pH 7 and 10, which prevails in most surface waters of the world, HCO_3^- predominates.

The extent of CO_2 exchange between the ambient air and some fresh waters cannot be determined by partial pressure differences alone. Many lakes with surface waters near neutrality are slightly supersaturated with CO_2 relative to the air. Alkaline bicarbonate lakes containing large amounts of HCO_3^- and CO_3^- are not in equilibrium with CO_2 of the air, and a net efflux of CO_2 to the atmosphere can occur throughout all or much of the year (e.g. Otsuki & Wetzel 1974).

In photosynthetically productive waters, consumption of CO_2 by phytoplanktonic algae and macrophytes can create a flux of atmospheric CO_2 into the water (Weiler 1974, Emerson 1975) as dissolved CO_2 concentrations decline and photosynthetically-induced concentrations of carbonate increase (e.g. Wood 1974, 1977, Talling 1976).

The amount of excess CO_2 required to maintain stability of bicarbonate in solution increases rapidly as the concentration of bicarbonate increases. If a solution of bicarbonate in calcium-rich hard waters in equilibrium with CO_2, H_2CO_3, and CO_3^- loses a portion of CO_2 required to maintain equilibrium, such as CO_2 assimilated in photosynthesis, the pH and carbonate concentration may increase to a level exceeding the solubility of $CaCO_3$. Calcium carbonate will precipitate (as marl) until equilibrium is reestablished. Excess CO_2 that is required to maintain large amounts of HCO_3^- in solution can also be lost to the atmosphere by photosynthetic removal, and result in massive precipitation of $CaCO_3$. As $CaCO_3$ nucleates, important nutrients (e.g. PO_4) and growth compounds (e.g. amino acids, vitamins) can simultaneously be coprecipitated or adsorbed and removed from zones of active assimilation (e.g. Otsuki & Wetzel 1972, 1973, Wetzel 1983a).

Atmospheric CO_2 is clearly the dominant source of inorganic carbon among emergent, floating-leaved (rooted), and freely-floating macro-phytes. A few species of freely-floating angiosperms (e.g., some duck-weeds, *Lemna*) utilize both atmospheric and aqueous carbon sources (Wetzel & Manny 1972, Ultsch & Anthony 1973). Heterophyllous aquatic angiosperms are presumably similarly adapted to assimilate both atmospheric CO_2 and aqueous inorganic carbon sources. Short-term atmospheric CO_2 enhancement (500 to 1000 ppm) has been shown to increase photosynthetic rates of three species of floating duckweeds (Loats *et al.* 1981).

At least three potential sources of CO_2 exist for plants with emergent foliage: sediment CO_2 absorbed directly by roots, sediment-derived CO_2 released into the canopy, and CO_2 which comes into the canopy from the above atmosphere. Studies with rice (*Oryza sativa* L.) have shown that most of the CO_2 fixed is derived from the atmosphere, with sediment-derived CO_2 supplying 0 to 12% of the carbon for net production. Direct absorption of CO_2 by the roots is 1 to 2% of the total (Murata *et al.* 1957, Mitsui & Kurihara 1962, Tanaka *et al.* 1966, Yoshida *et al.* 1974). It is not clear at present if rhizomatous species have similarly low root uptake. Where the sediment is in direct contact with the atmosphere, sediment derived CO_2 may be important to photosynthesis (Moss *et al.* 1961, Monteith *et al.* 1964). However, under wet conditions, the diffusion of CO_2 from the soil to the atmosphere is strongly inhibited.

Limited evidence suggests that the physiological processes in emergent foliage of aquatic plants are the same as in terrestrial plants. Most emergent species possess C_3 metabolism, and under conditions of high light and temperature these species exhibit high levels of photorespiration (Jones & Milburn 1978, Filbin 1980). Studies by McNaughton (1966, 1969) and McNaughton and Fullem (1969) present evidence that *Typha latifolia* possesses C_3 carbon metabolism, despite its high rates of photosynthesis and relatively low rates of photorespiration. C_4 species often found in freshwater habitats include *Cyperus esculentus, C. papyrus, Echinochloa crus-galli, Panicum* spp., and *Zizaniopsis miliacea*. In contrast to freshwater marshes, C_4 species *Spartina alterniflora, S. patens, S. townsendii,* and *Distichlis spicata* often dominate salt marshes.

Algae and submersed aquatic macrophytes require an abundant and readily available source of carbon for high sustained rates of growth. Abundant physiological evidence indicates that free CO_2 is the form of inorganic carbon most readily utilized by many algae and submersed vascular plants. Since RuBP carboxylase can utilize only CO_2 and requires high internal concentrations of CO_2 in the chloroplasts to maintain high rates of carbon fixation, the plants would be constantly reaction limited without some inorganic carbon augmentation. Although species of algae and higher plants differ in their half-saturation kinetics for CO_2 uptake, augmentation of CO_2 from the dehydration of HCO_3^- ions can provide an additional source of inorganic carbon from the water. The uptake of bicarbonate ions by submersed plants and its energy-consuming transport across membranes has been a topic of much study and controversy (cf. reviews of Raven 1970, 1981, Allen & Spence 1981, and Wetzel & Grace 1983). Many algae and submersed macrophytes, particularly the mosses and pteridophytes, utilize only free CO_2 (Bain & Proctor 1980). Other algae and aquatic vascular plants can utilize bicarbonate ions as an exogenous carbon source when free CO_2 is very low and HCO_3^- is abundant. There is little conclusive evidence that algae or higher aquatic plants assimilate CO_3^- directly as a carbon source. The recent monographs of Raven (1984) and Lucas and Berry (1985) summarize contemporary understanding of the physiology of carbon uptake in aquatic plants.

In some cases of high photosynthetic demand, the resupply of CO_2 from the carbonate species can be rate limiting. Below a pH of 8.5, the theoretical dissociation kinetics of the $CO_2-HCO_3^--CO_3^-$ complex, derived from pure solutions, agree very well with the concentrations of the ionic species found in natural waters (Talling 1973). Above pH 8.5, however, the total inorganic carbon dioxide concentrations fall considerably below values calculated on the basis of the apparent dissociation constants K_1' and K_2' of carbonic acid, as derived from measurements of alkalinity. Although the explanation for this phenomenon is not completely clear, it appears to be related to the presence of a non-carbonate, nonhydroxide alkalinity component arising from ionized silicate (Talling 1973).

Direct assimilation of bicarbonate ions is variable among planktonic algae, macroalgae, and submersed angiosperms. Bicarbonate utilization is an active process that includes the dehydration of HCO_3^- and is coupled to a stoechiometric excretion of hydroxyl ions. Although carbonic anhydrase activity is found in submersed plants, its activity is generally lower than that found in both terrestrial plants and in unicellular algae (Weaver & Wetzel 1980). In marine angiosperms that utilize HCO_3^-, carbonic anhydrase activity is apparently too low to contribute significantly to photosynthesis (Beer *et al.* 1980).

Since the affinity for HCO_3^- ions is lower than for CO_2, a higher concentration of HCO_3^- ions is required to saturate photosynthesis. The maximal rates of photosynthesis in HCO_3^- solutions are one-third to one-half those obtained at saturating concentrations of dissolved CO_2 in both freshwater and marine algae and submersed angiosperms (Lucas 1975, Beer *et al.* 1977, Lucas *et al.* 1978, Kadono 1980, Allen & Spence 1981, Raven 1984, Lucas and Berry 1985).

Equilibrium concentrations of dissolved CO_2, particularly in alkaline hardwater lakes with a pH $>$ 8.5, are inadequate to saturate photosynthesis in submersed plants. When waters become more productive, and in densely populated littoral zones of less productive lakes, pH is rapidly altered by metabolism on a diurnal basis (pH ranges from a low of 6 to a maximum of 10 or more in 24 hours), reducing carbon fixation or bicarbonate assimilation. Under stagnant conditions common to heavily colonized littoral zones of lakes, the shift to bicarbonate metabolism, as well as the increased pH, is often associated with severe reduction in CO_2.

Among planktonic diatoms, for example, successive CO_2 enrichments induced a three- to five-fold increase in algal biomass over that of unenriched controls (Jaworski *et al.* 1981). A certain degree of cellular adaptation occurred with successive generations exposed to CO_2 reductions. Inorganic carbon limitations could be counterbalanced by reducing the pH without any additions of total inorganic carbon; this increases the relative proportion of free CO_2. This effect often is observed under culture conditions in algae and has been found in the submersed angiosperm *Najas flexilis* as well (Moeller & Wetzel, in preparation).

Morphological adaptations among submersed leaves, stems, and some petioles include thin leaves (1 to 3 cell layers), reduced cuticle development, extreme reduction or elimination of mesophyll, and dense distribution of chloroplasts in epidermal cells. These adaptations all increase utilization and exchange of gases. Massive intercellular gas spaces in leaves, stems, and petioles facilitate rapid internal diffusion.

Limitation of photosynthesis by slow diffusion of CO_2 in water has received much attention, especially related to the alternate source of bicarbonate ions. Rate-limiting steps in the flux of CO_2 from the air to metabolic pathways of cells vary widely among aquatic environments depending upon differences in water chemistry, physical mixing, and rates of CO_2 utilization. CO_2 exchange between the air and water surface can be quantified using the 'stagnant film' model or more complex models (cf. review of Danckwerts 1970, Broecker 1974). Transfer rates depend upon the CO_2 gradient between the air and surface water, turbulent mixing of both air and water at the interface, function of wind speed controlling both air and water boundary layer thicknesses (and thus the transfer coefficients), and chemical enhancement by reaction of CO_2 with water and OH^- ions (cf. Wood 1974 for transfer coefficients from several sources). Transfer coefficients in well-stirred systems probably vary from 0.01 cm s^{-1} in lakes with high winds and whitecaps to less than 0.001 cm s^{-1} on calm days (Verduin 1975). The flux of CO_2 from the air may be too slow to support potential CO_2 demands by aquatic plants in nutrient-rich lakes. Production can be restricted by slow air-water transfer in softwater lakes, whereas bicarbonate sources may sustain productivity in hardwater situations.

In large submerged angiosperms and algae, and diffusion path is long (up to 50 μm). through the cells of the thallus or leaf, to which must be added another boundary layer of about 100 μm of unstirred water at the cellular surfaces; further, diffusion rates through the larger cells are much slower than among micro-algae (Steemann Nielsen 1947, Raven 1970, Smith & Walker 1980). Low velocity currents in the bulk fluid increase photosynthesis among submersed angiosperms by reducing the stagnant boundary layer, thus decreasing diffusion path lengths (e.g., Barth 1957, Westlake 1967, Riber & Wetzel 1987).

Stagnant layers around plant cells (algae) and tissues (macrophytes) clearly limit the assimilation of both CO_2 and HCO_3^- during photosynthesis (Browse et al. 1979, Smith & Walker 1980, Black et al. 1981, Prins et al. 1982a, 1982b). Internal diffusive resistance is much less than resistance associated with the stagnant boundary layer at the plant-water interface. Internal resistance apparently accounts for only a small fraction (ca. 5%) of total resistance to CO_2 assimilation in several species of Potamogeton. Turbulence reduces the thickness of the stagnant boundary layer and can promote photosynthesis at fixed levels of available carbon in aqueous systems (Westlake 1967). However, a residual boundary layer (ca. 10 μm thick) parallel to the plant surface remains even when plants are exposed to water movements or when planktonic algae sink. Mass transport across the residual layer is by molecular diffusion. Diffusion of CO_2 and HCO_3^- through these stagnant layers is an important rate-limiting process both to availability of CO_2 and to membrane transport of HCO_3^- ions. Mucilage sheaths surrounding algal cells, especially among blue-green algae, can further reduce the uptake rates of inorganic carbon (Chang 1980).

Bicarbonate ions as a carbon source supplementary to carbon dioxide has been demonstrated in a number of studies. Space does not permit detailed discussion of the results of all of these studies. Utilization of HCO_3^- as a carbon source under natural conditions has been reported for a number of freshwater and marine angiosperms (e.g., Raven 1970, Beer et al. 1977, 1979, Helder & Zanstra 1977, Lucas et al. 1978, Browse et al. 1979, Allen & Spence 1981, Beer & Wetzel 1981, Prins et al. 1980, 1982a, 1982b, 1982c), while other species were found to utilize mainly CO_2 (Wetzel 1969, Brown et al. 1974; Van et al. 1976, Moeller 1978, Winter 1978, Kadono 1980). In algae, especially larger macroalgae, the ability to utilize bicarbonate ions as a carbon source in photosynthesis is widespread, but not universal. The extensive work on this subject among mosses and angiosperms by Steemann Nielsen (1944, 1947), Ruttner (1947, 1948, 1960), and others (e.g., Bain & Proctor 1980; Raven & Beardall 1981) has shown that certain freshwater red algae and all assayed genera of freshwater submersed mosses utilize primarily free CO_2. These latter plants are restricted almost always to soft waters of relatively low pH, or to streams in which CO_2 concentrations are relatively higher. In two reported cases, the levels were 2.8×10^{-4} M free CO_2, which is 20 times the concentration in air (Bristow 1969, Browse et al. 1977).

In summary, evidence indicates that, among the diversity of concentrations and states of inorganic carbon in fresh and marine waters, there exists a large number of cases where free CO_2, even in equilibrium with the atmosphere, may be inadequate to sustained high levels of photosynthesis. In other cases, free CO_2 may be inadequate as a result of slow diffusion rates and chemical losses from the system. Possession of an affinity for bicarbonate transport with subsequent dehydration is a distinct adaptive advantage in most submersed plants.

The slower diffusion of CO_2 in water and the presence of massive internal gas lacunae can reduce loss of CO_2 from submersed angiosperms and facilitate refixation of respired and photorespired CO_2 (Carr 1969, Hough & Wetzel 1972, Hough 1974, Søndergaard 1979, Søndergaard & Wetzel, 1980). In softwater lakes, concentrations of total inorganic carbon in the water are very low; certain soft-water submersed angiosperms utilize CO_2 of the sediment interstitial water to supplement CO_2 assimilated from the water (Wium-Andersen 1971, Søndergaard & Sand-Jensen 1979, Wetzel et al. 1984). Uptake of CO_2 by root tissue for photosynthetic fixation by submersed plants of hardwater lakes, where concentrations of inorganic carbon were high, could not be demonstrated (Beer & Wetzel 1981).

In contrast to a relatively constant CO_2 compensation point for a given terrestrial species, values for submersed angiosperms appear to be quite variable, depending of prior growth conditions (Helder et al. 1974, Brown et al., 1974, Van et al. 1976, Lloyd et al. 1977, Hough & Wetzel 1978, Bowes et al. 1979, Salvucci & Bowes 1981; Barko & Smart 1981, Roelofs et al. 1984). Values are similar to those for certain unicellular algae (Tsuzuki & Miyachi 1981). The CO_2 compensation point is important as an indicator of the photosynthesis/photorespiration ratio. High values suggest significant photorespiration.

Photoperiod and temperature conditions of summer may decrease the CO_2 compensation point, whereas winter-like conditions may increase it (Salvucci & Bowes 1981). Correlated with these observations, photorespiration in Scirpus subterminalis is relatively low in summer and increases in winter (Hough 1974). Similarly, photorespiration and dark respiration are ten-fold greater in fall compared to summer as the annual submergent Najas enters senescence (Hough 1974). For Hydrilla and Myriophyllum, a low CO_2 compensation point is associated with higher rates of net photosynthesis (or an increased affinity for CO_2), decreased rates of photorespiration, and reduced oxygen inhibition of photosynthesis (Holiday & Bowes 1980, Salvucci & Bowes 1981). With a high CO_2 compensation point, the reverse is true (cf. Bowes 1985).

Preliminary evidence indicates that changes in the CO_2 compensation point may be related to the CO_2 concentration in the environment. For *Hydrilla*, low CO_2 levels (100 ppm gas phase) induced a low CO_2 compensation point, whereas high levels (2000 ppm gas phase) restored a high value (Salvucci & Bowes 1981).

Conditions conductive to accelerated photorespiration and reduced net photosynthesis can develop in the littoral zone, particularly among dense submersed plant populations with little turbulence (Wetzel 1965, 1975, Goulder 1980, Hough 1974). Many studies suggest that photo-respiration increases through the day as dissolved oxygen increases from photosynthesis, temperature increases, and CO_2 availability decreases (Wetzel 1975, 1983a).

Evidence indicates that freshwater and marine submersed angio-sperms possess the C_3 photosynthetic pathway (Hough & Wetzel 1977, Winter 1978, Browse *et al.* 1979, Valanne *et al.* 1982, critically discussed in Beer & Wetzel 1981, 1982a, 1982b), and thus there is a potential for photorespiratory CO_2 loss. A number of marine and freshwater species also exhibit C_4 acid formation, particularly malate, in the light (Beer *et al.* 1980, Browse *et al.* 1980, Holiday & Bowes 1980, Kremer 1981, Beer & Wetzel 1982b, Helder & Harmelen 1982). It has been suggested that these acids function to balance excess cation uptake, or as an anaplerotic source of compounds for the TCA cycle or amino acid synthesis (Browse *et al.* 1980). In *Hydrilla*, production and turnover of malate in the light undoubtedly contributes to its inducible low photorespiratory state (Holiday & Bowes 1980). However, malate may also accumulate in the dark for subsequent decarboxylation and refixation of CO_2 in the light analogous to that in terrestrial CAM plants (Keeley 1981, 1982, Beer & Wetzel 1981).

Some species, such as *Myriophyllum spicatum* (Salvucci & Bowes 1981) and certain seagrasses (Beer & Waisel 1979, Beer *et al.* 1980) possess low rates of photorespiration, yet exhibit no evidence of C_4 acid metabolism. It is possible that an inducible bicarbonate utilization mechanism exists in these species as has been demonstrated for unicellular green algae.

Efficient internal recycling of respired CO_2, utilization of bicarbonate, and augmentation of exogenous inorganic CO_2 by decarboxylation of malate are all mechanisms by which certain submersed species adapt to limited CO_2. The evolution of an extensive internal lacunal gas system in submersed angiosperms constitutes interrelated morphological adaptations to (a) enhanced efficiency of carbon fixation, (b) flexibility to withstand water movements, (c) positive buoyancy and positioning of leaves towards greater available light, and (d) an efficient transport system for oxygen diffusion to roots and rhizomes.

CHANGING WATER ENVIRONMENT WITH AQUATIC PLANT SUCCESSION

Interacting physical, chemical, and biological mechanisms that regulate autotrophic productivity in standing fresh waters follow several ontogenetic pathways as lake basins undergo succession and are ultimately obliterated as aquatic ecosystems are transformed into terrestrial ecosystems (cf. review of Wetzel 1983a). Many mechanisms that regulate photosynthesis in the water change during the general transition in lake ontogeny from an early dominance of allochthonous loading of nutrients and organic matter from the drainage basin to conditions of accelerated loading of organic matter and nutrient recycling from littoral and wetland plant productivity.

In early stages of lake ontogeny, particularly among lakes formed by glaciation, autochthonous and allochthonous productivity are severely limited by climatic conditions, especially cold temperatures. In intermediate stages of ontogeny, autochthonous productivity is first dominated by phytoplanktonic productivity, with some small contributions by littoral production (Figure 2). In later and terminal stages of lake development, however, a rapidly accelerating shift occurs to conditions of total dominance of productivity by littoral components.

The planktonic-dominated stages of lake development occur over relatively lengthy time periods unless disturbed by human activities. During these stages, inputs of organic sedimentation and deposition are nearly balanced by decomposition. In shallow basins and in lakes in which sedimentation has reduced basin depth sufficiently, massive production by littoral plants and contiguous wetland emergent macrophytes dominates the inputs of photosynthetically accrued organic matter. The increasing contributions of derivatives of lignified and cellulosic tissues from these higher plants result in less complete, and reduced rates of, decomposition (Wetzel 1979). This loading of more refractory organic materials leads to accelerated rates of sedimentation and favors conditions conducive to enhanced emergent macrophyte development in a cyclically reinforcing pattern (Figure 3).

Increased inorganic nutrient loading, particularly of phosphorus and combined nitrogen, is fundamental to initial eutrophication and to maintenance of high sustained productivity of the phytoplankton community. Low rates of productivity in oligotrophic lakes are maintained to a large extent by low inputs of inorganic nutrients from external

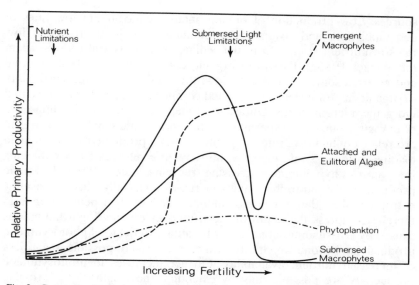

Fig. 2. Generalized relationship of primary productivity of submersed and emergent macrophytes, attached algae and metaphyton, and phytoplankton of lakes of increasing fertility of the whole lake ecosystem. (Modified from Wetzel & Hough 1973.)

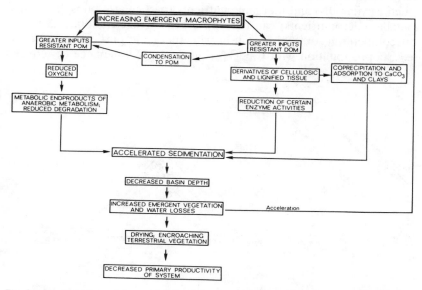

Fig. 3. Relationships of increased loading of relatively resistant organic matter from dominating emergent littoral and wetland flora to controls of decomposition rates, acceleration of emergent macrophyte development, and the ontogeny of lake ecosystems. POM and DOM = particulate and dissolved organic matter, respectively. (From Wetzel 1979.)

22

sources. Low production of organic matter, concomitant low rates of decomposition, and oxidizing hypolimnetic conditions result in low rates of nutrient release from the sediments in a cyclical causal system (cf. Wetzel 1983a). Dissolved organic compounds from both internal and external sources are usually low; the results are limited availability of organic micronutrients and limited complexing capability for essential inorganic micronutrients. Under eutrophic conditions, the loading rates of inorganic nutrients, especially of phosphorus and combined nitrogen, are relatively high. As rates of photosynthetic productivity and organic loading to lower strata increase, nutrients are released from sediments into anoxic hypolimnia, increasing recycling rates. Phytoplanktonic productivity of eutrophic lakes increases markedly. In increasingly eutrophic lakes, dense algal communities reduce light penetration, and thereby compress the depth of the trophogenic zone. Light limitations caused by self-shading set an upper boundary to the total photosynthetic productivity, beyond which further increases are not possible, regardless of increased nutrient availability. Further increases in photosynthetic productivty are possible only by extending the length of the growing season (e.g., in the tropics) or by increasing turbulence and frequency of exposure to available light (as in artificially mixed sewage lagoons). In calcareous hardwater lakes, availability of certain organic micronutrients (such as vitamins) and inorganic nutrients (particularly phosphorus and metallic micronutrients) can be suppressed by inactivation, i.e., chemical competition, or sedimentation with inorganic particulate materials (e.g., coprecipitation with $CaCO_3$) (Figure 4). Maintenance of reduced

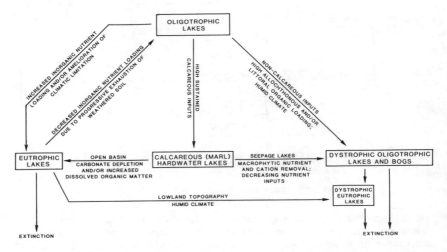

Fig. 4. Potential ontogeny and control mechanisms of the four main types of lakes. (From Wetzel 1983a.)

productivity in calcareous lakes by lowered nutrient availability depends upon high carbonate loading from the drainage basin. Calcareous loading may be reduced through long-term leaching or overcome by increased loading of dissolved organic matter. The result is a relatively rapid transition to eutrophic conditions. Under certain conditions, high cationic (e.g., Ca, Mg) loading also can be counterbalanced and reduced by cationic exchange mechanisms of littoral-wetland plants (e.g., bryophytes), which may result in the transition of calcareous lakes to acidic bogs (e.g. Glime *et al.* 1982).

Many lakes receive large amounts of dissolved, and to a lesser extent particulate, organic matter from allochthonous and surrounding wetland-littoral sources. These lakes, often called dystrophic, are commonly heavily stained as a result of the abundance of dissolved humic compounds of plant origin. Although phytoplanktonic productivity is usually low in highly stained lakes, productivity of the surrounding littoral vegetation is moderate to high. The surrounding macrovegetation can function as effective nutrient scavengers, and frequently reduces the nutrient loading that reaches the open water. In some cases, particularly in bogs, the vegetation can effectively shift the lake per se from an open to a closed basin, so that nutrient inputs to the lake are mainly from atmospheric precipitation.

Terminal stages in the successional sequence of lake ontogeny are dependent upon the types of macrovegetation in the littoral zone and wetland areas surrounding the basin. The development of the vegetation is, in turn, strongly influenced by prevailing climatic conditions (rainfall, humidity) of the region. Many fertile eutrophic lakes develop ever increasing amounts of emergent littoral vegetation, until littoral vegetation encroaches over the entire lake basin. The resultant *swamp conditions*, with standing water among the emergent vegetation, gradually succeed to *marsh* conditions with little or no standing water over the water-saturated sediments. Evapotranspiration from the vegetation can gradually exceed water income, permitting invasion of a terrestrial flora. Ecosystem productivity decreases in the transition to terrestrial conditions. Under climatic conditions of high rainfall and humidity, various types of *mire* ecosystems develop; in these systems, partially decayed plant organic matter (peat) accumulates in abundance. *Rheophilous mires* can develop from shallow drainage lake basins in which large amounts of organic matter accumulate in the depression. With time, the main water flow through the mire tends to be channelized to peripheral areas around the central portion of the organically filled basin. Eventually the mire is no longer under the influence of flowing groundwater. In these *ombrophilous mires*, water and nutrients are received primarily from rainfall. Salinity, buffering capacity, and pH of the water are low.

The acidity of the water often increases during mire development. Mosses, particularly *Sphagnum*, often develop in profusion in *bog mires*. Mosses have effective cation exchange mechanisms, in which divalent cations are retained with a commensurate release of H^+ ions and organic acids. *Sphagnum* and other acidophilic plants can develop above the mean water table (raised bog), yet still acquire sufficient water by capillarity. Bog mats and underlying peat accumulations can gradually encroach upon, and eventually eliminate, the open water of the lake remnant within the bog. Individual bog succession proceeds at rates and along developmental pathways that are controlled by a complex interplay of geomorphological, chemical, climatic, and biotic factors.

ACKNOWLEDGEMENTS

Supported in part by the U.S. Department of Energy (EY-76-S-02-1599). Contribution of the W. K. Kellogg Biological Station of Michigan State University.

REFERENCES

Allen, E. D. & Spence, D. H. N. (1981) The differential ability of aquatic plants to utilize the inorganic carbon supply in fresh waters. New Phytol. 87: 269—283.

Arber, A. (1920) Water Plants: A Study of Aquatic Angiosperms. Cambridge University Press, Cambridge. 436 pp.

Armstrong, W. (1978) Root aeration in the wetland condition. *In*: Hook, D. C. & Crawford, R. M. M. (eds.), Plant Life in Anaerobic Environments, Ann Arbor Science Publishers Inc., Ann Arbor, Michigan. pp. 269—297.

Bain, J. T. & Proctor, M. C. F. (1980) The requirement of aquatic bryophytes for free CO_2 as an inorganic carbon source: Some experimental evidence. New Phytol. 86: 393—400.

Barko, J. W. & Smart, R. M. (1981) Comparative influences of light and temperature on the growth and metabolism of selected submersed freshwater macrophytes. Ecol. Monogr. 51: 219—235.

Barth, H. (1957) Aufnahme und Abgabe von CO_2 und O_2 bei submersen Wasserpflanzen. Gewässer Abwässer 4(17/18): 18—81.

Beer, S. & Waisel, Y. (1979) Some photosynthetic carbon fixation properties of seagrasses. Aquatic Bot. 7: 129—138.

Beer, S. & Wetzel, R. G. (1981) Photosynthetic carbon metabolism in the submerged aquatic angiosperm *Scirpus subterminalis*. Plant Sci. Lett. 21: 199—207.

Beer, S. & Wetzel, R. G. (1982a) Photosynthesis in submersed macrophytes of a temperate lake. Plant Physiol. 70: 488—492.

Beer, S. & Wetzel, R. G. (1982b) Photosynthetic carbon fixation pathways in *Zostera* and three Florida seagrasses. Aquatic Bot. 13: 141—146.

Beer, S., Eshel, A. & Waisel, Y. (1977) Carbon metabolism in seagrasses. I. The utilization of exogenous inorganic carbon species in photosynthesis. J. Exp. Bot. 28: 1180—1189.

Beer, S., Shomer-Ilan, A. & Waisel, Y. (1980) Carbon metabolism in seagrasses. II. Patterns of photosynthetic CO_2 incorporation. J. Exp. Bot. 31: 1019—1026.

Bertani, A., Brambilla, I. & Menegus, F. (1980) Effect of anaerobiosis on rice seedlings: Growth, metabolic rate, and fate of fermentation productions. J. Exp. Bot. 31: 325—331.

Black, M. A., Maberly, S. C. & Spence, D. H. N. (1981) Resistances to carbon dioxide fixation in four submerged freshwater macrophytes. New Phytol. 89: 557—568.

Blotnick, J. R., Rho, J. & Gunner, H. B. (1980) Ecological characteristics of the rhizosphere microflora of *Myriophyllum heterophyllum*. J. Environ. Qual. 9: 207—210.

Bowes, G. (1985) Pathways of CO_2 fixation by aquatic organisms. *In*: Wetzel, R. G. (Ed.), Inorganic Carbon Uptake by Aquatic Photosynthetic Organisms. Amer. Soc. Plant Physiol., Rockville, Maryland. pp. 187—210.

Bowes, G., Holaday, A. S. & Haller, W. T. (1979) Seasonal variation in the biomass, tuber density, and photosynthetic metabolism of *Hydrilla* in three Florida lakes. J. Aquat. Plant Manage. 17: 61—65.

Bristow, J. M. (1969) The effects of carbon dioxide on the growth and development of amphibious plants. Can. J. Bot. 47: 1803—1807.

Bristow, J. M. (1974) Nitrogen fixation in the rhizosphere of freshwater angiosperms. Can. J. Bot. 52: 217—221.

Broecker, W. S. (1974) *Chemical Oceanography*. Harcourt, Brace, Jovanovich, New York.

Brown, J. M. A., Dromgoole, F. I. Towsey, M. W. & Browse, J. (1974) Photosynthesis and photorespiration in aquatic macrophytes. *In*: Bieleski, R. L., Ferguson, A. R. & Cresswell, M. M. (Eds.), Mechanisms of Regulation of Plant Growth, Bull. R Soc. New Zealand (Wellington), 12, pp. 243—249.

Browse, J. A., Dromgoole, F. I. & Brown, J. M. A. (1977) Photosynthesis in the aquatic macrophyte *Egeria densa*. I. ^{14}C fixation at natural CO_2 concentrations. Aust. J. Plant Physiol. 4: 169—176.

Browse, J. A., Brown, J. M. A. & Dromgoole, F. I. (1979) Photosynthesis in the aquatic macrophyte *Egeria densa*. II. Effects of inorganic carbon conditions on ^{14}C fixation. Aust. J. Plant Physiol. 6: 1—9.

Browse, J. A., Brown, J. M. A. & Dromgoole, F. I. (1980) Malate synthesis and metabolism during photosynthesis in *Egeria densa* Planch. Aquatic Bot. 8: 295—305.

Carr, J. L. (1969) The primary productivity and physiology of *Ceratophyllum demersum*. II. Micro primary productivity, pH, and the P/R ratio. Aust. J. Mar. Freshwat. Res. 20: 127—142.

Chang, T. P. (1980) Mucilage sheath as a barrier to carbon uptake in a cyanophyte, *Oscillatoria rubescens* D.C. Arch. Hydrobiol. 88: 128—133.

Crawford, R. M. M. (1978) Metabolic adaptations to anoxia. *In*: Hook, D. C. & Crawford, R. M. M. (eds.), Plant Life in Anaerobic Environments, Ann Arbor Science Publ. Inc., Ann Arbor, Michigan. pp. 119—136.

Dacey, J. W. H. (1981) Pressurized ventilation in the yellow waterlily. Ecology 62: 1137—1147.

Dale, H. M. (1981) Hydrostatic pressure as the controlling factor in the depth distribution of Eurasian watermilfoil, *Myriophyllum spicatum* L. Hydrobiologia 79: 239—241.

Danckwets, P. V. (1970) *Gas-liquid Reactions*. McGraw Hill, New York.

Dickerman, J. A. & Wetzel, R. G. (1985) Clonal growth in *Typha latifolia* L.: Population dynamics and demography of the ramets. J. Ecol. 73: 535—552.

Emerson, S. (1975) Chemically enhanced CO_2 gas exchange in a eutrophic lake: A general model. Limnol. Oceanogr. 20: 743—753.

26

Filbin, G. J. (1980) Photosynthesis, photorespiration and primary productivity in floating, floating leaved and emergent aquatic plants. Ph.D. Dissertation, Wayne State Univ., Detroit.

Gates, D. M. (1980) Biophysical Ecology. Springer-Verlag, New York. 611 pp.

Gessner, F. (1959) Hydrobotanik. Die Physiologischen Grundlagen der Pflanzenverbreitung im Wasser. II. Stoffhaushalt. Berlin, VEB Deutscher Verlag der Wissenschaften. 701 pp.

Glime, J. M., Wetzel, R. G. & Kennedy, B. J. (1982) The effects of bryophytes on succession from alkaline marsh to *Sphagnum* bog. Amer. Midland Nat. 108: 209—223.

Goulder, R. (1980) Day-time variations in the rates of production by two natural communities of submerged freshwater macrophytes. J. Ecol. 58: 521—528.

Grace, J. B. & Wetzel, R. G. (1981a) Phenotypic and genotypic components of growth and production in *Typha latifolia*: Experimental studies in marshes of differing successional maturity. Ecology 62: 789—801.

Grace, J. B. & Wetzel, R. G. (1981b) Habitat partitioning and competitive displacement in cattails (*Typha*): Experimental field studies. Amer. Nat. 118: 463—474.

Grace, J. B. & Wetzel, R. G. (1982a) Niche differentiation between two plant species: *Typha latifolia* and *Typha angustifolia*. Can. J. Bot. 60: 46—57.

Grace, J. G. & Wetzel, R. G. (1982b) Variations in growth and reproduction within two rhizomatous plant species: *Typha latifolia* and *Typha angustifolia*. Oecologia 53: 258—263.

Haslam, S. M. (1980) River Vegetation. Its Identification, Assessment, and Management. Cambridge Univ. Press, Cambridge. 154 pp.

Helder, R. J. & Harmelen, M. V. (1982) Carbon assimilation pattern in the submerged leaves of the aquatic angiosperm *Vallisneria spiralis* L. Acta Bot. Neerlandica 31: 281—295.

Helder, R. J. & Zanstra, P. E. (1977) Changes of the pH at the upper and lower surface of bicarbonate assimilating leaves of *Potamogeton lucens* L. Proc. Koninklijke Nederl. Akad. Wetenschappen, Ser. C. Biol. Med. Sci. 80: 421—436.

Helder, R. J., Prins, H. B. A. & Schuurmans, J. (1974) Photorespiration in leaves of *Vallisneria spiralis*. Proc. Koninklijke Nederl. Akad. Wetenschappen, Ser. C. Biol. Med. Sci. 77: 338—344.

Helder, R. J., Boerman, J. & Zanstra, P. E. (1980) Uptake pattern of carbon dioxide and bicarbonate by leaves of *Potamogeton lucens* L. Proc. Koninklijke Nederl. Akad. Wetenschappen, Ser. C. Biol. Med. Sci. 83: 151—166.

Holiday, A. S. & Bowes, G. (1980) C_4 acid metabolism and dark CO_2 fixation in a submersed aquatic macrophyte (*Hydrilla verticillata*). Plant Physiol. 65: 331—335.

Hough, R. A. (1974) Photorespiration and productivity in submersed aquatic vascular plants. Limnol. Oceanogr. 19: 912—927.

Hough, R. A. & Wetzel, R. G. (1972) A ^{14}C-assay for photorespiration in aquatic plants. Plant Physiol. 49: 987—990.

Hough, R. A. & Wetzel, R. G. (1977) Photosynthetic pathways of some aquatic plants. Aquatic Bot. 3: 297—303.

Hough, R. A. & Wetzel, R. G. (1978) Photorespiration and CO_2 compensation point in *Najas flexilis*. Limnol. Oceanogr. 23: 719—724.

Hutchinson, G. E. (1957) A Treatise on Limnology. I. Geography, Physics, and Chemistry. New York, John Wiley & Sons, 1015 pp.

Hutchinson, G. E. (1967) A Treatise on Limnology. II. Introduction to Lake Biology and the Limnoplankton. New York, John Wiley & Sons, 1115 pp.

Hutchinson, G. E. (1975) A Treatise on Limnology. Vol. III. Limnological Botany. New York, John Wiley & Sons. 660 pp.

Hynes, H. B. N. (1970) The Ecology of Running Waters. Toronto, University of Toronto Press. 555 pp.

Jaworski, G. H. M., Talling, J. F. & Heaney, S. I. (1981) The influence of carbon dioxide-depletion on growth and sinking rate of two planktonic diatoms in culture. Br. Phycol. J. 16: 395—410.

Jones, M. B. & Milburn, R. T. (1978) Photosynthesis in papyrus (Cyperus papyrus L.). Photosynthetica 12: 197—199.

Joshi, M. M. & Holllis, J. P. (1977) Interaction of Beggiatoa and rice plant: Detoxification of hydrogen sulfide in the rice rhizosphere. Science 195: 179—180.

Kadono, Y. (1980) Photosynthetic carbon sources in some Potamogeton species. Bot. Mag. Tokyo 93: 185—193.

Katayama, T. (1961) Studies on the intercellular spaces in rice. Crop Sci. Soc. Japan Proc. 29: 229—233.

Keeley, J. E. (1981) Isoetes howellii: A submerged aquatic CAM plant? Amer. J. Bot. 68: 420—424.

Keeley, J. E. (1982) Distribution of diurnal acid metabolism in the genus Isoetes. Amer. J. Bot. 69: 254—257.

Kremer, B. P. (1981) Aspects of carbon metabolism in marine macroalgae. Oceanogr. Mar. Biol. Ann. Rev. 19: 41—94.

Littlefield, L. & Forsberg, C. (1965) Absorption and translocation of phosphorus-32 by Chara globularis Thuill. Physiol. Plant. 18: 291—296.

Lloyd, N. D. H., Canvin, D. T. & Bristow, J. M. (1977) Photosynthesis and photorespiration in submerged aquatic plants. Can. J. Bot 55: 3001—3005.

Loats, K. V., Noble, R. & Takemoto, B. (1981) Photosynthesis under low-level SO_2 and CO_2 conditions in three duckweed species. Bot. Gaz. 142: 305—310.

Losee, R. F. & Wetzel, R. G. (1983) Selective light attenuation by the periphyton complex. In: Periphyton of Freshwater Ecosystems. R. G. Wetzel, Ed. Developments in Hydrobiology 17: 89—96.

Lucas, W. J. (1975) Photosynthetic fixation of [14]carbon by internodal cells of Chara corallina. J. Exp. Bot. 26: 331—346.

Lucas, W. J. & Berry, J. A. (Eds.) (1985) Inorganic Carbon Uptake by Aquatic Photosynthetic Organisms. Amer. Soc. Plant Physiol., Rockville, Maryland. 494 pp.

Lucas, W. J., Tyree, M. T. & Petrov, A. (1978) Characterization of photosynthetic [14]carbon assimilation by Potamogeton lucens L. J. Exp. Bot. 29: 1409—1421.

McNaughton, S. J. (1966) Ecotype function in the Typha community-type. Ecol. Mongr. 36: 297—325.

McNaughton, S. J. (1969) Genetic and environmental control of glycolic acid oxidase activity in ecotypic populations of Typha latifolia. Amer. J. Bot. 56: 37—41.

McNaughton, S. J. & Fullem, L. W. (1969) Photosynthesis and photorespiration in Typha latifolia. Plant Physiol. 45: 703—707.

Mitsui, S. & Kurihara, K. (1962) On the utilization of carbon in fertilizers through rice roots under pot experimental conditions. Soil Sci. Plant Nutr. 8: 226—233.

Moeller, R. E. (1978) Carbon-uptake by the submerged hydrophyte Utricularia purpurea. Aquatic Bot.. 5: 209—216.

Monteith, J. L., Szeic, G. & Yabuki, K. (1964) Crop phytosynthesis and the flux of carbon dioxide below the canopy. J. Appl. Ecol. 1: 321—337.

Morris, J. T. (1980) The nitrogen uptake kinetics of Spartina alterniflora in culture. Ecology 61: 1114—1121.

Moss, D. N., Musgrave, R. B. & Lemon, E. R. (1961) Photosynthesis under field conditions. III. Some effects of light, carbon dioxide, temperature, and soil moisture on photosynthesis, respiration and transpiration of corn. Crop. Sci. 1: 83—87.

Murata, Y., Osada, A. & Iyama, J. (1957) Physiological roles of carbon dioxide in plants. Agric. Hortic. 32: 11—14.

Otsuki, A. & Wetzel, R. G. (1972) Coprecipitation of phosphate with carbonates in a marl lake. Limnol. Oceanogr. 17: 763—767.

28

Otsuki, A. & Wetzel, R. G. (1973) Interaction of yellow organic acids with calcium carbonate in freshwater. Limnol. Oceanogr. 18: 490—493.

Otsuki, A. & Wetzel, R. G. (1974) Calcium and total alkalinity budgets and calcium carbonate precipitation of a small hard-water lake. Arch. Hydrobiol. 73: 14—30.

Payne, F. C. (1982) Influence of hydrostatic pressure on gas balance and lacunar structure in *Myriophyllum spicatum* L. Ph.D. Dissertation, Michigan State Univ. 109 pp.

Penhale, P. A. & Wetzel, R. G. (1982) Structural and functional adaptations of eelgrass (*Zostera marina* L.) to the anaerobic sediment environment. Can. J. Bot. 61: 1421—1428.

Prins, H. B. A. & Walsarie-Woff, R. (1974) Photorespiration in leaves of *Vallisneria spiralis*. The effect of oxygen on the carbon dioxide compensation point. Proc. Akad. van Wetensc. Amsterdam (Ser. C). 77: 239—245.

Prins, H. B. A., Snel, J. F. H. Helder, R. I. & Zanstra, P. E. (1980) Photosynthetic HCO_3^- utilization and OH^- excretion in aquatic angiosperms. Light-induced pH changes at the leaf surface. Plant Physiol. 66: 818—822.

Prins, H. B. A., Snel, J. F. H. & Zanstra, P. E. (1982a) The mechanism of bicarbonate utilization. *In*: Symoens, J. J., Hooper, S. S. & Compère, P. (eds.) Studies on Aquatic Vascular Plants. Royal Bot. Soc. Belgium, Brussels. pp. 120—126.

Prins, H. B. A., O'Brien, J. & Zanstra, P. E. (1982b) Bicarbonate utilization in aquatic angiosperms, pH and CO_2 concentration at the leaf surface. *In*: Symoens, J. J., Hooper, S. S. & Compère, P. (eds.) Studies on Aquatic Vascular Plants. Royal Bot. Soc. Belgium, Brussels. pp. 112—119.

Prins, H. B. A., Snel, J. F. H., Zanstra, P. E. & Helder, R. J. (1982c) The mechanism of bicarbonate assimilation by the polar leaves of *Potamogeton* and *Elodea*. CO_2 concentration at the leaf surface. Plant, Cell Environment 5: 207—214.

Raven, J. A. (1970) Exogenous inorganic carbon sources in plant photosynthesis. Biol. Rev. 45: 167—221.

Raven, J. A. (1981) Nutritional strategies of submerged benthic plants: The acquisition of C, N and P by rhizophytes and haptophytes. New Phytol. 83: 1—30.

Raven, J. A. (1984) Energetics and Transport in Aquatic plants. A. R. Liss, Inc., New York. 587 pp.

Raven, J. A. & Beardall, J. (1981) Carbon dioxide as the exogenous inorganic carbon source for *Batrachospermum* and *Lemanea*. Br. Phycol. J. 16: 165—175.

Riber, H. H. & Wetzel, R. G. (1987) Boundary-layer and internal diffusion effects on phosphorus fluxes in lake periphyton. Limnol. Oceanogr. 32: 1181—1194.

Roelofs, J. G. M., Schuurkes, J. A. A. R. & Smits, A. J. M. (1984) Impact of acidification and eutrophication on macrophyte communities in soft waters. II. Experimental studies. Aquatic Bot. 18: 389—411.

Rodewald-Rudescu, L. (1974) Das Schilfrohr *Phragmites communis* Trinius. *In*: Die Binengewässer 27. 302 pp.

Ruttner, F. (1947) Zur Frage der Karbonatassimiliation der Wasserpflanzen. I. Die beiden Haupttypen der Kohlenstoffaufnahme. Öst. Bot. Z. 94: 265—294.

Ruttner, F. (1948) Zur Frage der Karbonatassimilation der Wasserpflanze II. Das Verhalten von *Elodea canadensis* und *Fontinalis antipyretica* in Lösungen von Natrium-bzw. Kaliumbikarbonat. Öst. Bot. Z. 95: 208—238.

Ruttner, F. (1960) Über die Kohlenstoffaufnahme bei Algen aus der Rhodophyceen-Gattung *Batrachospermum*. Schweiz Z. Hydrol. 22: 280—291.

Sale, P. J. M. & Wetzel, R. G. (1983) Growth and metabolism of *Typha* species in relation to cutting treatments. Aquatic Bot. 15: 321—334.

Salvucci, M. E. & Bowes, G. (1981) Induction of reduced photorespiratory activity in submersed and amphibious aquatic macrophytes. Plant Physiol. 67: 335—340.

Sand-Jensen, K. (1978) Metabolic adaptation and vertical zonation of *Littorella uniflora* (L.). Aschers. and *Isoetes lacustris* L. Aquatic Bot. 4: 1—10.

Schwoerbel, J. & Tillmanns, G. C. (1964a) Konzentrationsabhängige Aufnahme von wasserlöslichem PO₄-P bei submersen Wasserpflanzen. Naturwissenschaften 51: 319—320.

Schwoerbel, J. & Tillmanns, G. C. (1964b) Untersuchungen über die Stoffwechseldynamik in Fliessgewässern. I. Die Rolle höherer Wasserpflanzen: Callitriche hamulata Kütz. Arch. Hydrobiol. (Suppl.) 28: 245—258.

Schwoerbel, J. & Tillmanns, G. C. (1964c) Untersuchungen über die Stoffwechseldynamik in Fliessgewässern. II. Experimentelle Untersuchungen über die Ammoniumaufnahme und pH-Änderung im Wasser durch Callitriche hamulta Kutz. und Fontinalis antipyretica L. Arch. Hydrobiol. (Suppl.) 28: 259—267.

Sculthorpe, C. D. (1967) The Biology of Aquatic Vascular Plants. St. Martin's Press, New York. 610 pp.

Smith, F. A. & Walker, N. A. (1980) Photosynthesis by aquatic plants: Effects of unstirred layers in relation to the assimilation of CO_2 and HCO_3^- and to carbon isotopic discrimination. New Phytol. 86: 245—259.

Søndergaard, M. (1979) Light and dark respiration and the effect of the lacunal system on refixation of CO_2 in submerged aquatic plants. Aquatic Bot. 6: 269—283.

Søndergaard, M. & Sand-Jensen, K. (1979) Carbon uptake by leaves and roots of Littorella uniflora L. Aschers. Aquatic Bot. 6: 1—12.

Søndergaard, M. & Wetzel, R. G. (1980) Photorespiration and internal recycling of CO_2 in the submersed angiosperm Scirpus subterminalis. Can. J. Bot. 58: 591—598.

Spence, D. H. N. (1982) The zonation of plants in freshwater lakes. Adv. Ecol. Res. 12: 37—125.

Steemann Nielsen, E. (1944) Dependence of freshwater plants on quantity of carbon dioxide and hydrogen ion concentration. Dansk Bot. Ark. 11: 1—25.

Steemann Nielsen, E. (1947) Photosynthesis of aquatic plants with special reference to the carbon sources. Dansk Bot. Ark. 12: 5—71.

Stocking, C. R. (1956) Vascular conduction in submerged plants. In: Ruhland, W. (ed.), Handbuch der Pflanzenphysiologie. Band 3. Pflanze und Wasser. Berlin, Springer-Verlag. pp. 587—595.

Talling, J. F. (1973) The application of some electrochemical methods to the measurement of photosynthesis and respiration in fresh waters. Freshwat. Biol. 3: 355—362.

Talling, J. F. (1976) The depletion of carbon dioxide from lake water by phytoplankton. J. Ecol. 64: 79—121.

Tanaka, A., Kawano, K. & Yamaguchi, J. (1966) Photosynthesis, respiration, and plant type of the tropical rice plant. Int. Rice Res. Inst. Tech. Bull. 7: 46.

Tessenow, U. & Baynes, Y. (1978) Redoxchemische Einflüsse von Isoetes lacustris L. im Litoralsediment des Feldsees (Hochschwarzwald). Arch. Hydrobiol. 82: 20—48.

Titus, J. E. & Adams, M. S. (1979) Coexistence and the comparative light relations of the submersed macrophytes Myriophyllum spicatum L. and Vallisneria americana Michx. Oecologia 40: 273—286.

Tsuzuki, M. & Migachi, S. (1981) Effects of CO_2-concentration during growth and of ethotyzolamide on CO_2 compensation point in Chlorella. FEBS Lett. 103: 221—223.

Twilley, R. R., Brinson, M. M. & Davis, G. J. (1977). Phosphorus absorption, translocation, and secretion in Nuphar luteum. Limnol. Oceangr. 22: 1022—1032.

Ultsch, G. R. & Anthony, D. S. (1973) The role of the aquatic exchange of carbon dioxide in the ecology of the water hyacinth (Eichhornia crassipes). Florida Sci. 36: 16—22.

Valanne, N., Aro, E. M. & Rintamaki, E. (1982) Leaf and chloroplast structure of two aquatic Ranunculus species. Aquatic Bot. 12: 13—22.

Van, T. K., Haller, W. T. & Bowes, G. (1976) Comparison of the photosynthetic characteristics of three submersed aquatic plants. Plant Physiol. 58: 761—768.

30

Verduin, J. (1975) Rate of carbon dioxide transport across air-water boundaries in lakes. Limnol. Oceanogr. 20: 1052—1053.

Weaver, C. I. & Wetzel, R. G. (1980) Carbonic anhydrase levels and internal lacunar CO_2 concentration in aquatic macrophytes. Aquatic Bot. 8: 173—186.

Weiler, R. G. (1974) Exchange of carbon dioxide between the atmosphere and Lake Ontario. J. Fish. Res. Bd. Can. 31: 329—332.

Westlake, D. F. (1967) Some effects of low-velocity currents on the metabolism of aquatic macrophytes. J. Exp. Bot. 18: 187—205.

Westlake, D. F. et al. (1980) Primary production. In: Le Cren, E. D. & Lowe-McConnell, R. H. (eds.), The Functioning of Freshwater Ecosystems. Cambridge, Cambridge Univ. Press, pp. 141—246.

Wetzel, R. G. (1964) A comparative study of the primary productivity of higher aquatic plants, periphyton, and phytoplankton in a large, shallow lake. Int. Rev. ges. Hydrobiol. 49: 1—64.

Wetzel, R. G. (1969) Factors influencing photosynthesis and excretion of dissolved organic matter by aquatic macrophytes in hard-water lakes. Verh. Int. Ver. Limnol. 17: 72—85.

Wetzel, R. G. (1975) Limnology. Saunders, Philadelphia. 743 pp.

Wetzel, R. G. (1979) The role of the littoral zone and detritus in lake metabolism. Arch. Hydrobiol. Beih. Ergebn. Limnol. 13: 145—161.

Wetzel, R. G. (1983a) Limnology. 2nd Edition. Saunders, Philadelphia. 860 pp.

Wetzel, R. G. (Ed.) (1983b) Periphyton of Freshwater Ecosystems. Dr. W. Junk Publishers, The Hague. Developments in Hydrobiology 17, 346 pp.

Wetzel, R. G. (1983c) Attached algae-substrata interactions: fact or myth, and when and how? In: Wetzel, R. G. (ed.), Periphyton of Freshwater Ecosystems. Developments in Hydrobiology 17, pp. 207—215.

Wetzel, R. G. & Grace, J. B. (1983) Aquatic Plant Communities. In: Lemon, E. R. (ed.), CO_2 and Plants: The response of plants to rising levels of atmospheric carbon dioxide. AAAS Selected Symposium 84. Westview Press, Inc., Boulder, Colorado. pp. 223—280.

Wetzel, R. G. & Manny, B. A. (1972) Secretion of dissolved organic carbon and nitrogen by aquatic macrophytes. Verh. Int. Ver. Limnol. 18: 162—170.

Wetzel, R. G. & Hough, R. A. (1973) Productivity and role of aquatic macrophytes in lakes: An assessment. Pol. Arch. Hydrobiol. 20: 9—19.

Wetzel, R. G., Brammer, E. S. & Forsberg, C. (1984) Photosynthesis of submersed macrophytes in acidified lakes. I. Carbon fluxes and recycling of CO_2 in Juncus bulbosus L. Aquatic Bot. 19: 329—342.

Winter, K. (1978) Short-term fixation of [14]carbon by the submerged aquatic angiosperm Potamogeton pectinatus. J. Exp. Bot. 29: 1169—1172.

Wium-Andersen, S. (1971) Photosynthetic uptake of free CO_2 by the roots of Lobelia dortmanna. Physiol. Plant. 25: 245—248.

Wood, K. G. (1974) Carbon dioxide diffusivity across the air-water interface. Arch. Hydrobiol. 73: 57—69.

Wood, K. G. (1977) Chemical enhancement of CO_2 flux across the air-water interface. Arch. Hydrobiol. 79: 103—110.

Yoshida, S., Coronel, V., Parao, F. T. & de los Reyes, E. (1974) Soil carbon dioxide flux and rice photosynthesis. Soil Sci. Plant Nutr. 20: 381—386.

METHODS OF EXPLORATION AND ANALYSIS OF THE ENVIRONMENT OF AQUATIC VEGETATION

H. L. GOLTERMAN, R. S. CLYMO, E. P. H. BEST & J. LAUGA

INTRODUCTION

Aquatic plants are found wherever there is liquid water and at least a modest flux of light. Sometimes they are conspicuous: 700 km^2 of head-high *Typha dominguensis* around the African L. Chilwa or 5000 km^2 of equally tall *Typha* and *Phragmites* in the Danube delta or the even larger Nile Sudd. Many freshwaters appear, at first sight, to lack plants. A filter and microscope will rapidly show that such waters usually do contain large populations of microscopic algae. If the populations are large enough they give the water colour. Even in the unpromising conditions of Antarctica there are seasonal lakes, frozen solid for most of the year, which contain an astonishingly active micro- and macro-flora. In contrast there are peat bogs, which consist of large masses of water 'gelled' by about 10% of organic matter. In the latter case the plants are so conspicuous that the system is rarely considered to be (what it actually is) a body of freshwater.

Macroscopic water plants may be grouped in several ways. A convenient one is to distinguish emergent taxa (such as *Phragmites* and *Typha*), floating-leaved but rooted ones (*Nuphar* and *Nymphaea*), free-floating ones with unattached roots (*Lemna, Eichhornia, Salvinia*) submerged rooted ones (*Myriophyllum, Isoetes*) and, lastly, the free floating rootless mosses (*Fontinalis*). Submerged rooted plants are sometimes called rhizophytes. Another group — haptophytes — is submerged and attached by modified organs other than roots. Macrophytes differ in conspicuousness: some form large floating mats (*Eichhornia*); others occur as scattered individuals of small stature in deep water and are therefore inconspicuous. One important feature of all but the emergent macrophytes is that the water supports most of the plant weight. Gas spaces may enhance this effect. It may well be that the structure of such plants is determined mainly by the need to transport gases and to photosynthesize, and only slightly by the need for structural support which is so important in terrestrial plants. Reviews of

31

J. J. Symoens (ed.), Vegetation of inland waters. ISBN 90—6193—196—7.
© *1988, Kluwer Academic Publishers, Dordrecht. Printed in the Netherlands.*

these and many other features of water plants are given by *inter alia* Sculthorpe (1971), Denny (1980) and Best (1982a).

The species-abundance and changes in most aquatic vegetation types are to a large extent controlled by chemical, physical and biotic factors. Aquatic vegetation can sometimes be a substantial nuisance — *Eichhornia, Salvinia* and *Elodea canadensis* are examples. The plants often decline as rapidly as they became abundant, affected by several, sometimes interacting factors.

In many cases the particular requirement of individual plant species for light, nutrients, substrate, a particular range of the temperature etc. in conjunction produces a distinct zonation.

Several examples illustrate some of the complexities.

The macrophytic vegetation of an upland stream flowing from a granitic catchment is almost completely different from that in a lowland chalk stream, which in turn is different from that fringing a large lowland lake in a catchment on soft sedimentary rocks. Water movement, water depth, and water clarity may partly explain the difference, but water and sediment chemistry are probably more important. Some macrophytes — *Lemna* is an obvious example — absorb solutes solely from the water in which they float. Others, such as *Typha* and *Phragmites australis*, have a large rhizome and root system in the sediment. They absorb solutes from the interstitial water mainly through their roots. Taxa such as *Vallisneria, Ceratophyllum demersum* and *Potamogeton* are intermediate in character (Denny 1972). The range of tolerance of chemical differences is enormous. Some species are restricted in their occurrence: *Isoetes* and *Lobelia* to oligotrophic, usually acid, water for example, while *Myriophyllum spicatum* is found in relatively eutrophic water. Others, such as *Phragmites australis*, have a wider range. One must, however, be cautious of interpreting such observations in terms of nutrient supply alone. For example, oligotrophic waters are usually clear whilst eutrophic waters are often turbid: the water types differ in light climate as well as in chemistry. In water rich in detritus or silt, plants with finely dissected leaves may be smothered by deposited particles. A dense growth of periphyton may have the same effect. *Phragmites australis*, which ranges in height from a few centimetres in acid nutrient-poor water to 7 metres or more in the Rhone and Danube deltas, is known to consist of a polyploid series. Other examples are given by Denny (1980). In general it is the combined effects of genotype and environment which control the occurrence of individual species.

In some cases a small difference in water chemistry is correlated with a large change in the macrophytic vegetation: Lambert (1951) noted that in tidal (but fresh) water *Scirpus lacustris* rather than *Typha angustifolia* was favoured. When long-lived perennials are involved one must beware of attributing the present distribution to present environmental conditions: Chapman (1964) records the occurrence of very small *Phragmites australis* plants at fairly high density at one small part of Coom Rigg bog. Examination of the peat showed that these were the last survivors of a reed swamp which surrounded a small lake 5000 years ago. A similar example is noted by Summerfied (1972).

Broad generalisations of these kinds are often based on observation of obvious factors such as water velocity, or less obvious but easily measured ones, such as pH, temperature, and electrical conductance of the water. But some very important factors, such as sediment chemistry, are not easily measured.

Finally, the strong correlation between the effects of concentrations of different substances and water movements, water depth (with its effect on light flux) and competition, prevent exact interpretation of the effects of single chemical factors on macrophyte growth. Such differences can only be made from controlled experiments — if at all.

Obviously, it is essential to define the objective(s) of work before measurements or experiments are begun. In the following we consider, therefore, a few of the consequent problems of sampling and analysis.

GETTING TO SAMPLING SITES

Wading is possible for short distances in waters up to 1 m depth. In temperate regions low temperatures during winter and in the tropics water-borne diseases will set constraints. Care should be taken not to sample where the sediments are disturbed by the sampler. Carrying equipment may be difficult. Collapsible (portable) rowing boats, of which the pneumatic type is probably the commonest, may be a good solution. They permit the transport of electrical measuring equipment for pH, oxygen concentration, conductivity and temperature, so that the variables can be measured *in situ* and often continuously. There is no practical limit to the number of samples than can be transported in a boat, although transport on land and the time required for the analyses will set limits.

With increasing site dimensions the boats have to be bigger and for large lakes large boats are needed for safety and speed. It must be realized that shallow lakes may have deeper parts and limnologists may drown just as easily as statisticians in a lake of average depth 10 cm. Safety measures are mandatory, particularly with small boats: life belts or jackets, somebody on the shore aware of the estimated time of return and so on. Radio contact is desirable.

A 'moon pool' i.e. a central hole (with sides of course) in the boat or raft makes the handling of sampling equipment far easier and the craft more stable.

Running waters can be sampled for dissolved compounds from the shore or from bridges. Small rivers can be waded. For large ones a boat with a powerful and reliable motor is essential.

MORPHOMETRIC AND HYDROLOGICAL VARIABLES

Water flow

An important factor for vegetation studies in rivers is water flow. The following variables and their relations must be considered:

Water height $(h$; normally in m);
Profile $(P$; normally in m^2);
Velocity $(v$; normally in m sec^{-1});
Flow rate $(Q$; normally in m^3 sec^{-2});

The profile (or cross-section at a given place) depends on the physical form of the river and water height. In hydrology water height is often called stage. Stage is the variable which controls the period and intensity of immersion of the vegetation. Velocity — or, for a given profile, flow rate — will control which vegetation may develop (Wetzel 1975, Dawson 1978). Measurement of these variables is often a specialized job; pitfalls are numerous. Often data can be obtained from hydrological surveys. Very often — especially for smaller streams — data must be obtained by the botanist himself. Some indications and guidelines follow here; for more details see Ven Te Chow (1964).

Stage

The term stage refers to the water-surface height at a point along the stream, measured above an arbitrarily fixed point. Readings are either taken at intervals by an observer or are recorded continuously. In use are staff, chain, tape and wire gauges plus pressure devices or transmitters and crest-stage indicators. A staff gauge is a graduated scale set in a stream and fastened to a wall, pier, bridge etc. The level is read with proper allowance for fluctuations and meniscus. In chain, tape and wire stages the stage is determined by lowering a weight to the water surface. The distance that the weight descends before touching the water is measured from a graduated scale, a reference bead etc. These gauges are easily installed and safe from damage by boats, flood-transported debris etc. It is often difficult to observe the moment of contact. The weight can be made a part of an electric circuit with the return through the water. The deflection of a galvanometer needle or an audible signal can be used as indicator of contact. In a float gauge a float is fastened to one end of a tape. The other end is mounted on a wheel and may drive the recorder directly. The stage may be converted to pressure by means of a cylinder and flexible diaphragm (pressure transmitter), however, the device is temperature-sensitive and needs careful calibration. The automatic instruments are based on floating gauges fitted to recorders, automatic printers etc. They give continuous records but their disadvantage is that they need protection from water-borne debris.

Velocity

Stream velocity can be measured by rotation or deflection of a mechanical device by floats or by chemical methods. Rotating current meters may be vertical or horizontal. Rotation around the vertical axle takes place by means of vanes or cups; around the horizontal axle by means of screw- or propeller-shaped blades. Examples of the first type are the Price (and derived) meters and of the second the Neyrpic (France) and Ott (F.R.G.) meters. Calibration is necessary and should be repeated, especially when they are wearing out. In humus-rich agressive waters and other agressive waters even stainless steel meters may corrode in a period as short as 6 months. Problems can arise if the current can reverse in the course of time, as may happen in artificial waterways.

Different floats can be used with different degrees of sophistication. They are discussed here in order of increasing reliability.

(1) Any object (cork, wood) floating on the water (but in good contact with the water) may be used ('Pooh Sticks', Milne 1928). Distance should be marked on the shoreline and the time between passing two marks should be noted. The float may easily be obstructed by branches, by bends in the river, or be deflected by wind.
(2) The object may be made to float at given depth by attaching a heavy weight to a cork by means of specific length of cable. These devices measure velocity at a single depth only — at the surface or at the depth of the weight respectively.
(3) Some sort of integration can be made by using a long rod (e.g. electric installation tube) closed at one end by a heavy stopper. If the lower end is near the bottom of the waterway the float probably approaches mean velocity. Floats are best in long and straight channels.

If the stream is not too large the sudden input of e.g. NaCl may be used to mark and define a certain water mass. At a fixed distance the passage of the salt can be detected by measuring the hump in a record of conductivity or of chloride concentration (measured with a silver electrode immersed in the river). Salt velocity or dilution may be measured. The same can be done with fluorescein; the method is extremely sensitive if a fluorescence meter is available, but a spectrophotometer may be used instead, though with reduced sensitivity. A preliminary test has to be carried out to estimate the amount to be introduced. Chemical-electric methods include oxygen polarography, the hot-wire anemometer, electro-voltage generation and supersonic wave. They require skilled operators.

Especially for low current velocities a new method 'Laser Doppler Velocimetry' is coming into operation. In Laser Doppler Velocimetry particles moving with the fluid (either natural or added) are illuminated with a focused Laser beam and become sources of scattered light. The velocity of these particles is determined from the change in frequency (Doppler shift) of the scattered light due to the movement of the particles when observed by a stationary detector. From the particles velocity the fluid velocity is inferred. A laser is used as a light source because it is easily focused and is coherent. Some effects of low current velocities are described by Westlake (1967) and Madsen & Sondergaard (1983). For methodological details see Durrani & Greated (1977) and Madsen & Warncke (1983).

Depth

Depth can be measured by means of a wading rod (small streams) cable and weight (larger streams, either from a boat or bridge — care should be taken to make sure that the line is vertical) or with an echo sounder. The last solution is the best for larger waters as it gives a continuous record and can be used to survey a whole lake quickly. Cheap equipment is nowadays available or can easily be borrowed. Normally the size of the stream determines what kind of equipment is needed.

Discharge

Discharge can be calculated as velocity times profile ($v \times P$: units $m^3 s^{-1}$). When stream velocity at a given observation point is measured sufficiently often, the stage-discharge relation can be found. This is a complex relation, which may be permanent, if channel characteristics are not changing; otherwise it is a varying relation which has to be recalibrated often (shifting control).

REMOTE SENSING (TELEDETECTION)

Information obtained from a distance is available in two forms: aerial photographs and digital data from satellites.

Panchromatic 'black and white' prints are commonest. They have relatively high resolution (often better than 1 m) and in many cases the same area has been recorded at several times over several decades. Individual prints are not expensive. Color prints are often made nowadays, but they are more expensive and may lack the resolving power of monochrome prints. But the color contrast may more than compensate for the disadvantages. 'False color' prints, of which those rendering the infrared visible are the most common, may be particularly valuable once the initially confusing change of color conventions is mastered. Such prints give information not observable at all by the unaided human eye.

The first non-military satellite designed to observe the Earth's surface — Landsat 1 — was put into orbit by the U.S. in 1972. Others followed and will continue to follow. The information provided by a satellite is not a (virtually instantaneous) photograph but a sequence of irradiance measurements at several wavelengths as the satellite scans the Earth's surface. It is thus more closely related to the production of a television picture by a scanning spot than to a photograph. The information received on Earth is stored on magnetic tape and it is these tapes which, after some editing, are copied and are the primary source that is sold. Coverage of the whole of the Earth's surface is available, repeated every few weeks, though cloud cover makes many of the scenes useless. The spatial resolution (pixel = smallest area analyzed) is about 80×80 m in Landsat images, and is thus very much less than that of most aerial photographs. The resolution of more recent satellites is better, but the limits are set more by political acceptability than by technology. The magnetic tapes can be read only by computer but it is also possible to obtain false color prints. They are produced by converting irradiance values into corresponding colors or shades of grey. The tape of a Landsat 'scene', or the corresponding print, is more expensive than a single aerial photograph and is of much lower resolution. But it does cover a much larger area, and the tape may be processed in a variety of ways to enhance or highlight chosen sorts of features. Aerial photographs and satellite images are, therefore, to a large extent complementary and not competitors.

Water quality and chlorophyll

Remote sensing of inland waters allows a synoptic view unobtainable by other means, and the first applications were in this field. Most of the work, however, has been applied to the seas and oceans. Several satellites have been designed primarily for this work. Observation of the way in which the flux of light at different wavelengths changes across a tract of water allows deductions to be made about the nature and concentration of the elements contained therein. The concentration of suspended matter can be estimated with mediocre accuracy, whilst for chlorophyll the problem is more complex. The following are some of the problems.

(1) The relation between reflectance and the concentration of suspended matter appears to be generally curvilinear (Munday & Alfoldi 1979). The turbidity of the water is caused by attenuation of light by both organic and inorganic components of the suspended matter. The relation between concentration and reflectance depends on the nature and amount of the surfaces of the suspended matter. The results from one site cannot, therefore, be generalised to all others.

(2) The measurement of chlorophyll requires that the sensor has precise performance in the following ways.

First, the chlorophyll 'particles' are usually less abundant but individually darker than the non-chlorophyll particles in the same environment. Precise results can be obtained only if the detector has a high resolution.

Secondly, chlorophyll absorbs light strongly over a narrow part of the spectrum only, while the rest of the suspended matter absorbs over a broad band of the spectrum. The detector response should be limited to the narrow band in which chlorophyll absorbs light.

Thirdly, the spatial variation of chlorophyll is at a smaller scale than that of suspended matter in general, so the detector must have a higher resolution than it needs for suspended matter in general.

These three constraints are at least additive and result in a relatively low signal : noise ratio for chlorophyll. Only high concentrations can be estimated with any certainty.

In spite of these restrictions some interesting results have been obtained for continental waters. In S.E. Australia the concentration of chlorophyll in lakes has been estimated, using multiple regression analysis (Carpenter & Carpenter 1983). The variables in their model were the Landsat spectral windows 4 (green) and 5 (red), the solar elevation, and the time at which the satellite passed. With logarithmically transformed pigment concentration as the dependent variable the square of the multiple regression coefficient was 0.9. The same article gives an extensive bibliography of the use of remote sensing in water quality studies. A critical account, based on the chromatic properties of water, is given by Bukata *et al.* (1983). These uses of remote sensing are still at the exploratory stage, but they do show great promise.

Wetlands, including marshes

Remote sensing is already in common use for the mapping and inventory of wetlands, though most published work deals with coastal wetlands. Soil type, sensed in this way, has been used as the basis for an inventory (Anderson *et al.* 1976) but only at a rather crude level. A more refined classification, specially designed for wetlands, has been used with some success in the U.S. (Cowardin *et al.* 1979).

False-color infrared aerial photographs have been used for an initial survey — reconnaissance — of emerging and floating vegetation (Carter 1977, 1982). But, for submerged vegetation, infrared sensing is not usually suitable because infrared radiation is so strongly absorbed by water that it penetrates only to a very shallow depth. Color aerial survey is probably best, but even this is seriously hindered by even small amounts of suspended matter. If only one survey is possible then it is probably best made in the autumn (if in temperate regions). Repetition of the survey at different seasons is much to be preferred as it improves discrimination and increases the reliability of identifications. With cover repeated over many years it is possible to follow the vegetation changes over large areas and long times, as has been done for the Great Dismal Swamp in Virginia, USA, between 1938 and 1952 (Garrett & Carter 1977).

The use of satellite data for mapping and inventory is still at the research stage. There is not, as yet, any nation-wide study of proven validity. For example, Gammon *et al.* (1979) compared the results obtained by aerial survey and by Landsat imaging of the Great Dismal Swamp. The comparison was disappointing: the proportion of correct classifications was low because of the small scale mixture of vegetation types, the relatively large pixel size, and the close spectral resemblance between different plant species. Considerably greater success was achieved in the mapping and inventory of peat covered wetlands in Ontario, Canada (Pala 1984) but it is still the case that satellite images allow the identification of wetlands only if they are of reasonable extent and homogeneity. Sensing from more spectral bands may improve our ability to discriminate between similar vegetation types. Narrow linear belts of vegetation, such as riverine and lake-shore zones cannot be distiguished on 80 × 80 m pixels. Only improved resolution will solve this problem. For most vegetation survey purposes aerial photographs, particularly infrared false-color ones, are still the most useful source of information about wetlands. This is likely to remain true for some years yet: low cost, high resolution, and the ability to choose the time and scale at which the survey is made will remain advantages of this method.

But satellite sensing is barely twenty years old. There are immense possibilities for development and these are being eagerly exploited. Newer satellites — Thematic Mappers — record in 7 spectral bands, rather than the 4 bands of the Landsat series. The pixel of the French satellite Spot is 10×10 m — a sixty-four-fold improvement on the Landsat pixel. Finally, the digital nature of the recordings allows and encourages computer-processing which can be used to enhance boundaries, to choose the best criteria for classification, to selectively enhance particular features, and to automate inventory. We may reasonably expect enormous improvements in satellite sensing, but aerial survey should retain its complementary place.

PLANT SAMPLING UNDER WATER

Rapid estimation of the distribution and abundance of submerged macrophytes has been limited for a long time by the available sampling methods. Many different overboard samplers have been developed, their success being dependent upon exhaustive sampling and homogeneous macrophyte growth patterns (Forsberg 1959). Conventional limnological methods are laborious in the study of water bodies with heterogeneous populations of submerged macrophytes. Sampling by Ekman dredge, grapnel or other overboard sampling methods in sparsely populated regions yields misleading results due to the distances between neighbouring plants.

However, although at present SCUBA diving is increasingly used for research on submerged plants, its application depends strongly on the transparency within the water body concerned and thus on the visibility for the SCUBA diver. For instance, in the sea and in oligotrophic freshwaters its use has obvious merits. In detritus-rich systems, such as estuaries or peaty, eutrophic lakes with consequent bad visibility, the opportunities for SCUBA diving are much more limited.

If the lake is small enough then orientation can be made on landmarks along the shoreline only. In larger lakes it may be necessary to use these in combination with aerial photography and/or topographical maps (Wile 1973). Aerial photography itself is intricate when used for submerged plants, and gives only limited information; it should always be backed up by groundtruth, preferably verification by SCUBA diving (Long 1979, Lachavanne & Wattenhofer 1975).

For insight into the species diversity and population density, the use of a transect based on depth with variable distance can be adopted. At each station SCUBA diver and boat operator choose a transect perpendicular to an innermost point on the shore. Orientation to this point is made using landmarks. Distance is estimated using topographic maps. The course of the transect can accurately be followed by aid of a diving compass and depth gauge. Depth and population density of each species are recorded on plastic sheet with a pencil. For the estimation of biomass of each species from density observations, the mean value for each population density index is multiplied by the average mature dry weight of an individual shoot. In choosing this method it should be borne in mind that it requires special expertise not only in the skills of SCUBA diving, but in taxonomy as well. Before population density can be estimated for a particular aquatic ecosystem, species diversity must be determined and the diver must be familiar with most of the species concerned and be able to make rapid taxonomic decisions. The few unknown species can easily be gathered for identification later on.

Relatively recently a recording fathometer has been used for the mapping of vegetation in the USA. Its use is limited to waters with good visibility and it has proved to give reliable information on plant biomass provided the species already has been identified (Maceina & Shireman 1980, Maceina *et al.* 1984).

In using the population density for biomass estimates as described, several factors should be taken into account, particularly the differences in seasonal growth patterns of the species, underground biomass and depth-dependent production of biomass.

The methods described are largely non-destructive. However, destructive methods are usually used. These include harvesting quadrats of different dimensions within macrophytic vegetation types. For the estimation of species diversity, random quadrat sampling has been shown by Sheldon & Boylen (1978) to be only about 60% as effective as SCUBA diving in enumerating species presence. On comparing data on population density collected by random quadrat sampling *vs* density-based estimates by SCUBA diving, they found that in general there was a highly significant correlation. However, a greater number of samples had to be taken in the random quadrat sampling method to give a set precision. Exceptions to this rule were densities lower than 1 plant m^{-2}, when random quadrat sampling yielded in general lower results compared with SCUBA diving, possibly due to the limited number of samples taken in the quadrat sampling method (Wood 1975, Sheldon & Boylen 1978; Wade & Bowles 1981). For the determination of growth rates of submerged plants, whole plants or parts of plants are tagged *in situ* at the onset of the growth season, and the tagged plants are hand-collected later on. This work can only be done by SCUBA diving (Jacobs 1979, Best 1982b).

When the aim of the study is to investigate the species distribution in relation not only to depth but also to type of substrate, then visual observation or photography of type and homogeneity of the sediments often combined with accurate coring of plant-colonized substrates is useful (Patriquin 1975). Concurrent with this, diver-operated transplantation of the plants can be carried out (Phillips 1976). Underwater photography has been used for investigations on the growth of submerged plants in general, and in particular on effects of pollution, recreation etc. (Zieman 1976).

Submerged plants often form a substantial substrate for periphyton and invertebrates. From above, quantitative sampling is virtually impossible, since part of the periphyton is usually loosely attached and many of the invertebrates flee from the sampler. For quantitative sampling of these groups of organisms, which live in close association with submerged macrophytes though with their own specific habitats, a large variety of methods has been devised by divers (Fager *et al.* 1966, Drew *et al.* 1976). The most generally useful is a diver-operated net with a large square or hoop-like opening at one side, the opening being large enough to easily enclose the target part of the vegetation. The choice of mesh-size depends on the size of the objects to be sampled. After the sample has been netted the open side of the net is closed by a drawstring previously threaded through an extra layer of gauze around the opening. The net is brought to the surface and emptied into a bucket filled with water. Animals and plants are subsequently separated by sieving. This apparatus and procedure is better than using a plastic bag or tube because the net causes much less water movement than the bag or tube. The animals may be identified and counted, and the activity of the periphyton may be measured. Sometimes those tasks are performed in the field, but it is more usual to do this in the laboratory. The activity of the macrofauna living on the macrophytes is nearly always measured in the laboratory because intricate handling is required.

By their presence and physiological activity, submerged plants affect their physical and chemical environment. Measurements of rate of photosynthesis, nutrient uptake and similar variables are often made in the laboratory, but the meaning of these data for field situations is often questionable. It is preferable to measure these processes *in situ* in enclosures or by installing monitors within the vegetation; this equipment is largely diver-operated (Lindeboom & De Bree 1982, Best & Dassen 1987).

As with terrestrial communities, there are several approaches to floristic description. Phytosociological systems have been devised (e.g. Den Hartog & Segal 1964). A more adaptable system such as that of Tansley (1964) may be preferred, or a simple annotated abundance list may be used (Best 1982b). More details are given in the chapter 'The Phytosociological Approach to the Description and Classification of aquatic macrophytic Vegetation (Best 1988).

CHEMICAL SAMPLING

Details of methods of chemical analysis of freshwater are given in the IBP manual No. 8 (Golterman, Clymo & Ohnstad 1978). A similar manual exists for the analysis of sediments and suspended matter (Golterman *et al.* 1983).

When trying to describe the chemical and physical environment it is first necessary to define the purpose of the study. Subsequently the number, frequency and location of the samples should be determined. Two extreme cases serve as examples.

(a) A process or a phenomenon is being studied. In this case it is essential that the chosen process is occurring at the sample location, even if that location is in other respects 'atypical' of the area as a whole. The results will be valid for the process under study and its influencing factors, but extrapolation to the whole lake is not possible.

(b) A lake or a given area must be studied. Now for each typical sub-area sufficient samples must be taken, the number of which depends on the desired precision and the variability of the site. In the first place, therefore, a map must be prepared to recognize how many different zones can be distinguished. Aerial photographs may be useful but if not available a first reconnaissance must be made by whatever means are available. The desired precision will then determine how many samples must be taken. For initial study of stable variables three samples in nominally identical locations may be sufficient; the mean can be calculated and the difference between highest and lowest value gives some idea about the precision. When 10 samples are taken a standard deviation or standard error can be calculated and statistical methods may be applied. Experience shows that it is often — but not always — the case that increasing from 5 to 10 samples gives little extra knowledge and that more than 10 samples is a waste of time and money.

The decision about the frequency of sampling is also difficult to make. The chemical variability during maximum plant density can often be assessed from samples at two or three different times, but to get a measurement exactly at the time of maximum growth rate will need far more samples. The timing of these is likely to be more efficient if prior knowledge of the pattern of variation exists. Sampling should now probably not be at regular intervals but timed to coincide with biomass formation. (Compare the sampling of suspended matter in rivers where the frequency should be greatest during periods of high flow rate.) For the chemical factors in the plant environment the frequency depends largely on the problem under study. To characterize the major elements or constituents four samples per year may often be sufficient — unless very rapid changes occur, e.g. as in tidal areas — while to study nutrient variations (N, P, Si) samples at 7 to 14 day intervals are indicated.

A first step in any detailed study should be an investigation of spatial and temporal variability. As a bare minimum three or more samples from different places taken as nearly as possible at the same moment and samples from one place at different moments (e.g. early morning, midday and sunset or late in the afternoon) are needed. Inadequate, invalid or even non-existent replication is one of the commonest deficiencies of published work.

Samples can be taken in one of three main ways.

(a) Cylinder samplers consist of a cylinder with caps which are held open while the cylinder is lowered to the desired depth. A 'messenger' weight is then slid down the supporting cable and causes the lids to spring closed. The sample thus trapped is then hauled to the surface. The commonest patterns are the Kemmerer, Ruttner and Friedinger types. The last is probably the best because the end caps are held open parallel to the length of the cylinder thus allowing water to flow through while the cylinder is being lowered.

(b) Gas-filled or evacuated flask samplers are sealed and then lowered to the desired depth. A 'messenger' then breaks the vacuum (Watt sampler) or a tug on a control string allows gas to escape (Dussart and Valas samplers). In all cases the flask fills with water from the sampling depth. These samplers are particularly useful for bacteriological and in situ work with tracers.

(c) Peristaltic pump samplers have the advantages that they cause little disturbance of stratification (if there is any) and produce large samples. They are particularly suitable for taking a "mixed" ("weighted") sample from different layers. The collecting flasks may be marked in advance for this purpose, the successive volumes in the flask being proportional to the corresponding volumes of the sampled layers in the lake.

Further details are given in Golterman, Clymo & Ohnstad (1978).

CORERS AND DREDGES

Corers are used to obtain sediment samples in which stratification is preserved and the sediment/water interface is relatively undisturbed. For sediment layers in shallow waters the 'push type' tube sampler is ideal, unless the sediment is too coarse. It consists of a Plexiglass

(Perspex) tube with an outlet valve at the top (any kind of valve will do) which is closed after penetration in the soil. The Jenkin sampler is the ideal sampler for sediment-surface sampling in deeper waters. The device allows plastic tubes to be driven into the sediment layer; the lids remain attached to the tube, so that one tube after another may be filled and brought to the laboratory. The Mackereth corer, driven by compressed air and controlled from the lake surface, allows long cores to be taken. It is a relatively expensive device and cannot take cores which are much shorter than its designed length. Cheaper and more flexible is the Livingstone corer in which a tube is driven by rods past a piston, held fixed at the top of the core by a wire attached to the boat.

Dredges or grab samplers (e.g. the Ekman or the Peterson Dredge) are used to sample the top few centimetres of bottom deposits. The samples are disturbed and the precise representation of depth in the sample is unknown. They provide a rapid means of characterizing the 'average' sediment composition over a limited range or depth — the so-called 'bulk sample'. A detailed critical evaluation of about 10 corers is given in Golterman *et al.* (1983) and of several sediment samplers, including the air-lift pump (which gives a very disturbed but large sample) by Elliott & Tullett (1978, 1983).

PHYSICAL VARIABLES

Temperature and light

Light and temperature are of interest in all vegetation studies and in the interpretation of, for example, pH measurements and solubility products. They are often measured at the same time that the water is sampled for chemical analysis. Both variables must be measured *in situ*; the methods are obvious and usually simple. Difficulties will be met when deciding what to measure, where to measure it, and when and what equipment to use. Both the light flux on the surface and its attenuation below the surface may be important, particularly for studies concerned with the growth of phytoplankton and submerged macrophytes.

Temperature

There are four types of apparatus in use.

(1) (*Reversing*) *thermometer*. If the water can be directly reached with a normal laboratory thermometer or if there is no serious delay in placing the thermometer in the sample itself (in the outlet tube of the pump for example), a normal laboratory thermometer is the

most common and simple instrument to use. Precision of 0.1 or 0.05 °C can easily be obtained. If direct access is impossible then a reversing thermometer may be used. After the thermometer has reached equilibrium at the required depth a messenger weight is sent down the cable and causes the thermometer to invert in its frame and separate the mercury thread from that in the main bulb, thus preserving the reading while the bottle is hauled up. A precision of 0.01 °C may be obtained.

(2) *Thermocouples.* These produce an electrical potential difference which is proportional to the difference in temperature between a reference junction and the sensing junction. The voltage is small and not easily amplified, though the difficulties are much smaller than they were before integrated circuits became common and cheap. They need a reference junction at known temperature.

(3) *Thermistors.* These are based on material which has a large negative temperature coefficient of resistance. They are relatively cheap. A precision of 0.1 °C is easily obtained.

(4) *Nickel or platinum resistance elements.* They are also based on a change in resistance. Compared with (3) they are more expensive, more accurate and more stable. A precision of 0.01 °C is possible.

The choice of equipment depends on purpose and availability. For general purposes a frequently calibrated thermistor is probably the most useful. If occasional more accurate measurements are needed a reversing thermometer will do very well. For the most accurate work — measurement of precise stratification for example — a nickel wire thermometer may be preferred.

Apparatus such as many O_2 electrodes includes a thermistor so that temperature is measured as well. In these cases it is necessary to make sure that the temperature of the water is being measured and not that of the cell.

Light and its attenuation

Measurement of light normally involves two aspects, the measurement of the total quantity and its spectral distribution.

Light, which is that part of the electromagnetic spectrum visible to the human eye, falls within the approximate limits 400 to 700 nm. This range is also roughly that of 'photosynthetically active (or available) radiation' or 'PAR', which falls between 350 or 390 and 700 nm depending on the definition used. Halldall (1967) has shown that light down to 310 nm may be active in photosynthesis in algae, but light of this short wavelength is rapidly attenuated by water.

Measuring light therefore implies integration over a particular range of the spectrum and often over time. The level of integration must be carefully chosen. Collecting complete spectra at several depths at frequent intervals is very expensive and for vegetation studies rarely necessary. Integration can best be done automatically by the sensor. The time integral of electric signals may be obtained electronically or by allowing the current to remove and deposit metal on the plate of an electrolytic cell (voltameter; see Westlake & Dawson 1965).

There are three types of target for measurements: incident quantity (irradiance and photon flux density), absorbed quantities (such as absorption spectra) and effective quantities (photosynthesis and illuminance). The choice of the target depends on the purpose of the work. If the problem requires an energy balance, irradiance will be needed. Irradiance in the PAR range may be best for production studies, as the units in these studies are normally those of energy flux.

Photochemical reactions (among which is photosynthesis) depend in nearly all cases on the number of quanta absorbed. The quantum of light is called a photon. It is therefore useful in some cases to measure not irradiance but photon flux density (PFD). There is a close physical relation between irradiance and PFD because the energy of a photon is inversely proportional to its wavelength. One chemical mole of molecules, ions or atoms contains about 6.02×10^{23} particles, so PFD is conveniently expressed in mol m^{-2} s^{-1}, sometimes called Einstein m^{-2} s^{-1} = E m^{-2} s^{-1}. Some authors prefer quanta m^{-2} s^{-1}. Figure 1 shows the spectral curves for irradiance and PFD, and the Table 1 shows some equivalences and typical values. Incident quantities such as irradiance and PFD are well-defined and unique; absorbed quantities, such as the absorption spectrum of chlorophyll a or of a whole leaf, are less well-defined.

Table 1. Some light quantities and units.

Quantity	Irradiance (= Irradiant flux density)	Photon flux density	*Illuminance
Units	$W\,m^{-2} = 0.1\ mW\,cm^{-2}$	$\mu mol\,m^{-2}\,s^{-1}$ $= \mu Einstein\,m^{-2}\,s^{-1}$ $= 6.02 \times 10^{17}$ $quantum\,m^{-2}\,s^{-1}$	$lx = lm\,m^{-2}$
Obsolete	$cal\,cm^{-2}\,min^{-1}$		foot candle $= lumen\,ft^{-2}$
Units	$692 = \dfrac{cal\,cm^{-2}\,min^{-1}}{W\,m^{-2}}$		$10.76 = \dfrac{ft\,candle}{lux}$
At 555 nm	$1.0\ W\,m^{-2}$	$= 4.15\ \mu mol\,m^{-2}\,s^{-1}$	$= 680\ lx$

* Illuminance units are given because they have been widely used but illuminance cannot be recommended as a measure in limnology.

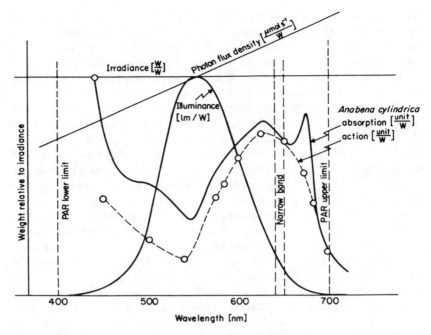

Fig. 1. 'Targets' for measurement: (a) incident (b) absorbed (c) effective. (a) Relative weight given to energy flux density as a function of wavelength for irradiance and photon flux density. (b) Absorption spectrum of whole *Anabaena cylindrica*. (c i) O_2 evolution (photosynthesis) action spectrum of *Anabaena cylindrica* for equal irradiance. (c ii) Illuminance: response of the light adapted young adult human eye. Both (b) and (c) are illustrative examples of whole classes of possible targets. Targets may be given specific spectral limits. For illustration a narrow band in the red and the broader 'photosynthetically active (or available) radiation', PAR, are shown. The definition of PAR is not agreed (see text). (From Golterman *et al.* 1978).

Measurement of total incident radiation

A series of thermocouple junctions alternate between the black and white areas of the sensor. As the black area becomes hotter, a combined potential difference is produced (typically between 5 and 15 mV in bright sunlight). This may be recorded by a potentiometer or it may be used to drive a sensitive current meter. A well-designed sensor has an almost level response between 400 and 2700 nm if glass-covered. In daylight PAR can be estimated as 0.50 times this irradiance, but it is preferable to use two instruments one of which has a filter which cuts out all wavelengths below 700 nm. PAR is then caluclated from the difference. Better still is to cover the sensor with filters which have, in combination, a level and small absorption within the PAR range and a sharp change to almost complete absorption outside it.

Other light sensors are the photomultiplier tube, the selenium barrier layer cell, the photodiode, silicon cell, and the CdS photoresistor. The energy-sensitivity of detectors other than thermopiles is markedly wavelength-dependent: Figure 2 shows examples. The selenium barrier layer photocell, which is the most commonly used cell, has a response fairly close to that needed for illuminance measurements, but this is of little use to the limnologist. On the other hand the electrical character-istics of these detectors are more favourable than are those of the thermopiles. Two solutions are possible.

(1) Use a filter or combination of filters which converts the spectral response to that needed for irradiance or photon flux density. The steps in the design are illustrated in Figure 3. Other solutions are given by Jerlov & Nygard (1969) and by Uphoff & Hergenrader (1976).

(2) Use filters which isolate narrow spectral bands and measure and calibrate with each separately. The filters are delicate. Evans (1969) gives details.

The CdS detector differs from others in that the property which varies with irradiance is the electrical conductance. Its spectral response is not easily transformed to irradiance or PFD.

The Si photocell is potentially useful; if the output is fed through a low (10—50 ohm) external resistor the response is linear, but if fed through a high resistor the response is logarithmic.

Photomultiplier tubes, which are in use in oceanography, are the

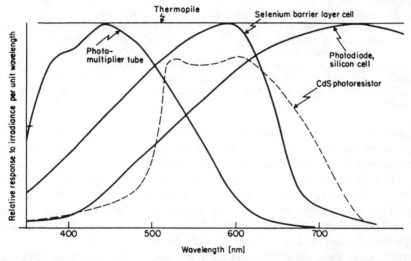

Fig. 2. 'Detectors': spectral sensitivity of various light detectors. These are typical for the type but *there is variation within types* i.e. do not assume that the curve here refers to your detector of this type. This is particularly true for photomultiplier tubes. The units are not comparable between detectors.

50

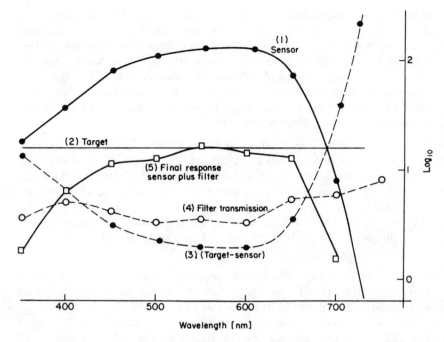

Fig. 3. Designing a sensor to measure irradiance in the PAR range. The example is that of Powell & Heath (1964) modified by Westlake & Dawson (1975), though the logarithmic plot used here is not theirs. First the sensor — a selenium barrier layer photocell (Megatron B) — is chosen and the spectral response (1) graphed with a logarithmic scale. Next the target — irradiance (2) — is graphed, again on a logarithmic scale. The vertical position is unimportant. Next the vertical distance (3) between target and sensor curves is plotted. This distance, a difference of logarithms, is the ratio target/sensor response and is the curve which must be *paralleled* by the filter transmission, again on a log scale. The filter curve (4) is for the combination of Cinemoid 'Steel Blue 17' and 'Pale Salmon 53' filters (Strand Electric). The overall response of sensor and filters is shown as (5), which has been shifted vertically to aid comparison with the target (2).

most sensitive of all detectors. They need high voltages and retain a memory of saturation, which occurs at very low irradiance. They are not a practical proposition for the non-specialist.

For detailed work on the photosynthesis of plants (or bacteria) absorption targets are attractive. Interactions with other chlorophylls and accessory pigments (including the photosynthetically inactive ones) may complicate matters and an action spectrum may be indicated. Photosynthesis action spectra in light limiting conditions are usually more level than absorption spectra, but are rarely known and are not measured *in situ*. The action spectra may well depend on other undefined physiological and environmental variables as well. Furthermore, near the surface of a water body during summer light is often non-limiting for photosynthesis so the target would be non-linear, and to different degrees for different plants. The effective target cannot in practice be defined for general limnological work when there are varying proportions of algae of different groups or a variety of vascular plants and varying environmental conditions.

It may be important to measure the amount of light absorbed by plants, particularly by phytoplankton, if primary production is to be calculated using a model. Examples for algae are given by Kirk (1976, 1980, 1981); for periphyton by Meulemans (1985) and Sand-Jensen & Sondergaard (1981); and for macrophytes by Best & Dassen (1987).

One effective target which is well defined is the response of the average photopic (which light adapted) young adult human eye. This is internationally agreed (Commission de l'Eclairage, 1924) and has recognised units. The target quantity is called illuminance, the units are lumen m^{-2} = lx (lx is the symbol for the unit 'lux'). Intensity, which is the property of the light source, and illuminance are not among the effective quantities and are not usually of interest to the limnologist. They are not commonly measured either, though the terms are commonly mistakenly used for irradiance or illuminance or some other measure similar to illuminance. Illuminance as a target can only be justified where the problem involves the human ability to see under water. Even then it is probably incorrect because the sensitivity of the dark-adapted (scotopic) eye differs from the photopic one used in the definition of illuminance. The sensitivity of the fish eye is considerably different from that of the human eye.

Though illuminance is indefensible as a limnological quantity many published works mention it or contain illuminance units measured with a sensor which does not have the illuminance spectral weighting. To complete the confusion this is often called 'light intensity'. The reason is that the selenium barrier layer photocell with which most of these measurements have been made, is cheap, robust and gives a large current, which can drive a meter directly. It can be calibrated in illuminance or even irradiance units for a source *with a particular spectral composition*, but the readings will be incorrect if the spectral composition of light falling on the detector changes, as it does if the detector is lowered deeper into water or amongst aquatic vegetation. This may lead to very large error: Tyler (1973) estimated that 600 or 700% is not unlikely. Table 2.3 of the IBP Manual gives some guidance.

For some purposes — calculation of heating and evaporation for example — absolute measurements are necessary. For productivity studies relative measurements may often suffice. Sometimes most measurements are relative but one of them is absolute; this one will at the same time be used to calibrate the whole set. This is often the case when spatial (horizontal or vertical) distribution must be measured. In limnological work light attenuation under water is often an important target. It can best be measured by lowering a sensor step by step. But because it is unlikely that during a series of measurements the light climate at the surface will remain constant, two matched sensors must be used one of which is kept in a standard position and the other moved to measurement positions. The meter may be switched rapidly between the two sensors and the movable sensor's reading recorded as a quotient of that of the fixed one. This may nowadays be done electronically with ease. In this way light attenuation with depth can easily be recorded; it must be kept in mind, however, that with increasing lake depth the distribution over the different parts of the spectra will change.

Light attenuation in water is caused by reflection at the surface both back into the air (albedo) and from below back into the waterbody, by scattering by particles and by absorption by the water itself and by the dissolved substances.

If no proper meter is available light attenuation can be estimated by recording the depth at which a white disk (Secchi disk, 25—50 cm diameter) just disappears. Measurements should be made preferably during the middle of a sunny day. It is usual to take the average of the depth at which the disk just disappears and just reappears. The results will be more reproducible, though different, if diving glasses or an observation box, both dipping just under water, are used.

In favorable conditions — calm water, bright light — irradiance at the Secchi disk depth is about 15% of that just below the surface. From this it follows, that the depth at which irradiance falls to 1% of that just below the surface is 2.5 times the Secchi disk depth and that the vertical attenuation coefficient ε_v is roughly 1.9/Secchi disk depth. These calculations assume however that the water column is homogeneous, and that the Secchi disk depth is related to irradiance. As there is little agreement about the size and pattern of markings on the disks, results are even more difficult to compare than need be.

The calculations should therefore be treated as rule of thumb estimates subject to errors of at least a factor 2. The method is much better than nothing however.

THE CHEMICAL VARIABLES

Chemical substances in the water can be divided into suspended material (or sediments) and dissolved compounds. The latter can be divided into gases, major, minor and trace elements or constituents, and organic compounds (Table 2). Of the gases, O_2 and CO_2 are measured

Table 2. Major, minor, trace elements, organic compounds, and gases normally found in natural waters.

Major elements		Minor elements	Trace elements and organic compounds	gases
Ca^{2+}	HCO_3^-	N (as NO_3^- or NH_4^+)	Fe, Cu, Co, Mo, Mn, Zn, B, V	O_2
Mg^{2+}	SO_4^{2-}	P (as HPO_4^{2-} or $H_2PO_4^-$)		CO_2
Na^+	Cl^-	Si (as SiO_2 or $HSiO_3^-$)		N_2
K^+	F^-			
H^+			organic compounds such as	
(Fe^{2+})			humic compounds, excretion	
(NH_4^+)			products, vitamins, other	
			metabolites	

Concentrations usually found

0.1—10 meq. l^{-1}	(< 1 mg l^{-1})	$\mu g\,l^{-1}$ (trace elements)	
(mean 2.4)	(Si sometimes higher)		

in most pollution and photosynthesis studies (including eutrophication studies). Nitrogen gas is seldom measured.

'Major' ions are generally present in the water and in living materials in relatively high concentration. They can normally be determined by classical titrations or precipitation reactions. Their sum largely determines the electrical conductivity. The 'minor' elements are often important determinants of the quantities of the selected biological species which an ecosystem can produce. They are normally measured by colorimetric procedures.

Titrations and precipitation procedures are normally specific or can easily be made to be so. Their calibration is absolute: e.g. 1 mg NH_3-N as base will neutralize 3.57 ml of 0.02M H^+.

Colorimetric reactions are normally not specific and interferences are to be expected and must be avoided by special measures which may differ from lake to lake. The extent of bias in the results will depend very much on knowledge of the possible presence of interfering substances.

It is common usage to speak about colorimetric methods if a filter colorimeter is being used and to use the term spectrophotometric procedures if a spectrophotometer is being used. This distinction is one of degree however: the same chemical reactions are involved in both cases.

Colorimetric and spectrophotometric methods depend on a calibration curve which gives the relation between quantities used and color developed. The relation is not necessarily a straight line, but it is convenient if it is. The final precision (and accuracy) of the method depends as much on the care with which the calibration curve has been prepared as on the care given to the sample.

If one studies a new environment, the following variables are normally the first to be measured: temperature, oxygen, conductivity, redox potential (pE) and pH. The information provided by these five variables is very valuable, especially when the variation in time and space are being assessed.

Conductivity gives information about the total salt content, the so-called salinity. It can normally be assumed that for freshwater with a pH in the range 4.5—10.5 then

$$C \simeq 0.01 \, k \quad \text{and} \quad C' \simeq 0.75 \, k$$

where: k = conductivity (μS cm^{-1}); C = sum of concentration of positive and negative ions (mmol l^{-1}); and C' = total salinity (mg l^{-1}).

Conductivity gives no information about which salts are present e.g. $Ca(HCO_3)_2$ or $NaCl$. In combination with the pH, however, a good guess at the nature of the ions can usually be made; waters containing $Ca(HCO_3)_2$ and in equilibrium with air usually have a pH greater than 8. A chloride titration (even a very simple one) may be extremely useful for a first classification of the chemical composition.

The redox potential gives a good description of the oxidation state of the water or bottom. It is a purely descriptive measure and it should not be given too much weight. Thermodynamically it is a well defined measure based on a reversible electrochemical reaction; in nature this is never the case. It is derived from unknown chemical reactions which essentially are not reversible. For detailed discussion see Golterman (1975) and Golterman et al. (1978). Temperature, oxygen, conductivity, pH and pE can be measured with reliable field equipment, but often not with great precision though sufficient for a first descriptive reconnaissance. The standard deviations are (roughly): temperature 0.1 °C; O_2 0.2 mg l^{-1}; conductivity 5 μS cm^{-1}; pH 0.1 unit.

Laboratory instruments provide a much smaller standard deviation: O_2 0.02 mg l^{-1} (Winkler titration); conductivity 0.5 μS cm^{-1}; pH 0.01 unit.

The equipment is described in Golterman et al. (1978, procedures 3.1, 3.2, 3.3 and 8.1). A few precautions should be noted however.

All these field instruments and instruments which are used for monitoring in the field should be inspected and cleaned regularly lest their response is changed by fouling and 'Aufwuchs'. It is essential that they are calibrated by independent means or equipment. O_2 probes, for example, should be calibrated regularly against the Winkler titration.

All these instruments have an analogue voltage and, increasingly commonly, a digital output. This may be recorded in the field or transmitted by telemetry. Such transmission of results by radio either continuously or on command may be useful for monitoring lakes at great distances or in relatively inaccessible areas. These signals can warn of sudden changes, but may be less suitable for recording small concentration variations, which for biological observations may be much more useful or important.

For several major elements (calcium, sodium, potassium, chloride) specific ion electrodes are available. They are similar to glass electrodes in that they produce a voltage proportional to the logarithm of the activity. Monovalent ions produce about 58 mV for a ten-fold change in concentration; divalent ions produce about 27 mV for the same change. A reliable high-impedance voltmeter is as necessary for such measurements as it is for pH measurements. A pH meter may be suitable, though calibration for divalent ions will be different from that marked on the scale. Sometimes it is necessary to add specific reagents to the sample, and for this reason such methods are often used in the laboratory only. Specific electrodes such as those for calcium and chloride may be useful for quick survey, provided that high accuracy is not desired.

The major elements

All major elements can be measured by classical titration or precipitation reactions. Calcium, for example can be estimated with the classical oxalate-precipitation reaction; the method needs a lot of practice but gives excellent precision and accuracy. Easier — but not more precise — are the procedures shown in Table 3, which are commonly in use:

Table 3. Common procedures for determining major elements.

Element	Method	IBP No. 8 procedure
Ca^{2+}	AAFS	4.3.4
	EDTA titration	4.3.1—4.3.3
Mg^{2+}	AAFS	4.3.4
	EDTA titration	4.3.1—4.3.3
Na^+	flame emission	5.7
K^+	flame emission	5.7
Cl^-	volumetric with Ag^+	4.6
	with potentiometric endpoint	4.6.3
	with indicator endpoint	4.6.2
	with conductometric endpoint	4.6.4
SO_4^{2-}	volumetric EDTA	
	indicator endpoint	4.7.3
	potentiometric endpoint	4.7.4
	turbidimetric	4.7.1
Total alkalinity (OH^-, HCO_3^- and CO_3^{2-})	acidimetric	3.4.2 or 3.4.3
Alkalinity (OH^-, CO_3^{2-})	acidimetric i.e. phenolphthalein	*idem*
Acidity	titration with $Ba(OH)_2$	3.6

The AAFS (absorption spectrophotometry) method requires expensive equipment, which demands proper maintenance and calibration. All acid-base titrations can be made with a suitable indicator — which must be selected carefully — or with a pH meter endpoint. Some chemical knowledge is needed; the HCO_3^- titration has a precise endpoint which depends on the concentration of the HCO_3^- present. If the full titration curve is being made, this information is provided automatically, but preset endpoints may be dangerous if a high precision is desired.

The minor elements

The minor elements occur in soluble form (NH_4^+, NO_3^-, o-phosphate, hydrolyzable phosphate i.e. organic- and polyphosphates, and silicate) and in particulate form (inorganic and organic particulate phosphate, organic nitrogen and inorganic silicate). The separation is achieved by filtration, normally through a 0.5 μm filter. This pore-size is arbitrary, but practical. For special cases 0.2 μm is sometimes used. The particulate forms are made soluble by one of the following digestions: hydrolysis in strong acid medium, with or without oxidation, or by fusion with alkali.

For the minor elements the following reactions are suggested.

Phosphate is measured as the blue color formed after complex-formation with molybdate (in the presence of antimony) and subsequent reduction, both in about 0.4M H^+. Ammonia is measured either as the yellow color formed with Nessler's reagent (the active component of which is mercury ion), or as indophenol blue. The first method can only be carried out after a distillation of the ammonia from an alkaline medium. This distillation is time consuming and may hydrolyze weakly bound organic nitrogen; it can be argued that such ammonia compounds are also easily hydrolyzable in water and are thus biologically available. The indophenol blue method is based on the blue complex formed by ammonia and phenol which are coupled under oxidation, e.g. with alkaline chlorine. The method is not yet standardized, many different modifications are still in use and interferences in natural waters are not yet well studied. Some organic nitrogen may react. Nitrate must first be reduced, either to NO_2^- or to NH_4^+. The former reaction is often effected with cadmium in some special form. Organic compounds may poison the reductant; the addition of internal standards is still necessary. Most organic compounds may be destroyed by boiling with $K_2S_2O_8$, but then some organic nitrogen may be oxidized to nitrate. Normally the precision is 2—3%, but with the different steps needed to achieve the reduction, it may easily increase to 5%. Silicate is measured, like phosphate, with molybdate; the molybdate complex is reduced again. The first step — the formation of the yellow molybdate complex — occurs in a weakly acid medium; the reduction step is carried out in strong acid medium or with specific organic acids in order to prevent the reduction of the phosphate complex. Arsenic interferes both in the phosphate and silicate procedures, but the concentrations are usually sufficiently low to be ignored. In special cases attention should be given to this interference.

The IBP Manual No. 8 gives for the three minor elements a choice of methods, often dependent on concentrations present, availability of equipment and complexity of the method itself.

Trace elements

The trace elements can be classified in one of two groups: those elements which occur naturally in water and are generally plant nutrients, and those which are potentially toxic even in very low concentrations and which have been widely distributed as a result of human activities. The elements of the first group are normally measured to obtain insight into the factors controlling biological characteristics, while elements in the second group are often monitored to detect threats to the health of human and other living things. Copper and zinc are examples of elements which may belong to both groups: they occur naturally and are plant nutrients, but the present concern is mainly related to their possible toxicity. Because of their low concentrations, analysis of these trace elements presents problems, which normally require special equipment and techniques. Contamination or losses can occur easily; dust in the laboratory is a serious problem because this interference is never reproduced in the blank. Standard additions and distilled water blanks should always be used. The validity of each method should be checked for each particular situation.

The trace elements normally occur as dissolved (both ionic and chelated), colloidal and suspended matter adsorbed to silt or as heavy metal oxides. The most common analytical method is atomic absorption flame spectrophotometry (AAFS), normally after a solvent extraction and chelation. AAFS methods do have some interferences. A development of flame emission spectrophotometry but at a temperature in the range 5000—10000 °C (produced by ICP — inductively coupled plasma) has even fewer interferences than AAFS and may prove useful because many elements can be measured simultaneously (or effectively so). The equipment is very expensive however. In the case of sediment and suspended matter, whatever the analytical method, the metals must first be extracted. For this purpose a destruction with several mineral acids is usually employed (Golterman *et al.* 1983). In specialized laboratories polarography (classical, cathode ray and anodic stripping voltammetry) is useful. There has been some discussion about which forms of the element can be measured using this method (Golterman *et al.* 1978).

A few colorimetric methods are in use (e.g. copper, iron and manganese) but normally the concentrations in the aquatic habitat are too low.

Typical sensitivities (mg l^{-1}) for some elements measured by AAFS are: Al 0.05—1.0; Cd, Co, Cu, Mn, Zn 0.001—0.01; Fe, Pb 0.001—0.05; Cr, Mo 0.0002—0.01.

Sampling, storage and sample digestion need some special attention. Completely non-metallic samplers should be used. Rubber parts should be avoided as rubber may contain considerable amounts of zinc.

Precision, bias, accuracy, sensitivity

All measurements are susceptible to errors. These may be erratic mistakes, systematic errors, and random errors. Mistakes may be instrumental or human. Sometimes they can be identified: for example a wrong flask number or a wrong pipette. It is wise to keep the equipment including flasks available on the laboratory bench until the results have been definitively calculated.

Systematic errors are related to the concept of bias; random errors are related to precision, reproducibility and dispersion. In rifle shooting it is possible to get several shots grouped close together, but a long way from the target. Such a group has a high precision or reproducibility (small dispersion) with small random errors. The group has however, a large bias. High precision may be accompanied by large bias if, for instance, a calibration curve is not properly made. Low precision with small bias may also occur, but is less dangerous: it is more obviously unsatisfactory. One may define accuracy^2 = bias^2 + precision^2 so that accuracy is the hypotenuse of a right triangle whose sides are bias and precision. The same accuracy may be obtained by small bias and large imprecision or by large bias and small imprecision.

Sensitivity has many definitions. One is the concentration giving a value three times that of the blank. In atomic absorption flame spectrophotometry sensitivity is often defined as twice the background noise (manufacturers of AAFS are tending to define sensitivity in terms of standard error or standard deviation of repeated blanks); the same must be done with high blanks in colorimetry (e.g. Nessler). The IBP manual No. 8 gives a more extensive discussion of these problems.

REFERENCES

Anderson, J. R., Hardy, E. E., Roach, J. T. & Witmer, R. E. (1976) A land-use and land-cover classification system for use with remote-sensor data. Prof. paper # 964, USGS, Reston, 1—28.

Best, E. P. H. (1982a) Effects of water pollution on freshwater submerged macrophytes. *In*: Varsney, C. K. (Ed.). Water Pollution and Management Reviews 1982. pp. 27—56.

Best, E. P. H. (1982b) The aquatic macrophytes of Lake Vechten. Species composition, spatial distribution and productivity. Hydrobiologia, 99: 65—77.

Best, E. P. H. (1988) The phytosociological approach to the description and classification of aquatic macrophytic vegetation. *In*: Symoens, J. J. (ed.), Handbook of Vegetation Science, Junk, 15 (This volume), pp. 155—182.

Best, E. P. H. & Dassen, J. H. A. (1987) Biomass, stand area, primary production characteristics and oxygen regime of the Ceratophyllum demersum L. population in Lake Vechten, The Netherlands. Arch. f. Hydrobiol. supl. 76: 347—367.

Bukata, R. P., Bruton, J. E. & Jerome, J. H. (1983) Use of chromaticity in remote measurements of water quality. Remote sensing of Environment, 13: 161—177.

Carpenter, D. J. & Carpenter, S. M. (1983) Modelling inland water quality using Landsat-data. Remote sensing of Environment, 13: 345—352.

Carter, V. (1977) Coastal wetlands: the present and future role of remote sensing. Proc. 11th Symp. on remote sensing of environment. Environ. Res. Inst. of Michigan. Ann Arbor, 301—323.

Carter, V. (1982) Applications of remote sensing to wetlands. Chapter 24, p. 284—300 in: Johannsen & Sanders (Ed.) 1982.

Chapman, S. B. (1964) The ecology of Coom Rigg Moss, Northumberland. 1. Stratigraphy and present vegetation. J. Ecol., 52: 299—313.

Commission de l'Eclairage (1924).

Cowardin, L. M., Carter, V., Golet, F. C. & LaRoe, E. T. (1979) Classification of wetlands and deep water habitats of the United States. FWS/OBS-79/31. Fish and wildlife Service, Washington D.C., 1—103.

Dawson, F. H. (1978) The seasonal effects of aquatic plant growth on the flow of water in a stream. Proc. EWRS 5th Symp. on Aquatic Weeds. pp. 71—78.

Den Hartog, C. & Segal, S. (1964) A new classification of the water-communities. Acta Bot. Neerl. 13: 367—393.

Denny, P. (1972) Sites of nutrient absorption in aquatic macrophytes. J. Ecol. 60: 819—829.

Denny, P. (1980) Solute movement in submerged angiosperms. Biol. Rev. 55: 65—92.

Durrani, T. S. & Greated, C. A. (1977) Laser systems in flow measurement. Plenum Press.

Drew, E. A., Lythgoe, J. N. & Woods, J. D. (1976) Underwater research. Academic Press. London, New York, San Francisco. 430 pp.

Elliott, J. M. (1977) Some methods for the statistical analysis of samples of benthic invertebrates. Freshwater Biological Assoc. Scientific Publication no 25. 160 pp.

Elliott, J. M. & Tullett, P. A. (1978) A bibliography of samplers for benthic invertebrates. FBA Occasional Publications. No 4. Windermere. 61 pp.

Elliott, J. M. & Tullett, P. A. (1983) A supplement to A bibliography of samplers for benthic invertebrates. FBA Occasional Publications Nr 20. Windermere. 000 pp.

Evans, G. C. (1969) The spectral composition of light in the field. 1. Its measurement and ecological importance. J. Ecol. 57: 109—125.

Fager, E. W., Flechsig, A. D., Ford, R. F., Clutter, R. I., & Ghelardi, R. G. (1966) Equipment for use in ecological studies using SCUBA. Limnol. Oceanogr. 11: 503—509.

Forsberg, C. (1959) Quantitative sampling of subaquatic vegetation. Oikos, 10: 233—340.

Gammon, P. T., Carter, V. & Rohde, W. G. (1979) Landsat digital classification of the vegetation of the Great Dismal Swamp with an evaluation of classification accuracy.

60

In: Satellite Hydrology, Proc., Pecora V. Symposium. American Water Resources Association, Minneapolis, Minesota.

Garrett, M. K. & Carter, V. (1977) Contribution of remote sensing to habitat evaluation and management in a highly altered ecosystem. Trans. 42d North Am. Wildlife and Natural Resource Conference. Wildlife Management Inst., Washington D.C., 56—65.

Golterman, H. L. (1975) Physiological Limnology. An approach to the physiology of lake ecosystems. Elsevier Scientific Publish. Company, Amsterdam. 489 pp.

Golterman, H. L., Clymo, R. S. & Ohnstad, M. A. M. (1978) *Is*: IBP Manual No. 8. Methods for physical and chemical analysis of freshwaters. Blackwell Scientific Publications. 2nd Ed. 124 pp.

Golterman, H. L., Sly, P. G. & Thomas, R. L. (1983) Studies of the relationship between water quality and sediment transport. Unesco. 231 pp.

Halldall, P. (1967) Ultraviolet action spectra in algology. Photochem. & Photobiol., 6: 445—460.

Jacobs, R. W. P. M. (1979) Distribution and aspects of the production and biomass of eelgrass, Zostera marina L., at Roscoff, France. Aquat. Bot. 7: 151—172.

Jerlov, N. G. & Nygard, K. (1969) Influence of solar elevation on attenuation of underwater irradiance. Inst. Fysisk. Oceanografi. Univ. Copenhagen, Rep. No. 4, 9 pp.

Johannsen, C. J. & Sanders, J. L. (Eds.) (1982) Remote sensing of resource management. Soil Conservation Society of America, Ankeny, U.S.A. 665 pp.

Lambert, J. M. (1951) Alluvial stratigraphy and vegetational succession in the region of the Bure Valley Broads. III. Classification, status and distribution of communities. J. Ecol. 39: 144—170.

Kirk, J. T. O. (1976) A theoretical analysis of the contribution of algal cells to the attenuation of light within natural waters. III. Cylindrical and spheroidal cells. New Phytol. 77: 341—356.

Kirk, J. T. O. (1980) Spectral properties of natural waters: contribution of the soluble and particulate fractions to light absorption in some inland waters of south-eastern Australia. Aust. J. Mar. Freshwater Res. 31: 287—296.

Kirk, J. T. O. (1981) Monte Carlo study of the nature of the underwater light field in, and the relationships between optical properties of turbid yellow waters. Aust. J. Mar. Freshwater Res. 32: 517—532.

Kodak Ltd (1972) Applied Infrared Photography. Doc # M. 28.

Lachavanne, J. B. & Wattenhofer, R. (1975) Contribution à l'etude des macrophytes du Léman. Commission Intern. pour la Protection des Eaux du Léman contre la Pollution, Conservatoire botanique de Genève. 147 pp.

Lindeboom, H. J. & de Bree, B. H. H. (1982) Daily production and consumption in an eelgrass (Zostera marina) community in saline Lake Grevelingen: discrepancies between the O_2 and ^{14}C method. Neth. J. Sea Res. 165: 362—379.

Long, K. S. (1979) Remote sensing of aquatic plants. Techn. Rep. U.S. Army Eng. Sta. A-79-2. Vicksbury, Miss., U.S.A. Army Eng. Waterw. Exp. Sta. Environm. Lab., 1979. 82 pp.

Maceina, M. J. & Shireman, J. V. (1980) The use of a recording fathometer for determination of distribution and biomass of Hydrilla. J. Aquat. Plant Manage. 18: 34—39.

Maceina, M. J., Shireman, J. V., Langeland, K. A. & Canfield, D. F. Jr. (1984) Prediction of submerged plant biomass by use of a recording fathometer. J. Aquat. Plant Manage. 22: 35—38.

Madsen, T. V. & Sondergaard, M. (1983) The effects of current velocity on the photosynthesis of Callitriche Stagnalis Scop. Aquatic Bot. 15: 187—193.

Madsen, T. V. & Warncke, E. (1983) Velocities of currents around and within submerged aquatic vegetation. Arch. Hydrobiol. 97: 389—394.

Meulemans, J. T. (1985) A method for measuring selective light attenuation within a periphyton community Arch. f. Hydrobiol. 109: 139—145.

Milne, A. A. (1928) The house at the Pooh corner. Chapter 6, pp. 91—107.

Munday, J. C. & Alfoldi, T. T. (1979) Landsat test of diffuse reflectance models for aquatic suspended solids measurement. Remote sensing of environment, 8: 169—183.

Pala, S. (1984) A Landsat-based method for the survey and inventory of peat resources over extensive regions. pp. 156—168. In: Robertson, R. A. (Ed). Remote Sensing in Peat and Terrain Recource Surveys. International Peat Society, Helsinki.

Patriquin, P. G. (1975) 'Migration' of blowouts in seagrass beds at Barbados and Carriacou, West Indies, and its ecological and geological implications. Aquat. Bot. 1: 163—189.

Phillips, R. C. (1976) Preliminary observations on transplanting and a phenological index of seagrasses. Aquat. Bot. 2: 93—103.

Powell, M. C. & Heath, O. V. S. (1964) A simple and inexpensive integrating photometer. J. exp. Bot. 15: 187—191.

Sabins, F. F. (1978) Remote sensing. Principles and Interpretation. Freeman & Co., San Francisco. 426 pp.

Sand-Jensen, K. & Sondergaard, M. (1981) Phytoplankton and epiphyte development and their shading effect on submerged macrophytes in lakes of different nutrient status. Int. Revue ges. Hydrobiol. 66: 529—552.

Sculthorpe, C. D. (1971) The biology of aquatic vascular plants. 2nd Ed. Edward Arnold, London. 610 pp.

Sheldon, R. B. & Boylen, C. W. (1978) An underwater survey method for estimating submerged macrophytic population density and biomass. Aquat. Bot. 4: 65—72.

Summerfield, R. J. (1972) Biological inertia — an example. J. Ecol. 60: 793—798.

Tansley, A. G. (1964) Introduction to Plant Ecology. Allen & Unwin. London. 260 pp.

Tyler, J. E. (1973) Lux vs. quanta. Limnol. Oceanogr. 18: 810.

Uphoff, G. D. & Hergenrader, G. L. (1976) A portable quantum meter-spectro-radiometer for use in aquatic studies. Freshw. Biol. 6: 215—219.

Ven te Chow (1964) Handbook of applied hydrology. Section 15: 1—41. McGraw-Hill Book Cy. New-York.

Wade, P. M. & Bowles, F. (1981) A comparison of the efficiency of freshwater macrophyte surveys carried out from underwater with those from the shore or a boat. Progress in Underwater Science 6: 7—11.

Wetzel, R. G. (1975) In: B. A. Whitton (Ed.): River Ecology. Blackwell Scientific Publications. Oxford. 725 pp.

Westlake, D. F. (1967) Some effects of low-velocity currents on the metabolism of aquatic macrophytes. J. Exp. Bot. 18: 187—205.

Westlake, D. F. & Dawson, F. H. (1975) In: Evans, G. C., Rackman, O. & Bainbridge, R., (Ed.). Light as an ecological factor II. 16th Symposium of the British Ecological Society, Blackwell, Oxford. 616 pp.

Wile, I. (1973) Use of remote sensing for mapping of aquatic vegetation in the Kawartha lakes. Remote sensing and water resources management. Proc. no. 17. Amer. Wat. Ass. 331—336.

Wood, R. D. (1975) Hydrobotanical methods. University Park Press. Baltimore. 173 pp.

Zieman, J. C. (1976) The ecological effects of physical damage from motor boats on turtle grass beds in Southern Florida. Aquat. Bot. 2: 127—141.

PHOTOSYNTHESIS OF AQUATIC PLANTS UNDER NATURAL CONDITIONS

MORTEN SØNDERGAARD

1. INTRODUCTION

The ability to capture photons and to use part of their energy to reduce inorganic carbon to organic carbon compounds is the most fundamental biological process on earth. This ability, photosynthesis (PS), along with a balanced assimilation of inorganic nutrients (e.g. phosphorus, nitrogen, potassium, etc.) creates growth, and is defined as primary production (PP).

The net primary production (NPP) of any single plant is an expression of the amount of matter available to consumption by other organisms. In an ecological context the amount of NPP and the rate of PS is often expressed in carbon units. Photosynthesis is the basis for all heterotrophic activity, including mankind. No wonder, the attempts to understand and to quantify these processes (PS, PP and NPP) have challenged and excited scientists to all times.

About 70% of the surface of our planet is covered with water and some 98% of the water-volume is saline. Only a small amount (0.01%) is present as freshwater in lakes and rivers. Fishery depends on the productivity of the sea, and human society depends totally on freshwater. The vegetation of aquatic systems and its metabolism has therefore attracted part of the research effort concerning the more quantitative aspects of photosynthesis. Not only due to its scientific value, but also due to its position in human interactions with the environment.

In an ecological context the basic questions are: How much organic matter is produced and at what rate? Which biotic and abiotic factors are controlling the amount and the instantaneous rates? How and how much of the primary production is transported to higher trophic levels in the ecosystem? The search for answers to these questions has been the guideline for some major currents in aquatic ecology during the last two decades.

Numerous are those investigations claiming to measure photosyn-

63

J. J. Symoens (ed.), Vegetation of inland waters. ISBN 90—6193—196—7.
© *1988, Kluwer Academic Publishers, Dordrecht. Printed in the Netherlands.*

thetic rates or primary productivity (terminology as in Wetzel 1983). However, doubt has always been raised to what extend the measurements represent true rates. Here, it might be useful to recall the controversy in marine ecology and the seemingly discrepancy between zooplankton (secondary) production and primary production (Pomeroy 1974, Williams 1981a).

The measurement of photosynthesis in aquatic plants do involve both methodological problems and questions on the interpretation no matter what method is choosen. Since there are several updated and excellent reviews on primary production and photosynthesis in aquatic ecosystems (Westlake 1963, 1980, 1982, Bunt 1975, Likens 1975, Boynton *et al.* 1983, Wetzel 1983), I intend in this chapter to focus more on physiology and methods than to present and summarize quantitiative data from the literature. I hope to demonstrate that such an approach can be useful.

2. PRIMARY PRODUCERS AND THEIR ENVIRONMENT

The scientific use of the term aquatic plant covers a wide range of taxonomic groups and life forms. Three major groups can be classified: the emergent, the submerged and the plants with floating leaves.

The emergent macrophytes are present in both freshwater and saline environments with waterlogged soils. The emergent vegetation encircles the open water and creates a border (not very sharp) to the true terrestrial ecosystem. This group of plants is dominating large marsh areas with the watertable at or just below the surface and will in many places dominate at shallow depths (< 2 m) in lakes and rivers (Hutchinson 1975, Westlake 1975).

Viewed on a global scale the most important element of the aquatic vegetation is the plants living permanently submerged (with the exception of the tidal flora). The group contains a diverse flora from the giant kelp (*Macrocystis pyrifera*) to the smallest cyanophycean and prasinophycean species (< 1.0 μm, Johnson and Sieburth 1982).

The third group of plants is the floating leaved vegetation. These plants often dominate in sheltered and slow flowing freshwater environments. Especially in the subtropic and tropic areas they can constitute the most important group of aquatic plants and create trouble. A total plant cover will make re-oxygenation of the water difficult and influence fish production; further, the boat traffic will have problems (Sculthorpe 1967). No single review can cover the ecological and productivity aspects of all aquatic plant groups and their special environments. In this chapter, I will restrict myself to the treatment of submerged plants.

The reason for this restriction is motivated by the fact that emergent and floating leaved macrophytes to some extent resemble terrestrial plants in their photosynthetic performance. The CO_2 source is the atmosphere, and light is not to any great extent influenced by water and the inorganic nutrients are in most instances supplied from the sediments (Wetzel 1979).

The submerged vegetation can somewhat arbitrarily be divided into macrophytes and microphytes based on size. Such a grouping has a functional basis, since different strategies for growth have developed and the metabolism of each species will to some extent depend on size. Macrophytes will normally contain more supportive and structural tissue than uni- and multicellular microphytes (algae). Moreover, a rooted submerged macrophyte is fixed in its place and not as exposed to water circulation and to such fast fluctuations in quantum flux density as is the phytoplankton. The submerged macrophytes have a distinct vertical zonation, where some species are always found in shallow water and others in deep water (Levring 1966, Spence 1967). The microphytes inhabit both the open water (phytoplankton), and cover submerged surfaces (periphyton). The latter group is described by special terms as epiphytes, epipelic and epipsammic communities (Wetzel 1983). All these growth sites give rise to special photosynthetic performance and create special methodological problems, as will become clear later on.

The rate of photosynthesis of aquatic plants is controlled by two main factors: (1) The physiological state of the organisms themselves; and (2) Abiotic and biotic factors in the environment. In the following paragraphs I will introduce some physicochemical variables with direct influence on photosynthesis and then give a short introduction to major aquatic plant categories. The introduction will focus on photosynthetic behavior linked with environmental variables.

2.1. *Physico-chemical variables controlling photosynthesis*

Photosynthesis under natural conditions is controlled by a series of interacting environmental variables. The basic variables are light, temperature, water movements, CO_2, pH, O_2 and nutrients. Allelopathy and the effect of special compounds as cyclic nucleotides might also be involved.

2.1.1. Light

No single environmental variable has been studied so intensely as light. This is of course logic recalling the known relationships between light and photosynthesis. In the aquatic environment, this relationship is further reinforced by the attenuation of light in water.

The mean attenuation coefficient from z_1 to z_2 meters depth of any wavelength or range of wavelengths is defined as:

$$\bar{K} = \frac{\ln\left(E_d(z_1) / E_d(z_2)\right)}{z_2 - z_1} \quad (m^{-1})$$

The logaritmic attenuation of light and different $\bar{K}$ values for light of different wavelength creates an environment with a special light climate at each depth. A detailed treatment of light in water and some ecological implications can be found in Jerlov (1976), Højerslev (1979) and Spence (1975).

In the context of phototrophs the word light is a little too broad and not well defined. What is really of interest is the photosynthetic active radiation (shortened PAR) from 350 to 700 nm. The unit of expression should be in quanta, as the light harvesting chlorophyll-protein complex and ultimately the reduction center responds to discrete quanta (Björkman 1981). The unit of quantum flux density (QFD) is $\mu E\ m^{-2}\ sec^{-1}$ ($= 6.02 \times 10^{17}$ quanta $m^{-2}\ sec^{-1}$). Approximate conversion factors among QFD and other measurements of radiation is presented by Harris (1978) and Richardson et al. (1983).

All photoautotrophic plants respond in a predictable way to an increasing flux of quanta (Figure 1). At low QFD the rate of photosynthesis is controlled by photochemical processes, although not necessarily linear at very low densities. This part of the light response is temperature independent. The slope of the curve and the light compensation point (LCP = net photosynthesis is zero) is dependent on the organisms in question, their adaptation and physiological state. The LCPs have a very wide range, from 0.2 to 55 $\mu E\ m^{-2}\ sec^{-1}$ in submerged macrophytes (Spence & Chrystal 1970, Van et al. 1976). The minimum range of QFD allowing growth of micro-algae is from 1 to 30 μE $m^{-2}\ sec^{-1}$ in a 12 : 12 L/D cycle (Richardson et al. 1983). The response to light by different microalgal groups is to some extent genotypically fixed, e.g. green-algae seem to require higher intensities than other algal groups (Richardson et al. 1983). However, adaptations can and do occur as shown by Spence & Chrystal (1970) for Potamogeton polygonifolius.

At increasing QFD the photosynthesis ultimately becomes saturated (P_{max}) and eventually decrease at exposure to very high light intensities. This theoretical response curve is shown in Figure 1. Note that photoinhibition did not occur in Littorella, not even at 800 $\mu E\ m^{-2}$ sec_{-1} (not shown). The level of P_{max} in conditions where no resource limits photosynthesis is, for any organism, temperature dependent. The level of P_{max} is controlled by enzyme reactions and feed-back mecha-

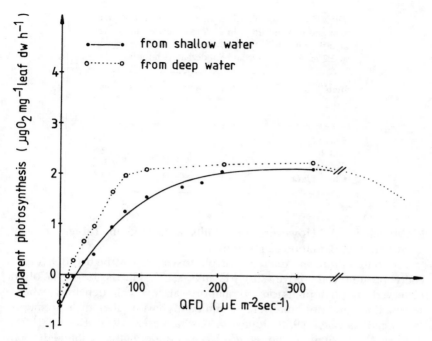

Fig. 1. Apparent photosynthesis of *Littorella uniflora* collected June 10 at 0.2 and 2.2 m depth in Lake Kalgaard. A theoretical photoinhibition which did not occur in *Littorella* is indicated. (After Bonde & Søndergaard, unpublished).

nisms in the photosynthetic cycle. The P_{max} is not fixed for any plant, but will depend on the previous growth conditions. Plants grown at high QFD will normally have a higher photosynthetic capacity (higher P_{max}) and become light saturated at higher light intensities than plants grown at low light. (Björkman 1981. Harris 1978). This adaptation is well known from both phytoplankton and macrophytes in nature (Spence & Chrystal 1970, Sand-Jensen 1978, Falkowski & Owens 1980).

Adaptation to low QFD in macrophytes is often seen to involve changes in the specific leaf area and a reduced dark respiration (Spence & Chrystal 1970, Sand-Jensen 1978). Shade adapted plants also have a higher pigment concentration than plants growing at higher QFD and can have a relative enrichment of chlorophyll *b* or *c* to chlorophyll *a* (Table I). Apparently, this change in pigment composition and concentration increase the photon capturing efficiency at low QFD (Falkowski & Owens 1980, Björkman 1981, Falkowski *et al.* 1981, Richardson *et al.* 1983, Bonde & Søndergaard, unpubl.). Questions have been raised whether aquatic macrophytes generally respond to shade by alteration of the chlorophyll *a : b* ratio (Barko & Filbin 1983, Wiginton &

Table I. An example of change in pigment composition and concentration in *Littorella* related to growth at different QFD. S = growth at 0.2 m, D = growth at 2.2 m. After Bonde and Søndergaard, unpubl.

Item	Chlorophyll ($\mu g\ mg^{-1}$ leaf dw.)	
	S	D
Chl. *a*	4.2	6.8
Chl. *b*	1.4	2.7
Total Chl.	5.6	9.5
Chl. *a* : *b* ratio	3.1	2.5

McMillan 1979). However, the results presented in Table I clearly show that this adaptation can be found.

Photoinhibition, the decrease of photosynthesis at high QFD (especially pronounced above 200 $\mu E\ m^{-2}\ sec^{-1}$, Harris 1978) is often observed in phytoplankton populations at low cell densities, where selfshading do not interfere significantly with the measurements (Steemann Nielsen 1958, Jónasson & Mathiesen 1959, Rodhe 1965, Talling 1965, Stadlemann *et al.* 1974, Søndergaard & Sand-Jensen 1979a, among numerous papers). To my knowledge photoinhibition has never been observed in submerged macrophytes.

In his review on photoinhibition, Harris (1978) concluded that light intensity, spectral composition and duration of exposure all influence the degree of inhibition observed in a particular species and that difference in susceptibility among different species occur. The physiological mechanisms involved in photoinhibition are not fully understood, but both photosystems seem to be affected (Harris 1978).

In a situation with free mixing of the water, the QFD reaching an alga will be fluctuating, and the relative duration of exposure to high QFD will as a consequence be short. In natural conditions the suppression of photosynthesis at the surface might be of minor importance. The dependence of exposure time is illustrated in Figure 2. Recently, Walsh & Legendre (1983) have found that fluctuating light significantly (up to 30%) enhance photosynthesis at QFD below saturation compared with a constant light level. The fluctuations used were in accordance with fluctuations predicted by sea surface waves and the results can probably be extended to fluctuations in a mixed water column.

These experiments support the conclusion that the exposure of phytoplankton in a fixed depth and especially at the surface can give erroneous and too low photosynthetic rates. Primary production is underestimated. It also must lead to a reexamination of primary pro-

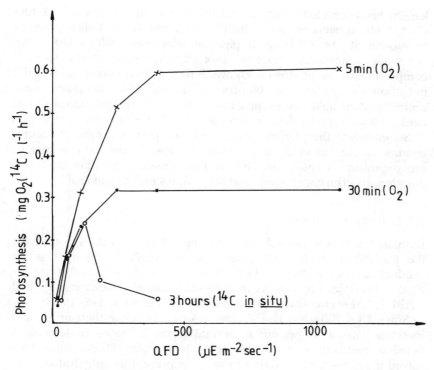

Fig. 2. Comparison of photosynthesis of *Ceratium hirundinella* measured at different light intensities and increasing time of incubation. Redrawn after Harris (1978).

duction models and calculations from incubators, where light fluctuations are deliberately kept to a minimum (Rodhe 1965, Vollenweider 1965, Fee 1969, Talling 1970, Gargas *et al.* 1976). The 'afternoon depression' of photosynthesis so often observed in planktonic assemblages (see Harris 1978) might merely appear to be an effect of incubation method, which enhances photoinhibition, photorespiration, and CO_2 depletion (high pH). Hysteresis observed in both algae (Gallegos *et al.* 1980) and in submerged macrophytes (Thyssen 1980) is probably linked with environmental variables and is not a function of any circadian rhythm, although it can be observed in phytoplankton as cellular fluorescence changes in (Vincent 1980).

Returning to the spectral composition of light. Chromatic adaptation is an often claimed feature, especially of marine benthic algae (Levring 1966) and in blue-green algae (Jeffrey 1981).

The scientific foundation for the old paradigm of 'pure chromatic adaptation' (viz. the red algae are growing in deep water due to their ability to efficiently harvest the dominating photons of green wavelength) has been challenged several times and, most recently by Dring (1981). His conclusion, with which I agree, was that red algae respond to growth at low QFD by a pigment adaptation, which looks like chromatic adaptation. However, biological response to the spectral composition of the photons is known. Jeffrey (1981) concluded that the induction of synthesis of biliproteins in red and blue-green algae optimizes their light-harvesting capacity in the green and orange-to-red light dominating at depth in the sea and in many freshwaters.

Summarizing the physical prerequisite for photosynthesis at natural conditions: the flux of photons should be viewed from the positions of the organisms in question, that is the density (light intensity) and duration, spectral composition and fluctuations in time and space.

2.1.2. Inorganic carbon

Inorganic carbon in the unhydrated form — CO_2 — is the substrate for the RuBP-carboxylase in C_3-plants with 3-PGA as the first stable product (Latzko & Kelly 1980). Probably HCO_3^- is the substrate for PEP—carboxylase present as the dominant first carboxylase in C_4 and CAM (CAM = crassulacean acid metabolism) (Kluge & Ting 1978).

Since CO_2 diffusion is 10^4 times slower in water than in air and inorganic carbon is present in several forms in water, it deserves a detailed treatment as a possible control of photosynthetic rates. Dissolved inorganic carbon (DIC) in water is present as unhydrated CO_2, and H_2CO_3, HCO_3^- and CO_3^-. The distribution among the three important forms, CO_2, HCO_3^- and CO_3^- is regulated by pH, temperature and ionic strength (Rebsdorf 1972, Stumm & Morgan 1981) The equilibrium chemistry of the complex carbonate system is treated in details by Stumm & Morgan (1981).

In the marine environment pH and DIC is rather constant at about 8.3 and 2.2 mM, respectively, and create a steady situation for DIC supply in most situations. The reverse happens in freshwaters where large variations occur among different localities and in time and space within a given site (Wetzel 1983). At air equilibrium the concentration of CO_2 is low, about 14 and 17 μM in the sea and freshwater at 15 °C, respectively (Stumm & Morgan 1981). In the sea the ratio of HCO_3^-/CO_2 is constant at about 160, a situation favoring organisms able to supply CO_2 from the HCO_3^- pool. The ratio is very variable in lakes and rivers. In acidic waters the ratio is < 1 and in the pH range from 7 to 8.5, the ratio ranges from 2 to 150. In most situations, HCO_3^- will dominate the DIC pool.

The low concentration of CO_2 and its slow diffusion raises the question of CO_2 as a limiting factor for photosynthesis, if HCO_3^- cannot be used. CO_2 depletion in freshwater can take place in situations with low DIC (low alkalinity) and high plant biomass. The depletion is due to assimilation and the simultaneous increase in pH (Schindler 1971, Schindler & Fee 1973, Talling 1976, Van *et al.* 1976).

It is known that all submerged aquatic plants can utilize CO_2 (Raven 1970, 1974), some plants can also utilize HCO_3^- (HCO_3^- utilization is treated in Section 4.2.1.), and none uses $CO_3^=$ directly (Raven 1970). Raven 1970 and Black *et al.* (1981) have stated that it is possible that the rate of photosynthesis under natural conditions can be limited (controlled) by the rate at which inorganic carbon enters the cell. A simple model to illustrate the total resistance to CO_2 uptake is given by Raven (1970):

$$PS = \frac{C_0}{\dfrac{1}{k} + \dfrac{1}{p}}$$

where PS = the rate of photosynthesis per area and time, $1/k$ is the reaction (carboxylation) resistance, $1/p$ is the diffusion resistance, and C_0 is the bulk phase CO_2 concentration at the outer surface of the cell wall. A detailed treatment is given by Raven (1970), Smith & Walker (1980), and Black *et al.* (1981). Summarizing the conclusion: CO_2 diffusion is a major factor limiting photosynthesis under natural conditions, the reaction resistance is without importance in macrophytes, but might be in microphytes. The major influence of diffusion resistance is not simply a function of low CO_2, but due to the size of the unstirred layers surrounding all aquatic plants, that is the length of the diffusion path.

Returning to the response of photosynthesis and the CO_2 concentrations. The half-saturation constant for CO_2 assimilation $K_{1/2}(CO_2)$ can be used to analyse CO_2 limitation. In submerged macrophytes the range is between 0.07 and 0.3 mM (Van *et al.* 1976, Smith & Walker 1980, Allen and Spence 1981, Sand-Jensen 1983). At optimum photosynthetic condition of light and temperature and at air equilibrium concentrations of CO_2, macrophytes that utilize only CO_2 will be CO_2 limited. However, an ability to use HCO_3^- will increase the photosynthetic rates. Microalgae have apparently a higher affinity for both CO_2 and HCO_3^- than macrophytes (Allen & Spence 1981). This difference is to some extend reflected in their CO_2 compensation points, which can be very low ($< 1 \mu M$, Bidwell 1977, Lloyd *et al.* 1977a, Allen & Spence 1981). Most macrophytes have higher CO_2 compensation points ranging from about 2 to 25 μM (Søndergaard 1979, Bain & Proctor 1980, Allen & Spence 1981). However, it must be emphasized that the CO_2 compensation is only of theoretical interest in plants able to utilize HCO_3^-.

The difference between micro- and macrophytes can perhaps be related to glycollate excretion in many algae (Harris 1978). In higher plants glycollate is metabolized to CO_2 (photorespiration) and with an efficient recycling this can give rise to lower compensation points (Søndergaard 1979, Søndergaard & Wetzel 1980).

In a study on the ability of different phytoplankters to deplete CO_2 in lakes, Talling (1976) showed that different species had different capabilities. This was later confirmed by Maberly & Spence (1983). Talling's experimental studies implied DIC limitation to photosynthesis. Indirect evidence of inorganic carbon as a controlling factor for photosynthesis under natural conditions were also presented by Nygaard (1968) for phytoplankton in a soft water lake. Shapiro (1973), Talling (1976), Jaworski et al. (1981) and Maberly & Spence (1983) have drawn attention to the differences in ability to deplete DIC as a factor in seasonal succession patterns and competition of phytoplankton.

From the experimental evidence, it is clear, that the concentrations of CO_2 at saturation are much higher than the CO_2 concentrations present. In macrophytes the range is from 0.25 to 1.5 mM (Van et al. 1976, Allen & Spence 1981, Sand-Jensen 1983), and probably lower in microphytes, which seems to have a higher CO_2 affinity (Allen & Spence 1981). The dominance of HCO_3^- do not counteract low CO_2 availability for freshwater macrophytes, as the $K_{1/2}$ for HCO_3^- is high (0.7 to 23 mM). In some unicellular algae and marine macrophytes the affinity for HCO_3^- uptake is, however, higher (low $K_{1/2}(HCO_3^-)$, Allen & Spence 1981, Sand-Jensen & Gordon, 1984) and do not necessarily imply limitation. Allen & Spence (1981) in their treatment of freshwater plants concluded, that DIC must be considered as a factor controlling photosynthesis under natural situations.

DIC limitations have been shown to occur in situ for some macrophytes (Van et al. 1976, Sand-Jensen & Søndergaard 1978, Søndergaard & Sand-Jensen 1979b).

If the here mentioned examples and statements are rare exceptions or wrong, conclusions concerning CO_2 as a rate limiting resource must await further investigations and analysis. However, there is not doubt that special physiological, morphological and anatomical adaptations have evolved to compensate the low CO_2 supply and the high HCO_3^-/CO_2 ratio found in most aquatic systems (Talling 1976, Søndergaard & Sand-Jensen 1979b, Allen & Spence 1981, Keeley 1981). The concentration of DIC can control and limit the current rate of photosynthesis in situ. Along with Schindler & Fee (1973) and Sand-Jensen (1983), I would like to emphasize, that this does not mean that growth (biomass accumulation) is limited by DIC. A major point for ecologists.

2.1.3. Water movement

One reason, it is so difficult to reach conclusions on CO_2 limitation, is the presence of the stagnant boundary layer around any organism in water (Leyton 1975). The transport of DIC through the layer is by molecular diffusion and the mass transport is thus dependent on the resistance in the layer, e.g. the thickness of the layer. A decrease in thickness by water movement or movement by the organism will decrease the resistance to CO_2 transport and would look like an increased concentration of CO_2, if the photosynthesis is diffusion limited.

Water flow velocities in experiments with macrophytes are easily controlled. However, the real problem is to measure and define the waterflow in natural situations. Due to their small size, the problem is not so severe in phytoplankton (Raven 1970), but how should an experiment with phytoplankton and epiphytes be conducted to simulate the true diffusion situation?

The effect of flow velocity on the photosynthetic rates of submerged macrophytes has been investigated by Westlake (1967), Smith & Walker (1980), and Madsen & Søndergaard (1983). In Figure 3 the effect is exemplified in an experiment with *Littorella uniflora*. In nature, *Littorella* is supplied with CO_2 from the sediment interstitial water via diffusion in the lacunal system of the roots, stem and leaves (Søndergaard & Sand-Jensen 1979b). There is of course no flow around the roots, and the $K_{1/2}$ (the half saturation constant) was about 1.9 mM CO_2. In an experiment with a very moderate mixing of water around the roots, the apparent $K_{1/2}$ decreased to about 0.9 mM. This clearly illustrates the role of the boundary layer and mixing.

In macrophytes the unstirred boundary layer have been calculated to be several hundred microns (μm) even in a mixed situation (Raven 1970, Walker et al. 1979, Smith & Walker 1980, Madsen 1984). This means that the diffusion resistance of CO_2 is a major (the major) controlling process in photosynthesis (Raven 1970, Smith & Walker 1980, Black et al. 1983, Madsen 1984). True response to inorganic carbon can, as a concequence, only be investigated in experiments with a constant and fast flow around the plants, and if the CO_2 fixation can be discribed by Michaelis—Menten kinetics (Smith & Walker 1980). Judgement of carbon limitation in nature is thus very difficult.

Small unicellular algae (e.g. *Chlorella*) have a very thin (5—10 μm) unstirred layer under well stirred conditions (Raven 1970). Performance of carbon fixation, which can be discribed by Michaelis—Menten kinetics, would imply that the rate of photosynthesis should not be diffusion limited. However, many phytoplankton species are large, and are not placed in a well-stirred experimental set up, thus in most natural

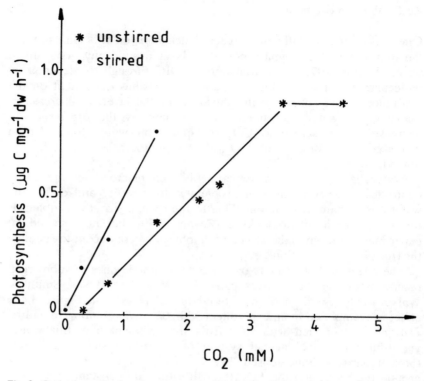

Fig. 3. Carbon uptake by the roots and photosynthesis of *Littorella uniflora* during stirred and unstirred incubations. Modified from Søndergaard and Sand-Jensen (1979).

situations the stagnant layer is probably thicker. Sinking and motility will create water movements around each cell, but knowledge on these relationships is absent. It only seems possible to conclude that we do not know if the photosynthetic rate of phytoplankton is diffusion limited under natural conditions.

2.1.4. Oxygen

No aquatic plants have so far been shown to possess C_4-photosynthesis (Osmond *et al.* 1982). Photorespiration, the light-mediated release of previously reduced CO_2, is thus an inherent process in aquatic plants unless compensated by special mechanisms. To what extent the photorespiratory carbon oxidation cycle will be of quantitative importance depends on the CO_2 and O_2 concentrations (partial pressure) at the carboxylation site. In terrestrial plants a two-to-threefold increase of the CO_2 concentration can prevent the oxidation of ribulosebiphosphate. Any CO_2-concentrating mechanism will favor plants in a natural selection.

Photosynthesis of aquatic plants is affected by oxygen. An increased concentration in the external medium decreases the apparent photosynthetic rate (Brown *et al.* 1974, Hough 1974, Burris 1977, Van *et al.* 1976, Søndergaard 1979). Similarly, the CO_2 compensation point is O_2 sensitive in many freshwater and marine macrophytes (Lloyd *et al.* 1977b, Søndergaard 1979). Seasonal variations in sensitivity have been shown to occur and probably can be induced by temperature (Bowes *et al.*, 1978). In *Hydrilla*, high temperature and shift in day length induce a lowering of the CO_2 compensation point, which might be of competitive value (Bowes *et al.* 1978).

Photorespiration might be counteracted by high intracellular CO_2 concentrations. The transport of HCO_3^- into the cells and a high carbonic anhydrase activity can be viewed as such a process. The light dependent HCO_3^- uptake in *Chara spp.* do increase the internal CO_2 concentration (Lucas 1979, 1980) and favor the carbon reduction function of the RuBP — carboxylase/oxygenase complex.

The normal C_3-fixation of CO_2 (Calvin cycle) was discovered in algae, and the RuBP — carboxylase activity is positively correlated with maximum photosynthesis in many plankton species (Glover & Morris 1979). Many phytoplankton species have oxygen sensitive photosynthesis (Burris 1981, Birmingham *et al.* 1982), although CO_2 concentration mechanisms have been observed (Badger *et al.* 1978, Miller & Colman 1980).

Low CO_2 compensation points do not necessarily imply low photorespiration. The elevated CO_2 compensation points in C_3-plants compared with almost zero in C_4-plants is a function of the glycollate—oxidase enzyme and the decarboxylation of glycine. The release of glycollate found in many algae (se Harris 1978) is partly related to photorespiration, but no CO_2 release occurs and the CO_2 compensation point remain low. Net photosyntesis, viewed from the point of the algal cell is lowered, however, the bacterial assimilation of glycollate returns some of the carbon to a particulate form.

Aquatic plants are often exposed to situations, which enhance photorespiration, both on a seasonal and a diel scale. Photorespiration is controlled by O_2, CO_2, light intensity and temperature (Zelitch 1971). A combination of low CO_2 and high values of the other variables enhance photorespiration. In water, a higher temperature and high QFD increase photosynthesis, deplete CO_2 and increase O_2. This is most pronounced in situations with high biomass and low alkalinity (e.g. Talling 1976). The midday/afternoon depression of photosynthesis claimed to be a general feature might to some extend be explained by enhanced photorespiration (Hough 1974, Harris 1978). There seems no reason to expect photorespiration to be lower in aquatic plants than in their terrestrial counterparts.

2.1.5. Inorganic nutrients

The supply of inorganic nutrients, especially nitrogen and phosphorus, is the key factor controlling the general level of primary productivity and community production in aquatic systems. Factors as grazing, sedimentation, water mass stability etc. all affect growth and current biomass. However, on a global scale, nutrients control volume and area based photosynthesis both in the sea and in freshwaters (Bunt 1975, Likens 1975).

The availability of different inorganic nutrients and trace elements (micronutrients) influences the physiological state and the current rate of photosyntheis in aquatic plants, as known for terrestrial plants (e.g. Salisbury & Ross 1978). A few examples can exemplify the statement. Phosphorus is normally a growth limiting factor in freshwater environments. The content in macrophytes can be very different and varies with place, time, species and organ (Hutchinson 1975), and has been used as an indicator of P limitation. And along with other elements as a way to assay the nutrient status of natural waters (Gerloff 1973). Schmitt & Adams (1981) have shown that the photosynthetic rate of shoot tips of *Myriophyllum spicatum* is dependent of the tissue concentration of P, up to a limit, at which no further increase in P increases the photosynthesis. If such behavior is a general phenomenon in aquatic plants and eventually can be extended to other key nutrients, the experiments by Schmitt & Adams (1981) have expanded the general relationship between growth and the critical tissue concentration published by Gerloff & Krombholz (1966). The tissue concentration found to saturate photosynthesis was about 0.3% P compared with 0.07% for growth. A comparison with the natural phosphorus concentrations of *Myriophyllum* in Lake Wingra, made Schmitt & Adams (1981) conclude that earlier statements of phosphorus sufficiency may be misleading viewed in a physiological photosynthetic context.

Similar comparisons with natural phytoplankton can be difficult to assess, since a separation of algae from detrital particles is very difficult, if not impossible. A direct *in situ* measurement of algal phosphorus (and other nutrients) is thus uncertain at best and probably dubious in most cases. Experiments with algal cultures have shown, that the supply of nutrients of course influence the growth rate (e.g. Nalewajko & Lean 1980). How any limitation influences the photosynthetic performance of single organisms is, however, largely unknown.

The effect of micronutrients (mostly metals as Fe, Mn, Zn, Cu, Mo, V etc.) on phytoplankton photosynthesis has received some attention.

One rationale for this interest is the low solubility and the ability of the metals to chelate with organic compounds. A series of cases have been described, where the addition of one or several micronutrients can increase the short-term (hours) carbon fixation in natural phytoplankton populations (Goldman 1972, Wetzel 1972). Often the effect is most pronounced when the element (e.g. Fe) is added with a weak complex-former such as EDTA. However, special situations with photosynthetic enhancement observed in lakes in bedrock areas can be explained by low concentrations. This must probably be considered as single and unique phenomenons. Most intension have been adressed to the physiological availability, which can be reduced by complexing with organic matter or adsorption onto (into) $CaCO_3$ in a colloidal or particulate form (Wetzel 1983). There seems to be no evidence to suggest a significant influence on photosynthesis considering the aquatic environment on a wider scale. The influence of inorganic nutrients on species composition, competition, biomass development (growth) and the general level of productivity is well known (e.g. Vollenweider 1968). Any direct effect on the photosynthesis of individual plants under natural conditions is only known for a few special cases (limitation or poisoning). The complex interactions of an infinite number of variables make it very difficult to render general conclusion and to assay the effects of single factors. Experimental work with cultures might help in this respect, but shortcomings by extrapolation to nature seems more likely to occur than the contrary. This, however, should not influence our attempt to elucidate photosynthetic behavior in nature.

2.1.6. Miscellaneous

Photosynthesis in aquatic plants is not only affected and controlled by physicochemical variables, but can also be influenced by the presence of other organisms and specific organic products. Examples of both enhancement and inhibition are known.

Auxotrophy is common among algae. The term is used to characterize organisms, which do not have the ability to synthesize one or several essential vitamins, but are forced to rely on uptake from the environment. The three water-soluble vitamins include B_{12}, thiamine and biotin. In special cases the photosynthetic rates of phytoplankton can be significantly increased by adding one of the vitamins (Wetzel 1965, 1966, Hagedorn 1971). The subject has been reviewed by Provasoli & Carlucci (1974).

A newcomer in our knowledge concerning regulation of carbon fixation is cyclic adenosine-3,5-monophosphate (cAMP). Francko (1983) has reviewed this new and interesting field of 'biochemical ecology', and among other things focused on the stimulating and inhibitory functions of cyclic nucleotides in aquatic plants, (see Franko 1984). Although our knowledge is growing, it seems a little premature to make general conclusions on significance for photosynthesis in nature.

The metabolic regulation of photosynthesis by dissolved effector molecules is an almost unknown part of aquatic ecology. Allelopathy is a subject, which has received attention. Allelopathy is the adverse influence of one organism on another mediated through chemical agents. It has been demonstrated, that extracts from algae (e.g. blue-green algae) inhibit the development of other algae (Keating 1978, Chan et al. 1980). However, only a few specific compounds have been identified and investigated, although many speculative effects have been put forward. Looking at the effect on photosynthesis as such, the clearest example has been presented by Wium-Andersen et al. (1982). Two sulfur compounds extracted from Characean species have been chemically identified (Anthoni et al. 1980) and were shown to severely inhibit the carbon fixation of an epiphytic diatom and natural phyto-plankton. In nature, the Characeans are without epiphytes and often occur in monotypic stands. This is most probably due to the allelopathic effect of the sulfur compounds (Wium-Andersen et al. 1982).

2.2. Primary producer communities

In the preceding sections it has been made clear that different plant communities have to adapt and respond to the environment in different ways. The crude division into two major categories, microphytes and macrophytes is not a satisfactory basis to analyse and explain the behavior and physiological adaptations of each groups. The size of an organism do influence its potential to fast changes in metabolism and photosynthetic performance, but this is not the only significant feature. The ability to stay in a fixed position in the water column is also of importance.

2.2.1. Phytoplankton

Due to eddy currents and turbulence in the open water, the phyto-plankton has to photosynthesise and grow in gradients of nutrients and fluctuating light. The changes of light can be very fast. Although a lot of algal groups and species are motile, the capability of movement is limited compared with the wind forced mixing. Anyway, both horizontal and vertical differences in algal density and community composition are often found. Diel migration can in special cases be observed. There is no a priori homogeneous distribution, not even in isothermal and 'mixed' water masses (e.g. Kristiansen & Mathiesen

1964, Berman & Rodhe 1971). Buoyancy by means of gas vacuoles is found in blue-green algae and is a very efficient way to avoid sedimentation and stay in the photic zone. Accumulation of oil is another way to avoid sinking, this is found in *Botryococcus*.

The life in the open water creates several specific problems.

1. As most algae have a specific density larger than one, they eventually sink out of the photic zone. Sinking into or below a thermocline or chemocline prevents a return to the surface and light. Eddy currents prevent sinking at the maximal theoretical speed (Wetzel 1983). Extrusions from the algae might to some extend slow down sinking and help the algae to follow turbulence. Changes in buoyancy are partly controlled by the physiological state of the cell. Nutrient limitation, e.g. silica depletion can influence the sinking rate of diatoms (Jewson *et al.* 1981).

2. In a mixed water column the algae are exposed to fast changes in QFD and light quality. They have to be adapted to manage both high and low photon densities. During spring and autumn overturn in dimictic lakes and deep mixing in the oceans, the pelagic population will travel between the photic and aphotic zone. During summer stratification, if present, the mixing depth is from several meters in lakes to several tens of meters in the oceans and many coastal areas. Exposure to low quantum flux density is thus the situation with the highest probability. The reversible photoinhibition can be viewed as a protection against high intensities occurring with a low frequency and during short time. Adaptation to changes in the light regime can be very fast. Intracellular photosynthetic pigments as chlorophyll and accessory pigments can change within 15 minutes (Falkowski 1980) and a response in chloroplast structure, size and orientation to changing light conditions can take place within hours (Harris 1978). A cell division as proposed by Steemann Nielsen *et al.* (1962) is not necessary in all cases.

   The small size gives a high potential for growth, with doubling times between a few hours (3) and 7 days. A fast physiological response as changes in the photosynthetic characteristics is possible. natural community was reflection of the light environment of the previous week.

3. The pelagic is often a low nutrient environment. The small size, sinking, motility and water turbulence break down any depleted microzone around the cells. The diffusion path is short. One adaptation to low availability of nutrients is the high affinity uptake systems (Nalewajko and Lean 1980, McCarthy 1980) and the use of membrane-bound enzymes as alkaline phosphatase (Pettersson 1980).

4. Low K_{CO_2} and $K_{HCO_3^-}$ values have been found in planktonic algae (Allen & Spence 1981) and prevent inorganic carbon from limiting growth. However, the photosynthetic rate can be influenced.

5. The high surface area to volume ratio and the life in water seem to create leachage problems. Phytoplankton releases extracellular organic carbon (EOC) as an integrated part of their photosynthetic performance (Mague et al. 1980, Riemann et al. 1982). The relative amount is variable, but is probably higher in oligotrophic than in more productive areas (Anderson & Zeutschel 1970, Thomas 1971). The subject is further treated in section 3.3.

2.2.2. Macrophytes

Most submerged macrophytes are anchored to or rooted in the bottom. For the anchored species as marine macroalgae, and free floating macrophytes, the water is the sole nutrient source. Rooted angiosperms can and do obtain a major part of their nutrients from the sediment and inorganic carbon via leaves and stems (Huebert & Gorham 1983). Special cases exist for uptake of CO_2 via the roots (see Section 4.2.2.).

1. Most angiosperm macrophytes have an extensive development of gas filled lacunae (Sculthorpe 1967). The lacunae give buoyancy to the shoot-systems and aid a fast internal diffusion of gases (O_2, CO_2). Depending on growth form, macrophytes can grow toward the surface and develop a leaf-canopy in strata with good light conditions. If phytoplankton and epiphyte shading are insignificant, the macrophytes are exposed to a light climate, with rather low fluctuations with time. Some macrophytes live at several meters' depth and can adapt to low QFD. Other species only possess a limited ability to adapt (Spence & Chrystal 1970). Colonization with epiphytes decreases the QFD reaching the leaf surface (Sand-Jensen 1977, Sand-Jensen & Søndergaard 1981, Sand-Jensen & Borum 1984). The apparent light compensation point is increased and the plants have to decrease their depth of colonization.

2. The slow diffusion of CO_2 and HCO_3^- in water and the unstirred layer is an important factor controlling in situ photosynthesis. Along with light, perhaps the most important. Dense epiphyte communities enhance the effect by competition for CO_2 and by increasing the thickness of the layer. Wave action, water flow and turbulence must be regarded as important factors, as they reduce the resistance of CO_2 diffusion.

3. The high HCO_3^-/CO_2 ratio is often circumvented by HCO_3^- utilization. The affinity for HCO_3^- seems rather fixed in marine species living at constant and high HCO_3^- concentration.

In freshwater macrophytes the affinity is more plastic, and probably can be induced by the environment (Sand-Jensen & Gordon, 1984). The availability of DIC, which can change with season might be involved. A lowering of the CO_2 compensation point along with a very low light compensation point can be seen as the physiological answer to counteract DIC shortage during much of the day (Van et al. 1976, Bowes et al. 1978).

2.2.2. Epiphytes

Epiphytes are part of a mixed heterotrophic-autotrophic community on substrates of stone, plants, and sediment in the photic zone. The vertical distribution on the substrate is restricted.

1. The algae in a dense community are faced with diffusion problems as macrophytes and exposed to large diel fluctuations in oxygen concentrations (Revsbech & Jørgensen 1983) and probably nutrients as well. The close juxtaposition for autotrophs and heterotrophs makes an efficient internal community cycling of elements likely (Wetzel & Allen 1970).
2. Light is attenuated very fast in these populations. For epipsammic communities perturbation of the substrate by waves or animals can influence the light climate. Benthic diatoms are adapted to these fluctuations (Rasmussen et al. 1983).

3. METHODS

The photosynthetic process consists of several sub-processes. A simplified model of these sub-processes is presented in Figure 4. The most important are:

1. The uptake and reduction of CO_2 and production of O_2
2. Photorespiration (PR) if present; use of O_2 and release of CO_2
3. Dark respiration (R)
4. Any extracellular loss of immediate photosynthetic products (EOC).

In the ideal situation the exchange of CO_2 and O_2 should occur simultaneously. However, CO_2 and O_2 do not have to be exchanged at a fixed ratio = the photosynthetic quotient.

Primary production, synonymous with the net uptake of CO_2 is also called apparent photosynthesis (APS). True photosynthesis is APS + PR + R. The primary goal for the production ecologist is to be able to measure how much biomass there is produced in time and space and available to heterotrophic production. An array of methods and tech-

82

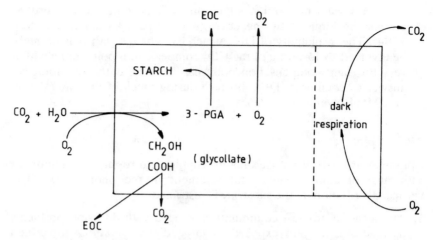

Fig. 4. A simple model of the pathways of O_2 and CO_2 during photosynthesis. The release of glycollate as EOC or CO_2 is dependent on the enzymatic systems of the species in question.

nical devices have been developed to measure photosynthesis in nature and in laboratory experiments. In theory, any substance linearly related to the photosynthetic rate can be applied. Measurements of CO_2 (HCO_3^-) and O_2 are most widely applied; even $^3H—H_2O$ uptake has been used as well (McKinley & Wetzel 1977). The most direct and best method is to measure the CO_2 budget. This can be done directly on CO_2 (or $^{14}CO_2$ which will be considered in Section 3.3.) or indirectly through pH and alkalinity. Oxygen exchange is another well proven method. If the accumulation of carbon biomass is the main purpose, problems arise when the O_2 values are converted to carbon.

3.1. *Enclosures*

The basic approach in the majority of studies on primary production in natural communities is to use some kind of enclosure and measure the metabolism inside the enclosure. This can be done in a small bottle with a pelagic community or in large enclosures surrounding macrophytes or other benthic communities.

The use of enclosures is unavoidable if not the total community of a system wants to be included (e.g. the upstream — downstream oxygen method of Odum (1957)). The use of enclosures is also inherent in many methods. It is very difficult to use the radiocarbon method without containment (Hesslein *et al.* 1980).

Measurements in open systems can be conducted in situations, where the photosynthesis is dominated by the autotrophic component of interest. The diffusion of gases (CO_2 or O_2) from or to the measuring points and the air-water exchange must, however be taken into account. Physical mixing introducing new populations and water masses with new physico-chemical properties might blur the results. 'Drogue' measurements in the oceans (e.g. Burney et al. 1982) have been used to compensate for the two problems mentioned. The most obvious situation in which to use the free water approach, is in situations without water mixing and atmospheric contact, that is below or in thermo- and haloclines. Free water methods have been discussed and recommended by Hall and Moll (1975) and Verduin (1975) and applied in ecophysiological studies by Talling (1976). What is wrong with containment? A series of theoretical and practical problems occur:

Phytoplankton: 1. Most enclosures change the light regime; UV and general light attenuation can be of special importance in surface waters (Hobson & Hartly 1983). 2. The free mixing is hindered, both on a large and a small scale. 3. Sampling needs not to be representative; photosynthesis in 100 to 300 ml is often used to recalculate to km^2 and $10^6 m^3$. 4. The algae are exposed to constant light. 5. Lack of gas exchange with the atmosphere. 6. 'Bottle effects'. 7. Incubation time. Items 2, 3, 4, and 5 have been treated previously.

Does containment itself stress the organisms and to what extent do bottle size and incubation time interfere with the results? Conflicting results have been presented in the literature (Venrick et al. 1977). Recently, Gieskes et al. (1979) in samples from the Atlantic Ocean observed an accelerated death of algae in small bottles (30—300 ml) incubated 12 hours in light compared with large (3.8 l) bottles. Photosynthesis was up to ten times higher in the larger bottles. Other investigators have not been able to demonstrate 'bottle effects' in short-term incubations (Lewis 1974, Chr. Ursin, pers. comm.; Søndergaard, unpubl.).

The duration of incubation might be a problem. Prolonged incubation time can stimulate bacterial growth and change the ratio of photosynthesis and respiration. Sedimentation or aggregation of the phytoplankton can take place (Venrick et al. 1977). Stimulation of carbon fixation due to long incubation time is shown in Figure 5. During the second day of continuous incubation in a 5 l bottle, the carbon fixation increased 2.5 times although the total quantum fluxes recieved over the two days were identical. The release of EOC, bacterial uptake of EOC and bacterial secondary production also increased (Riemann & Søndergaard 1984). The observed enhancement can be a time-dependent bottle effect. No doubt, a prolonged containment does have effects; however, the influence of bottle size seems uncertain.

Macrophytes 1: Containment will change the natural flow and influence the unstirred layer. 2. Interruption of sediment exchange if

84

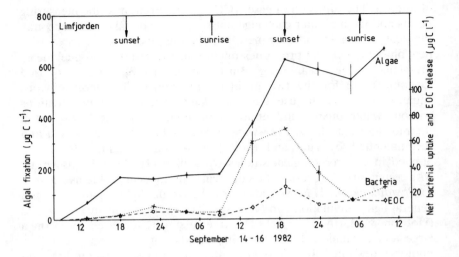

Fig. 5. Cumulative carbon fixation by algae and the carbon flux into bacteria and an extracellular carbon pool (EOC) during two diel cycles in a coastal area in Denmark. The incubation was performed in a 5 l bottle and subsamples were taken at 5 hours intervals. Note enhancement on day two.

rooted macrophytes are bottled. 3. Interference by epiphytes and sediment metabolism.

The most serious effect of containment is a change of carbon diffusion across the unstirred layer. In most situations an increase in thickness and a lowered photosynthetic rate will occur. The exchange with sediments is probably only a problem in species with isoetid growth form where the gas exchange takes place across the roots. If these species are removed from the sediment their natural CO_2 source is interrupted. Any attempt to measure photosynthesis as gas exchange only across the leaves with e.g. *Lobelia* in the rooted position will lead to a dramatic underestimation (Sand-Jensen & Søndergaard 1978, Sand-Jensen *et al.* 1982) as experienced by Nygaard (1958).

The influence on the measurements by other phototrophs and heterotrophs is trivial. However, special care must be taken if any sediment with a very active microflora is present in the enclosure.

Epiphytes and microbenthos: Briefly, the same problems arise as for the former two components. Technical problems are perhaps even more prominent working with photosynthesis of epiphytes and microbenthic algae than with plankton and macrophytes. Different approaches have been presented by Allen (1971) and Riber *et al.* (1984) on epiphytes and Rasmussen *et al.* (1983) on the microbenthos.

3.2. The oxygen method

The use of oxygen changes to measure primary production in the sea was introduced by Gaarder & Gran (1927). Since then, the method has been applied to many systems and situations. The use of reliable O_2-electrodes have expanded its application to continuous measurements of gas exchange (Kelly et al. 1974, Harris & Piccinin 1977, Sand-Jensen 1983, among numerous papers). With the development of microelectrodes the versatility have increased and made studies of O_2 exchange in microsites possible (Revsbech & Jørgensen 1983).

The measurements are based on the changes in oxygen concentration in transparent and opaque bottles or other devices. Net community production is the increase of O_2 in the light bottles compared with the initial concentration. Respiration is the decrease of O_2 in the dark bottles and gross production is the difference between light and dark bottles. Since the plankton comprises both photoautotrophs and microheterotrophs (bacteria and zooplankton), the oxygen technique always measure the community metabolism. Only in situations where the autotrophic components are totally dominating, the measurements approximate the photosynthetic oxygen exchange.

In plankton communities the respiration of microheterotrophs often is a large fraction of the community respiration. Williams (1981b, 1982) has found from 40 to 60% of the respiration to be due to microheterotrophs (<3 μm) in both marine and freshwater. Similar values have been observed by Schwærter & Søndergaard (unpublished) in eutrophic lakes. Such results have to be considered in the evaluation of the O_2 method.

The oxygen method implies equal respiration rates in light and dark. Photorespiration does not occur in darkness, however, it is a matter of controversy if the dark respiration is suppressed or enhanced in light. Several studies have shown that phytoplankton respiration is dependent on their former light history (Ganf 1974, Stone & Ganf 1981). Indications of a higher respiration in light are indirect, but strong (Gibson 1975, Stone & Ganf 1981). The opposite view was put forward by Mangat et al. (1974) and Canvin et al. (1976) for terrestrial angiosperms.

One major drawback of the oxygen method is its sensitivity. In cases with low photosynthetic rates, any significant changes can only be detected using very long incubation times. In oligotrophic systems, synthetic illumination must be used, since the normal daylight period is not long enough. However, the versatility of the method, the easy application of continuous measurements, and the possibility to measure respiration, makes it a very useful tool. One of its forces is in free water measurements, both in rivers, lakes and in the sea.

Besides the sensitivity, one has to mention problems with conversion to carbon units. The photosynthetic quotient (O_2/CO_2 = PQ) ranges from 0.8 to about 2 and depends among other things on the nitrogen source. If the plants uses NO_3^- the PQ is high, with NH_4^+ it is lower. A mean value of 1.2 is often applied, but by no means universally (Harris 1978).

Some submerged aquatic macrophytes show CAM features e.g. *Isoetes lacustris* and *Littorella uniflora* (Keeley and Bowes 1982, Madsen 1985) where O_2 and CO_2 exchange is not coupled. In the dark, there is a net uptake of CO_2 without an O_2 production (see Figure 11). During the day, the decarboxylation of accumulated malate and CO_2 fixation in the Calvin-cycle releases oxygen, but does not involve a corresponding CO_2 uptake.

The oxygen method can be used to measure photosynthesis of macrophytes, epiphytes and microbenthos. The restrictions mentioned above are still valid. Macrophytes containing large volumes of lacunae (intercellular gas space) might create problems in closed systems. At equilibrium, about 30 times more oxygen is present in a unit of air than in a unit of water. Incubation in a closed system without stirring leads Hartman & Brown (1967) to suggest that O_2 accumulated in the lacunae and only entered the medium after a lag period. Later experiments have clearly shown this to be due to the experimental conditions. Under stirred conditions the lacunae and the medium equilibrate and no lag occurs (Westlake 1978, Kelly *et al.* 1981). It must, however, be emphasized that if the measurements are made in a closed system, oxygen increase in 1 ml of lacuna gas space equals the increase in 30 ml of water. Knowledge of lacunal volume and stirring is necessary in achieving correct results.

3.3. *The radiocarbon method (^{14}C)*

Since the introduction of the ^{14}C-method to measure primary production in aquatic ecosystems (Steemann Nielsen 1952) the overwhelming majority of studies on planktonic populations have used this method. Although there have been improvements in several technical aspects of the method, its basic and simple approach have not changed.

The fundamental assumption can be explained by the simple formula:

$$\frac{^{12}CO_{2\,(uptake)}}{^{12}CO_{2\,(present)}} = \frac{^{14}CO_{2\,(uptake)}}{^{14}CO_{2\,(present)}}$$

Applying an isotope discrimination factor of 1.05, knowledge of $^{14}CO_2$ (added) and measurement of $^{12}CO_2$ (present) and the $^{14}CO_2$ uptake, the $^{12}CO_2$ uptake over time can easily be calculated.

What is actually measured by the ^{14}C-method, net or gross photosynthesis or something in-between? This question has been a matter of debate and controversy since 1952. Recent contributions to the debate have been published by Hobson *et al.* (1976) and Dring & Jewson (1982). The history of the ^{14}C-method, and the controversy surrounding it has been excellently reviewed by Peterson (1980). The most unambiguous answer to the question above is: the accumulation of ^{14}C-labelled organic matter in particles and in extracellular products. A distinction among the two categories is not necessary (Schindler *et al.* 1972, Theodòrsson & Bjarnason 1977).

A model for ^{14}C flux in natural plankton should at least include the pathways and pools shown in Figure 6. Experiments with macrophytes, epiphytes and the microbenthos include identical pathways, although grazing on macrophytes in most cases can be neglected and the loss of EOC seems trivial (Søndergaard 1981). The outcome of an experiment is the net accumulation of ^{14}C in each of the compartments.

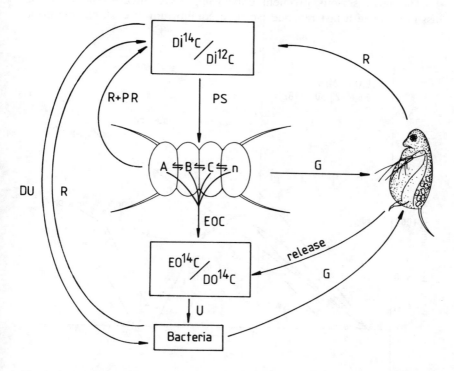

Fig. 6. A minimum model to be used in studies with the ^{14}C-method in natural pelagic systems. PS = photosynthesis; PR = photorespiration; R = respiration; A, B, C, n = intracellular carbon pools and substances; EOC = extracellular organic carbon; DOC = dissolved organic carbon; G = grazing; U = uptake of DOC and EOC; DU = dark uptake (anaplerotic) of CO_2.

88

We do not know, how incubation time affects the results. Summation of short-time incubations (2—3 hours) normally show higher values than a long time incubation (one day) (e.g. Vollenweider & Nauwerck 1961). This could imply a movement from gross community towards net community production. The result of a comparison of apparent carbon fixation in a plankton community by means of ^{12}C and ^{14}C indicates, that the ^{14}C values are higher than net community exchange of $^{12}CO_2$ (Figure 7). The result emphazise that with the ^{14}C-method most respiratory components are lacking, a major weakness of the method.

Another important factor in ^{14}C measurements in pelagic systems can be the release of extracellular products. Both high and low release values have been reported (reviewed by Fogg 1983). However, most measurements have only included net EOC values and not the amount assimilated by bacteria during incubation (see Figure 6).

The most serious problem concerning EOC measurements is the assumption of a fast reached isotopic equilibrium in each carbon pool.

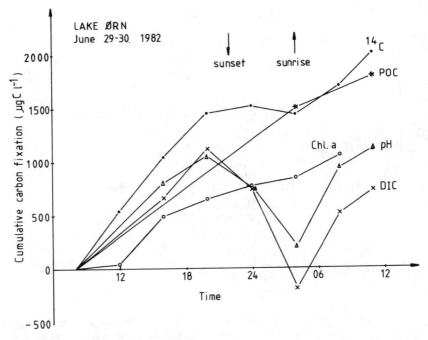

Fig. 7. Cumulative carbon fixation in a pelagic community measured by several different methods during a diel cycle in eutrophic Lake Ørn, Denmark. Chlorophyll a converted to carbon using a factor 30.

Storch & Saunders (1975), Sellner (1981) and Jensen *et al.* (1985) have shown that this assumption should be handled with great care. The use of ^{14}C in algal cultures has shown that the specific activity of the EOC only approximated the specific activity of the DIC pool after 34 to 40 hours in continuous light (Jensen *et al.* 1985). A serious underestimation of EOC release in nature might be present in most investigations.

If the EOC products are not measured, the carbon fixation obtained with the ^{14}C-method can be severely underestimated especially in oligotrophic areas (Fogg 1983). An example from a eutrophic and an oligotrophic lake in Denmark can illustrate this. In Lake Hylke and Almind, the carbon fixation by algae and the subsequent release and bacterial uptake of EOC was measured every three days in May 1983. A size-fractionation method was used to separate algae and bacteria (Larsson & Hagström 1982, Jørgensen *et al.* 1983). The integrated daytime values are shown in Table II. About 2.1 mg C l^{-1} was fixed in the surface (20 cm depth) of Lake Almind and about 0.7 mg C was found as EOC products. Using common particulate primary production estimates (activity in particles > 0.2 µm) would have resulted in an underestimation of 50%. In eutrophic Lake Hylke the underestimation would only have been 5%. A large part of the fixed carbon was not present in the algae, but was recovered in bacteria. Using a mean bacterial growth yield of 0.6, the 'gross release of EOC' was 56 and 37% of the 'gross' fixation in Almind and Hylke, respectively. The inverted commas are used to imply that the expressions do not equal gross photosynthesis. The ^{14}C-method measures the net accumulation of labelled organic products.

The results can also be biased by zooplankton grazing on phytoplankton and bacteria. Some of the ^{14}C is respired during incubation. Removing the phytoplankton grazers can give a higher estimate as shown in Figure 8. The results were obtained in a diel study, where larger zooplankters (> 200 µm) were removed before incubation.

The ^{14}C-method has also been used in studies on macrophyte photosynthesis. Although it should be easy to compare apparent photosynthesis measured as $^{12}CO_2$ and $^{14}CO_2$ exchange, only a few experiments have been published (Titus *et al.* 1979, Søndergaard & Wetzel 1980). Titus *et al.* (1979) observed a reasonable agreement between $^{12}CO_2$ and $^{14}CO_2$ net uptake. However, the experiments by Søndergaard & Wetzel (1980) using intact leaves implied that the ^{14}C-method in the beginning of an experiment would result in values aboves APS. Recent studies using simultaneous $^{12}CO_2$ and $^{14}CO_2$ measurements in water-flow systems have confirmed the last observation (Figure 9). The faster rate of $^{14}CO_2$ uptake can be explained if the assimilated $^{14}CO_2$ do not participate or only partly participate in dark respiration and

Table II. Time integrated values of carbon fixation and release of extracellular organic carbon by the phytoplankton in two Danish lakes, May 1983[1].

Lake	Yearly primary production (g C m^{-2})	Carbon fixation in May 1983 (mg C l^{-1})	Compartment (mg C l^{-1})			Underestimation without EOC[2] %	'gross' fixation[3] (mg C l^{-1})	'gross' EOC release[4] %
			Algae	bacteria	EOC			
Almind	60	2.1	1.0	0.4	0.7	50	2.4	56
Hylke	300	6.2	4.4	1.5	0.3	5	7.0	37

[1] Method: size fractionation (Jørgensen et al. 1983), [2] Calculation: EOC × 100/(Algae + Bacteria), [3] Assuming a bacterial growth yield of 0.6, [4] The respired EOC is included in the calculations: [(Bacteria × 1.67) + EOC]/[Algae + (Bacteria × 1.67) + EOC].

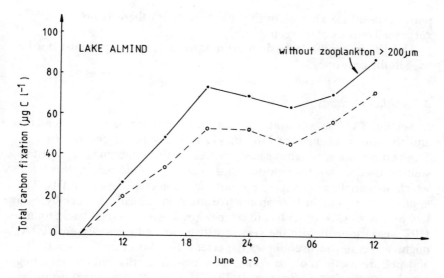

Fig. 8. Cumulative total carbon fixation during a diel study in oligotrophic Lake Almind, Denmark. Parallel samples were incubated. Zooplankters > 200 μm were removed from one of the bottles. ^{14}C-method.

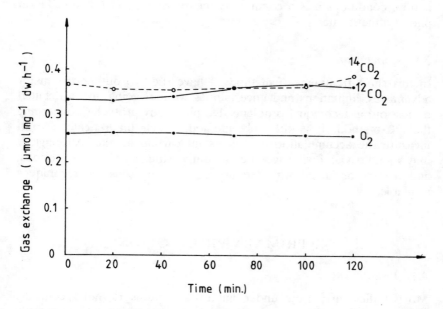

Fig. 9. Simultaneous measurements of photosynthetic rate in *Batrachium* using $^{14}CO_2$, $^{12}CO_2$ and O_2. With the permission of Lisbeth D. Hansen, Botanical Institute, (unpublished data).

photorespiration. That is, during the first hours there is no, or only a very small outflux of inorganic $^{14}CO_2$

The ^{14}C-method is a basic tool in aquatic ecology, but must not be used as the only one.

3.4. *Other methods*

In Section 3.1. I have included a short note on free water measurements and the use of changes in DIC or pH to measure system metabolism. The two methods, which basically represent the same thing, can just as well be used in bottles or other closed systems. One major drawback, which also applies to oxygen, but not the ^{14}C-method, is sensitivity. The application *in situ* in low productive areas is difficult. Especially as a low uptake of CO_2 often has to be measured against a high background. DIC measurements with the very sensitive infrared gas analyser (IRGA) might increase the usefulness. Especially in laboratory studies IRGA and pH are useful tools as continuous measurements without sampling can be applied (e.g. Browse 1979). The newly developed sensitive CO_2 electrodes will aid these measurements. The major advantage of direct CO_2 measurements compared with the O_2 method is that no conversions and assumptions on PQ are involved. However, under natural conditions it is a community measurement and not specific on photosynthetic rates.

3.5. *Conclusion*

In this short treatment of methods, I have tried to outline the major advantages, limitations and uncertainties without going into a detailed discussion on technical problems. Despite many problems concerning the ^{14}C-method, it is the only method, which theoretically should measure the accumulation of the instant carbon uptake. Although it may sound trivial, I will suggest comparative studies to be taken serious, and to be conducted with the use of the new sensitive techniques available.

4. PRIMARY PRODUCTION

4.1. *Productivity*

Most studies carried out under 'natural conditions' do not specifically measure photosynthetic rates, but are measurements of community metabolism. The probability to be net or gross community productivity, something between or approximate photosynthetic rates (net or gross) will depend on methodology and the community structure. In this context, photosynthesis and primary productions will be considered identical terms.

The majority of published values of phytoplankton productivity have been established with the use of the ^{14}C-method. Most measurements have been short-time incubations ($<$ one day). The data from marine environments have mostly been obtained using deck incubators and in lakes in bottles placed at fixed depths.

It is common to express the results per volume or area in order to make easy comparisons among systems. Often the data are integrated over time to give an impression of the yearly productivity. Table III has

Table III. Phytoplankton productivity in different environments.*

Location	g C m^{-2} year^{-1}	g C m^{-2} day^{-1} (mean)
Oceans	50	0.1
Coastal	100	0.3
Upwellings	300	1.0
Lakes: Oligotrophic	$<$100	$<$0.5
Mesotrophic	100—200	0.2—1.0
Eutrophic	200—1500	1.0—8.0

* Compiled after numerous sources the most important of which are: Bunt (1975), Likens (1975), Boynton *et al.* (1983), Wetzel (1983).

been compiled for comparative purposes. To extract the essence, I have omitted all details, exceptions due to latitude and altitude, season and day-to-day fluctuations. In an elaborate analysis of each system all these important fluctuations must, however, be included. Globally, there is a low productivity per area (and volume) in places with a low input of nutrients, and vice versa (Boynton *et al.* 1983). Perhaps the most striking feature is the rather narrow range from 50 to 500 g C m^{-2} year^{-1}. Exceptions are available, but the overwhelming majority of surface waters fall within this range.

A presentation per volume would widen the range from a factor 10 to above a factor of 100. In eutrophic lakes the light attenuation is high and net primary production is restricted to the upper few meters or decimeters if the water is hypereutrophic. In oceanic waters the photic zone can extend over 150 m. As shown by Mathiesen (1971), it can be of importance in evaluating eutrophication not only to compare the area productivity, but also the productivity per volume at the best depth (G 24 max.).

From a physiological viewpoint, the data based only on area and volume are of limited value. They reflect to a certain degree the accumulation of biomass and the total activity. More concern in future studies should also be adressed to an expression per biomass. The assimilation number = fixed carbon per chlorophyll *a* and light might prove a better way of expression if any adaptive responses are under investigation.

A generalizing global Table of the productivity of macrophytes, epiphytes and microbenthos cannot be compiled due to their very heterogenous distribution and partly due to lack of comparative data. Information on the productivity of specific sites and vegetational types can be found in the Tables published by Westlake (1963, 1982) and Wetzel (1983). Some of the difficulties in comparing macrophyte productivity data have been reviewed by Westlake (1965).

The most reliable measurements on the primary productivity of macrophytes have been carried out using different methods to estimate the biomass turnover (Westlake 1982, Wetzel 1983). Many studies on single species have been conducted in enclosures, which change the very important flow velocity factor. The relation to true *in situ* photosynthetic rates are mostly unknown in these experiments.

Potential photosynthetic rates of numerous species of submerged macrophytes are available from the literature. The data presented in Table IV are just a minor part. Comparing the values in Table IV directly must be done with reservation, since most investigators have used different methods and experimental conditions. In the selection of the data, I have used the maximum values to get as close to the photosynthetic potential as possible.

For the angiosperms, the observed range of photosynthetic rates is very narrow, especially when expressed per chlorophyll *a*. The range widens dramatically including the marine macrophytes. The photosynthetic rates found in some macroalgae (e.g. *Ulva lactuca*) are 3 to 4 times higher than in the angiosperms. The reason for this difference is not obvious.

Most photosynthetic rates occurring under natural conditions are probably much lower than the rates shown in Table IV. Sand-Jensen (1983) concluded that APS in nature would be approximately 10 and 6% of the maximum attainable rates for *Callitriche* and *Sparganium*, respectively, due to a suboptimal supply of CO_2. In any case, the 'natural' rates would enable the plants to grow rapidly. Sand-Jensen (1983) further concluded, that the carboxylation capacity was in excess in nature. The amount of RuBP-carboxylase in *Sparganium* did not show seasonal fluctuations in accordance with the photosynthetic rates (C. Prahl, personal communication) The APS was a function of environmental variables.

The productivity of aquatic ecosystems are set by the general

Table IV. Some examples of photosynthetic rates in submerged macrophytes.

Species	μg C mg^{-1} dw. h^{-1}	μmol C mg^{-1} chl. *a* h^{-1}	Reference
Littorella uniflora	5.4	65	1
Callitriche stagnalis	21.5	90	3, 4*
Myriophyllum spicatum	8.0		8,
Sparganium simplex		114	4*
Scirpus subterminalis	4.9		2
Elodea canadensis	17.2		1
Fontinalis antipyretica	14.9		1
Ceratophyllum demersum	2.2—4		10**
Ceratophyllum submersum	0.6		10**
Potamogeton trichoides	17.7		9
Potamogeton crispus		89	4*
Potamogeton pectinatus		126	4*
Utricularia purpurea	14.5		6
Egeria densa	13.7	100	7
Zostera marina		30	5*
Fucus vesiculosus		51	5*
Ceramium rubrum		312	5*
Ulva lactuca		310	5*
Cladophora glomerata		46—520	11

[1] Søndergaard (1979), [2] Søndergaard & Wetzel (1980), [3] Madsen & Søndergaard (1983), [4] Sand-Jensen (1983), [5] Sand-Jensen & Gordon (1984), [6] Moeller (1978), [7] Browse *et al.* (1977, 1979), [8] Titus & Stone (1982), [9] Horbach, Horning & Weise (1974), [10] Søndergaard (1981), [11] Wallentinus (1978).
* probably optimum conditions, O_2 values recalculated with PQ = 1.
** low waterflow, very high resistance to CO_2 diffusion.

nutrient level and supply to each system. The fluctuations of the photosynthetic rates by each group of primary producers and individual organisms are set by both intrisic and external factors. In any specific situation one factor e.g. light might limit the photosynthetic rate, but this can change very rapidly. For each organism or group of organisms one has to look at the time-scales of changes by important environmental factors and make comparison with the time-scales of organism responses. Choice of method and aim should always be judged from such view, which is discussed by Harris (1978) concerning phytoplankton.

4.2. *Special photosynthetic features*

It has been common knowledge during long time that the supply of inorganic nutrients (N, P and Si) has major impact on the productivity of aquatic ecosystems. The general level of biomass is controlled by nutrients and not by photosynthesis. During recent years it has also become evident that the availability of inorganic carbon can be of importance for the primary production and the competition among different plant species and groups. A shortage of DIC is most likely to take place in soft waters and especially if they are rich in nutrients, supplied by human activity. In very productive sites a high pH lower the physiological availability of DIC. Talling (1976) and recently Maberly & Spence (1983) have stressed the importance of DIC uptake abilities in such situations.

In Section 2.1.2. it was stated that a low $K_{1/2}(CO_2)$ (high affinity) or utilization of HCO_3^- as an inorganic carbon source could be one explanation for the ability to have high photosynthetic rates at low CO_2 and high pH situations. Other mechanisms as root uptake of CO_2, CAM photosynthesis or just a general efficient use of the available inorganic carbon could also be of importance.

Two special photosynthetic features will be treated here: (1) The utilization of HCO_3^-; and (2) Inorganic carbon uptake via CAM. Both of these physiological types have developed among marine and freshwater angiosperms and the former among phytoplankton. If carboxylation in the dark (CAM) has any quantitative significance among micro- and macroalgae is unknown at present.

4.2.1. Bicarbonate utilization

In freshwater, submerged macrophytes are distributed among acid, neutral, and permanent alkaline waters in a predictable way (Iversen 1929). Shortly, he observed that species with the elodeid growth form (e.g. *Elodea, Potamogeton, Myriophyllum spicatum*) dominated in alkaline lakes. Acid, soft water lakes with clear water were dominated by the isoetids: *Lobelia dortmanna, Isoetes lacustris*, and *Littorella uniflora*. A mixed population could be found in lakes with a pH fluctuating around 7. Iversen (1929) concluded that pH was the controlling environmental variable.

Later, physiological experiments by Ruttner (1947) and Steemann Nielsen (1947) showed the reason for the distribution not to be pH as such, but rather due to the ability of many elodeid species to make use of the large HCO_3^- — pool in alkaline waters. Similar observations and experiments have been done with freshwater phytoplankton and similar conclusions can be taken (Maberly & Spence 1983).

Studies on HCO_3^- utilization and its physiological explanation (s) are numerous and there is a growing list of plant species categorized as HCO_3^- users and non-users (Raven 1970, Beer *et al.* 1977, Bain & Proctor 1980, Kadono 1980, Allen & Spence 1981, Raven *et al.* 1982, Maberly & Spence 1983, Sand-Jensen 1983). As emphazised by Allen

& Spence (1981) and Maberly & Spence (1983), the use of the two distinct categories (users and non-users) is somewhat misleading. An apparent non-user of HCO_3^- (e.g. *Chlorella pyrenoidosa*) can be induced to utilize HCO_3^- if cultured in media with $2\,mM\ HCO_3^-$ (Steemann Nielsen & Jensen 1958). This is probably also true for some macrophytes (Kaj Sand-Jensen, personal communication). There seems to exist a continous range and some plasticity in the HCO_3^- utilization ability. The tested phytoplankters are the most efficient carbon extractors compared with macrophytes, which again are very different (Maberly & Spence 1983).

Since the experiments conducted by Steemann Nielsen (1947) several theories have been suggested to give a physiological explanation of HCO_3^- utilization. The earlier theories were reviewed by Raven (1970). At present, two alternatives are in focus (it might turn out that this is not alternatives). Lucas (1980) suggests an active uptake of HCO_3^-, whereas Prins and co-workers (1979, 1980, 1982b) from their excellent experimental evidence suggest an active efflux of protons. The protons generate CO_2 from HCO_3^- at the leaf surface. Acid and alkaline sites (H^+ and OH^- efflux) can be observed in bands on single *Chara* cells (Lucas & Smith 1973) or be spatially arranged between the upper and lower sides of single leaves; the so called polar species (Prins *et al.* 1980).

The polar model suggested by Prins *et al.* (1982b) is visualized in Figure 10. The upper part of the figure shows continuous measurements of pH at the upper and lower side of a polar leaf e.g. *Potamogeton lucens*. In the beginning of a light period the pH increases at both sides. After about 10 to 15 minutes the pH on the lower side decreases to values lower than the stirred solution and on the upper side the pH further increases. The explanation for the observed time course in HCO_3^- uptake is a short lag in the efflux of H^+ and OH^-; the efflux seems to be induced by light and can be suppressed by high CO_2 tension. Initially, CO_2 is taken up from both sides and pH consequently increases. The initiation of a further increase on the adaxial side, which gives the biphasic curve, is due to a OH^- release concomitant with the abaxial H^+ efflux. The biphasic time course of pH is used by Prins as a criterion to categorize a plant as an HCO_3^- — user.

In the lower part of Figure 10 the basic proton flux model is presented as a leaf cross section including some chemical characteristics and a theoretical indication of pH and CO_2 gradients in the unstirred layers. Prins *et al.* (1982b) suggested the HCO_3^- conversion into CO_2 to take place in plasmalemma invaginations and the intracellular surplus of OH^- to be balanced by a unidirectional flow of K^+ from the lower to the upper side of the leaf. A pH lowering can be measured in the unstirred layer, consequently some external conversion of HCO_3^- must

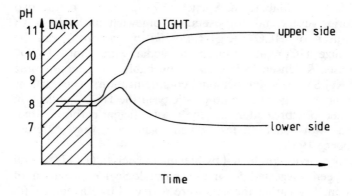

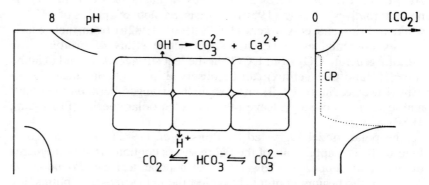

Fig. 10. The measurement of pH on the upper and lower side of a leaf from a polar 'HCO$_3^-$ — user' (above) and a simple model of the mechanism of HCO$_3^-$ — utilization (below). pH and concentrations of CO$_2$ in the unstirred layers indicated. CP = the CO$_2$ compensation point concentration. The figures are based on the results found by Dr. Prins and co-workers (1982a, b).

take place as depicted in the figure. At or in the chloroplasts the concentration of CO$_2$ probably approximate the compensation point and the influx of CO$_2$ is driven by the CO$_2$ gradient. OH$^-$ is released on the upper leaf side, pH increase, and HCO$_3^-$ is converted to CO$_3^-$. The precipitation of CaCO$_3$ can now take place.

The proton pump theory nicely fits the results of Sand-Jensen (1983) that a high pH in the medium decreases the photosynthetic rate due to the buffering capacity of the created CO$_3^-$. The use of buffers in photosynthetic experiments with macrophytes should be avoided and used with great care in phytoplankton studies.

Present knowledge of HCO$_3^-$ utilization by micro-algae and especially macrophytes is extensive, although gaps still exist. Future studies should take the ecophysiological approach and investigate if and how much plants in nature make use of HCO_3^- and not just look at a potential ability. Seasonal and environmental control of the capacity to convert HCO_3^- to CO_2 (or assimilation of HCO_3^-) should also be considered. The physiological ability to make the large HCO_3^- pool available must be viewed as an elegant way to counterbalance the negative effects and make use of the physics of the unstirred layer.

4.2.2. CAM in macrophytes

Until the surprising discovery by Keeley (1981) that the submerged macrophyte *Isoetes howellii* showed CAM features (night-time accumulation of malic acid), this special dark-fixation of CO_2 by β-carboxylation was entirely viewed as a mechanism of succulents to conserve water in a drought climate (Osmond *et al.* 1982). Later experiments with *Isoetes* have confirmed the first observations and shown that a net gain of CO_2 takes place in the dark (Keeley & Bowes 1982). Screening for diel fluctuations in titratable acidity have revealed CAM to be a likely carbon fixation pathway in several *Isoetes* species including *I. lacustris* (Keeley 1982), *Littorella uniflora* (Boston & Adams 1983), and *Crassula aquatica* (Keely & Morton 1982). The species *Hydrilla verticillata* (Holaday & Bowes 1980) and *Scirpus subterminalis* (Beer & Wetzel 1981) have a limited capacity for overnight acid accumulation.

Recent gas exchange studies in our laboratory by Tom V. Madsen have confirmed a substantial net carbon uptake in the dark by *Isoetes lacustris* and *Littorella uniflora*. *Lobelia dortmanna*, have neither acid accumulation nor a net carbon uptake in the dark. However, a lowered CO_2 dark respiration points to a weak CAM behavior (Madsen 1985).

The CAM performance of *Littorella* leaves in a laboratory experiment is shown in Figure 11. The time course of CO_2 uptake is somewhat different from the one observed in *Isoetes howellii* by Keeley & Bowes (1982). They found a rather short lag in the initial dark CO_2 fixation followed by decreasing uptake rates during the rest of the dark period. In *Littorella*, a net CO_2 uptake first appeared after about 2 hours, stayed at a constant rate for about 6 hours and then declined. The ratio of net CO_2 fixation in the light and the dark was about 3 : 1.

Simultaneous oxygen measurements showed a constant dark respiration and the uncoupling of oxygen and carbon metabolism. The observed time course can be interpreted as an initial induction period followed by an accumulation of malic acid in the vacuoles. The decreasing rate after about 8 hours in the dark is probably due to shortage of acid storage capacity or exhaustion of phosphoenol pyruvate generated through glycolysis. The stored acid functions as an internal CO_2 source utilized during the day.

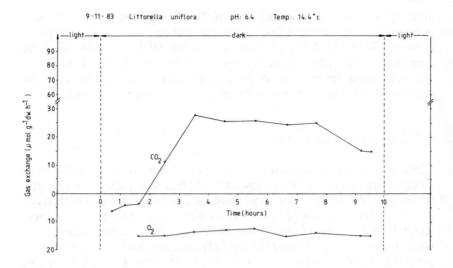

Fig. 11. Gas exchange of *Littorella uniflora* in light and dark. Note net uptake of CO_2 in the dark. After Madsen (1985).

It is ecologically interesting to notice CAM in two out of three species often totally dominating the submerged vegetation in lakes low in DIC (Sand-Jensen & Søndergaard 1979). The three species rely totally on root uptake of CO_2 (they do not use HCO_3^-) to meet their inorganic carbon requirements (Søndergaard & Sand-Jensen 1979b). The unique presence of mycorrhiza in *Lobelia* and *Littorella* further underline their special ecology (Søndergaard and Lægaard 1977). As an analog to the water conserving function of CAM in succulents, the CAM development can be viewed as a conservation of carbon. CAM enables *Isoetes* and *Littorella* to utilize the limited CO_2 resources throughout a diel cycle.

The slower growth of *Lobelia* compared with the other species (Sand-Jensen & Søndergaard 1978) and its often less quantitative development can perhaps be explained by the weak CAM. However, a large ratio of rootsurface to photosynthetic tissues might counteract the weak CAM. The large root area makes the plant able to explore larger volumes of the interstitial water with its high CO_2 concentrations.

The CAM-feature is of importance for the quantitative success of *Isoetes* and *Littorella*. However, the exact importance in nature is at present unknown and needs further studies.

The presence of CAM in other submerged macrophytes and perhaps microalgae should be looked for in future research.

5. CONCLUDING REMARKS

Photosynthesis by the submerged vegetation in aquatic ecosystems is controlled by an array of variables changing in time and space. Each group of organisms have their special environmental requirements and physiological behavior. The microscopic algae in open waters and the rooted macrophytes is exposed to very different growth conditions and show special and sometimes unique photosynthetic features to solve their problems. However, many common problems are also solved in different ways. This is what makes aquatic ecology and the physiological ecology of aquatic vegetation such an exciting and 'living' science.

In have tried to extract some important items from the literature on productivity and photosynthesis in aquatic plants. What is viewed as important can only be a subjective judgement. It strikes me that our knowledge, especially concerning net primary production in nature is based on values partly (mostly?) collected from point measurements without a proper time scale. And with methods, which we know, that are just a bad imitation of nature. I do not want to mistrust all values, but it looks as if too much time has been spent on data collection without challenging the old methods. In future research more attention should perhaps be paid to free-water measurements combined with the traditional 'enclosure' experiments.

In experimental aquatic ecology, the physiological approach has proved its value and should probably continue to be fruitfull. Questions I have raised and left open in the preceding sections were not treated that way with the purpose of confusing or frustrating newcomers but, on the contrary, to stimulate new thinking in a field that has produced many useful results for mankind. We all depend on a wise exploitation of aquatic ecosystems.

ACKNOWLEDGEMENTS

I wish to acknowledge the support by several colleagues during writing this review. Especially, I appreciate the help of Tom V. Madsen, L. Møller Jensen, Steen Schwærter and Lisbeth D. Hansen. Part of the study was supported by The Danish Natural Science Research Council (11—1816).

REFERENCES

Allen, E. D. & Spence, D. H. (1981) The differential ability of aquatic plants to utilize the inorganic carbon supply in fresh waters. New Phytol. 87: 269—283.

102

Allen, H. L. (1971) Primary productivity, chemoorganotrophy and nutritional interactions of epiphytic algae and bacteria on macrophytes in the littoral of a lake. Ecol. Monogr. 41: 97—127.

Anderson, G. C. & Zeutschel, R. P. (1970) Release of dissolved organic matter by marine phytoplankton in coastal and off shore areas of the northeast Pacific Ocean. Limnol Oceanogr. 15: 402—407.

Anthoni, U., Christophersen, C., Madsen, J. Ø., Wium-Andersen, S. & Jacobsen, N. (1980) Biologically active sulphur compounds from the green alga *Chara globularis*. Phytochemistry 19: 1228—1229.

Badger, M. R., Kaplan, A. & Berry, J. A. (1978) A mechanism for concentrating CO_2 in *Chlamydomonas reinhardtii* and *Anabaena variabilis* and its role in photosynthetic CO_2 fixation. Carn. Inst. Was. Yearb. 77: 251—261.

Bain, J. T. & Proctor, M. C. F. (1980) The requirement of aquatic Bryophytes for free CO_2 as an inorganic carbon source: Some experimental evidence. New Phytol. 87: 269—238.

Barko, J. W. & Filbin, G. J. (1983) Influence of light and temperature on chlorophyll composition in submersed freshwater macrophytes. Aquat. Bot. 15: 249—255.

Beer, S., Eshel, A. & Waisel, Y. (1977) Carbon metabolism in seagrasses. I. The utilization of exogenous inorganic carbon species in photosynthesis. J. exp. Bot. 28: 1180—1189.

Beer, S. & Wetzel, R. G. (1981) Photosynthetic carbon metabolism in the submerged aquatic angiosperm *Scirpus subterminalis*. Plant Sci. Lett. 21: 199—207.

Berman, T. & Rodhe, W. (1971) Distribution and migration of *Peridinium* in Lake Kinneret. Mitt. Int. Ver. theor. angew. Limnol. 19: 266—276.

Bidwell, R. G. S. (1977) Photosynthesis and light dark respiration in freshwater algae. Can. J. Bot. 55: 809—818.

Birmingham, B. C., Coleman, J. R. & Colman, B. (1982) Measurement of photorespiration in algae. Plant Physiol. 69: 259—262.

Björkman, O. (1981) Responses to different quantum flux densities. *In*: Lange, O. L. *et al*. (Eds.), Physiological plant ecology I. Responses to the physical environment, pp. 57—107. Springer-Verlag, Berlin.

Black, M. A., Maberly, S. C. & Spence, D. H. N. (1981) Resistance to carbon dioxide fixation in four submerged freshwater macrophytes. New Phytol. 89: 557—568.

Bonde, G. & Søndergaard, M. (In prep.) Photosynthetic adaptation of the submerged macrophyte *Littorella uniflora* (L.) Aschers.

Boston, H. L. & Adams, M. S. (1983) Evidence of crassulacean acid metabolism in two North American isoetids. Aquat. Bot. 15: 381—386.

Boynton, W. R., Hall, C. A., Falkowski, P. G., Keefe, C. W. & Kemp, W. M. (1983) Phytoplankton productivity in aquatic ecosystems. *In*: Lange, O. L., Nobel, P. S., Osmond, C. B. & Ziegler, H. (Eds.), Physiological plant ecology IV. Ecosystem processes: Mineral cycling, productivity and man's influence. pp. 305—327. Springer-Verlag, Berlin.

Bowes, G., Holaday, A. S., Van, T. K. & Haller, W. T. (1978) Photosynthetic and photorespiratory carbon metabolism in aquatic plants. *In*: Hall, D. O. Coombs, J. & Goodwin, T. W. (Eds.), Photosynthesis 77. pp. 289—298. The Biochem. Soc., London.

Brown, J. M. A., Dromgoole, F. I., Towsey, M. W. and Browse, J. (1974) Photosynthesis and photorespiration in aquatic macrophytes. *In*: Bielski *et al*. (Eds.), Mechanism of regulation of plant growth. Bull R. Soc. New. Zealand, 12: 243—249.

Browse, J. A. (1979) An open-circuit infrared gas analysis system for measuring aquatic plant photosynthesis at physiological pH. Aust. J. Plant Physiol. 6: 493—498.

Browse, J. A., Dromgoole, F. I. & Brown, J. M. A. (1977) Photosynthesis in the aquatic macrophyte *Egeria densa*. I. $^{14}CO_2$ fixation at natural CO_2 concentrations. Aust. J. Plant Physiol 4: 169—176.

Browse, J. A., Dromgoole, F. I. & Brown, J. M. A. (1979) Photosynthesis in the aquatic macrophyte *Egeria densa*. III. Gas exchange studies. Aust. J. Plant Physiol. 6: 499—512.

Bunt, J. S. (1975) Primary productivity of marine ecosystems. *In*: Lieth, H. & Whittaker, R. H. (Eds.), Primary productivity of the biosphere. pp. 169—183. Springer — Verlag, New York.

Burney, C. M., Davis, P. G., Johnson, K. M. & Sieburth, J. McN. (1982) Diel relationships of microbial trophic groups and *in situ* dissolved carbohydrate dynamics in the Caribbean Sea. Mar. Biol. 67: 311—322.

Burris, J. E. (1977) Photosynthesis, photorespiration and dark respiration in eight species of algae. Mar. Biol. 39: 371—379.

Burris, J. E. (1981) Effects of oxygen and inorganic carbon concentrations on the photosynthetic quotient of marine algae. Mar. Biol. 65: 215—219.

Canvin, D. T., Lloyd, N. D. H., Fock, H. & Przybylla, K. (1976) Glycine and serine metabolism and photorespiration. *In*: Burris, R. H. and Black, C. C. (Eds.), CO_2 metabolism and plant productivity. pp. 161—176. Univ. Park Press, Baltimore.

Chan, A. T., Anderson, R. J., Le Blanc, M. J. & Harrison, P. J. (1980) Algal plating as a tool for investigating allelopathy among marine algae. Mar. Biol. 59: 7—13.

Dring, M. J. (1981) Chromatic adaptation of photosynthesis in benthic marine algae: An examination of its ecological significance using a theoretical model. Limnol. Oceanogr. 26: 271—284.

Dring, M. J. & Jewson, D. H. (1982) What does ^{14}C uptake by phytoplankton really measure? A theoretical modelling approach. Proc. R. Soc. Lond. B. 214: 351—368.

Falkowski, P. G. (1980) Light-shade adaptation in marine phytoplankton. *In*: Falkowski, P. G. (Ed)., Primary productivity in the sea. pp. 99—119. Plenum Press, New York.

Falkowski, P. G. & Owens, T. G. (1980) Light-shade adaptation. Two strategies in marine phytoplankton. Plant Physiol. 66: 592—595.

Falkowski, P. G., Owens, T. G., Ley, A. C. & Mauzerall, D. C. (1981) Effects of growth irradiance levels on the ratio of reaction centers in two species of marine phytoplankton. Plant Physiol. 68: 969—973.

Fee, E. J. (1969) A numerical model for the estimation of photosynthetic production, integrated over time and depth, in natural waters. Limnol. Oceanogr. 14: 906—911.

Fogg, G. E. (1983) The ecological significance of extracellular products of phytoplankton photosynthesis. Bot. Mar. 26: 3—14.

Francko, D. A. (1983) Cyclic AMP in photosynthetic organisms: Recent developments. Adv. Cyclic Nucleotide Res. 15: 97—117.

Francko, D. A. (1984) Phytoplankton metabolism and cyclic nucleotides. I. Nucleotide induced perturbations of carbon assimilation. Arch. Hydrobiol. 100: 341—354.

Gaarder, T. & Gran, H. H. (1927) Investigation of the production of plankton in the Oslo Fjord. Rapp. et Proc. Verb. Cons. Int. Explor. Mer. 42: 1—48.

Gallegos, C. L., Hornberger, G. M. & Kelly, M. G. (1980) Photosynthesis — light relationships of a mixed culture of phytoplankton in fluctuating light. Limnol Oceanogr. 25: 1082—1092.

Ganf, G. G. (1974) Rates of oxygen uptake by the planktonic community of a shallow equatorial lake (Lake George, Uganda). Oecologia 15: 17—32.

Gargas, E., Nielsen C. S. & Lønholdt, J. (1976) An incubator method for estimating the actual daily plankton algae primary production. Water Res. 10: 853—860.

Gerloff, G. C. (1973) Plant analysis for nutrient assay of natural waters. EPA environmental Health Effects Res. Ser. EPA-RI-73-001. 65 pp.

Gerloff. G. C. & Krombholz, P. H. (1966) Tissue analysis as a measure of nutrient availability for the growth of angiosperm aquatic plants. Limnol Oceanogr. 11: 529—537.

Gibson, C. E. (1975) A field and laboratory study of oxygen uptake by planktonic blue-green algae. J. Ecol. 63: 867—880.

104

Gieskes, W. W. C., Kraay, G. W. & Baars. M. A. (1979) Current ^{14}C methods for measuring primary productions: gross underestimates in oceanic waters. Neth. J. Sea Res. 13: 58—78.

Glover, H. E. & Morris I. (1979) Photosynthetic carboxylating enzymes in marine phytoplankton. Limnol. Oceanogr. 24: 510—519.

Goldman, C. R. (1972) The role of minor nutrients in limiting the productivity of aquatic ecosystem. In: Likens, G. E. (Ed.) Nutrients and eutrophication: the limiting nutrient controversy. Spec. Symp. Amer. Soc. Limnol. Oceanogr. 1: 21—38.

Hagedorn, H. (1971) Experimentelle Untersuchungen über den Einfluss des Thiamins auf die natürliche Algenpopulation des Pelagials. Arch. Hydrobiol. 68: 382—399.

Hall, C. A. S. & Moll, R. (1975) Methods of assessing aquatic primary productivity. In: Lieth, H. & Whittaker, R. H. (Eds.), Primary productivity of the biosphere. pp. 19—53. Springer-Verlag, New York.

Harris, G. P. (1978) Photosynthesis, productivity and growth: The physiological ecology of phytoplankton. Ergebn. Limnol. 10. 1—171.

Harris, G. P. & Piccinin, B. B. (1977) Photosynthesis by natural phytoplankton populations. Arch. Hydrobiol. 80: 405—457.

Hartman, R. T. & Brown, D. L. (1967) Changes in internal atmosphere of submersed vascular hydrophytes in relation to photosynthesis. Ecology 48: 252—258.

Hesslein, R. H., Broecker, W. S., Quay, P. D. & Schindler, D. W. (1980) Whole-lake radiocarbon experiment in an oligotrophic lake at the experimental lake area, Northwestern Ontario. Can. J. Fish. Aquat. Sci. 37: 454—463.

Hobson, L. A., Morris, W. J. & Pirquet, K. T. (1976) Theoretical and experimental analysis of the ^{14}C technique and its use in studies of primary production. J. Fish. Res. Board Can. 33: 1715—1721.

Hobson, L. A. & Hartly, F. A. (1983) Ultraviolet irradiance and primary production in a Vancouver Island fjord, British Columbia, Canada. J. Plank. Res. 5: 325—332.

Holaday, A. S. & Bowes (1980) C_4 acid metabolism and dark CO_2 fixation in a submersed aquatic macrophyte (Hydrilla verticillata). Plant Physiol. 65: 331—335.

Horbach, W., Horning. L. & Weise, G. (1974) Untersuchungen des CO_2-stoffwechsels submerser Höherer Wasser-Pflanzen im fliessgewässer unter Einsatz eines fahrbaren Infrarot-Gasanalysator-Labors. Int. Revue ges. Hydrobiol. 59: 17—29.

Huebert, D. B. & Gorham, P. R. (1983) Biphasic mineral nutrition of the submersed aquatic macrophyte Potamogeton pectinatus L. Aquat. Bot. 16: 269—284.

Hough, R. A. (1974) Photorespiration and productivity in submersed aquatic vascular plants. Limnol. Oceanogr. 19: 912—927.

Hutchinson, G. E. (1975) A treatise on limnology. Vol. 3 Limnological Botany. John Wiley, New York. 660 pp.

Højerslev, N. K. (1979) Daylight measurements appropriate for photosynthetic studies in natural sea waters. J. Cons. int. Explor. Mer. 38: 131—146.

Iversen, J. (1929) Studien über die pH-Verhältnisse dänischer Gewässer und ihren Einfluss auf die Hydrophyten-Vegetation. Bot. Tidsk. 40: 277—333.

Jaworski, G. H. M., Talling, J. F. & Heaney, S. I. (1981) The influence of carbon dioxide on growth and sinking rate of two planktonic diatoms in culture. Br. Phycol. J. 16: 395—410.

Jeffrey, S. W. (1981) Responses to light in aquatic plants. In: Lange, O. L. et al. (Eds.) Physiological plant ecology I. Responses to the physical environment. Vol. 12 A, pp. 249—276. Springer-Verlag, Berlin.

Jensen, L. M., Jørgensen, N. O. G. & Søndergaard, M. (1985) Specific activity. Significance in estimating release rates of extracellular dissolved organic carbon (EOC) by algae. Verh. intern. Ver. theor. angew. Limnol, 22: 2893—2897.

Jerlov, N. G. (1976) Marine Optics. Elsevier Oceanography Series, 14. Elsevier, Amsterdam.

Jewson, D. H., Rippey, B. H. & Gilmore, W. K. (1981) Loss rates from sedimentation, parasitism, and grazing during the growth, nutrient limitation, and dormancy of a diatom crop. Limnol. Oceanogr. 26: 1045—1056.

Johnson, P. W. & Sieburth, J. McN. (1982) In-situ morphology and occurrence of eucaryotic phototrophs of bacterial size in the picoplankton of estuarine and oceanic waters. J. Phycol. 18: 318—327.

Jónasson, P. M. & Mathiesen, H. (1959) Measurement of primary production in two Danish eutrophic lakes, Esrom sø and Furesø. Oikos 10: 137—167.

Jørgensen, N. O. G., Søndergaard, M., Hansen, H. J., Bosselman, S. & Riemann, B. (1983) Diel variation in concentration, assimilation and respiration of dissolved free amino acids in relation to planktonic primary and secondary production in two eutrophic lakes. Hydrobiol. 107: 107—122.

Kadono, Y. (1980) Photosynthetic carbon sources in some *Potamogeton* species. Bot. Mag. Tokyo 93: 185—194.

Keating, K. J. (1978) Blue-green algal inhibition of diatom growth: Transition from mesotrophic to eutrophic community structure. Science 199: 971—973.

Keeley, J. E. (1981) Isoetes howellii: A submerged aquatic CAM plant? Amer. J. Bot. 68: 420—424.

Keeley, J. E. (1982) Distribution of diurnal acid metabolism in the genus *Isoetes*. Amer. J. Bot. 69: 254—257.

Keeley, J. E. & Bowes, G. (1982) Gas exchange characteristics of the submerged aquatic crassulacean acid metabolism plant, *Isoetes howellii*. Plant Physiol. 70: 1455—1458.

Keeley, J. E. & Morton, B. A. (1982) Distribution of diurnal acid metabolism in submerged aquatic plants outside the genus *Isoetes*. Photosynthetica 16: 546—553.

Kelly, M. G. Hornberger, G. M. & Cosby, B. J. (1974) Continuous automated measurement of rates of photosynthesis and respiration in an undisturbed river community. Limnol. Oceanogr. 19: 305—312.

Kelly, M. G., Moeslund, B. & Thyssen, N. (1981) Productivity measurement and the storage of oxygen in the aerenchyma of aquatic macrophytes. Arch. Hydrobiol. 92: 1—10.

Kluge, M. & Ting, I. P. (1978) Crassulacean acid metabolism. Ecological studies 30. Springer-Verlag, Berlin 209 pp.

Kristiansen, J. & Mathiesen, H. (1964) Phytoplankton of the Tystrup-Bavelse lakes, primary production and standing crop. Oikos 15: 1—43.

Larsson, U. & Hagström, A. (1982) Fractionated phytoplankton primary production, exudate release and bacterial production in a Baltic eutrophication gradient. Mar. Biol. 57: 57—70.

Latzko, E. & Kelly, G. J. (1980) Carbon metabolism: Chloroplast capability and the uncertain fate of CO_2. Progress in Botany 42: 58—79. Springer, Berlin.

Levring, T. (1966) Submarine light and algal shore zonation. *In*: Bainbridge, R. *et al.* (Eds.), Light as an ecological factor. pp. 305—319. Blackwell, Oxford.

Lewis, W. M. Jr. (1974) Primary production in the plankton community of a tropical lake. Ecol. Monogr. 44: 377—409.

Leyton, L. (1975) Fluid behavior in biological systems. Oxford Uni. Press, Oxford. 235 pp.

Likens, G. E. (1975) Primary production of inland aquatic ecosystems. *In*: Lieth H. & Whittaker, R. H. (Eds.), Primary productivity of the biosphere. pp. 185—202. Springer-Verlag, New York.

Lloyd, D. H., Canvin, D. T. & Culver, D. A. (1977a) Photosynthesis and photorespiration in algae. Plant Physiol. 59: 936—940.

Lloyd, N. D. H., Canvin, D. T. & Bristow J. M. (1977b) Photosynthesis and photorespiration in submerged aquatic vascular plants. Can. J. Bot. 55: 3001—3005.

Lucas, W. J. (1979) Alkaline band formation in *Chara corallina*. Plant Physiol. 63: 248—254.

Lucas, W. J. (1980) Control and synchronization of HCO_3^- and OH^- transport during photosynthetic assimilation of exogenous HCO_3^-. *In*: Spanswick, R. M., Lucas, W. J. & Dainty, J. (Eds.), Plant membrane transport: Current conceptual issues. pp. 317—327. Elsevier, Amsterdam.

Lucas, W. J. & Smith, F. A. (1973) The formation of alkaline and acid regions at the surface of *Chara corallina* cells. J. exp. Bot. 24: 1—14.

Madsen, T. V. & Søndergaard, M. (1983) The effects of current velocity on the photosynthesis of *Callitriche stagnalis* Scop. Aquat. Bot. 15: 187—193.

Madsen, T. V. (1984) Resistance to CO_2 fixation in the submerged aquatic macrophyte *Callitriche stangnalis* Scop. J. Exp. Bot. 35: 338—347.

Madsen, T. V. (1985) A community of submerged aquatic CAM plants in Lake Kalgaard, Denmark. Aquat. Bot. 23: 97—108.

Mague, T. H., Friberg, E., Hughes, D. J. & Morris, I. (1980) Extracellular release of carbon by marine phytoplankton, a physiological approach. Limnol. Oceanogr. 25: 262—279.

Mangat, B. S., Levin, W. B. & Bidwell, R. G. S. (1974) The extent of dark respiration in illuminated leaves and its control by ATP levels. Can. J. Bot. 52: 673—681.

Maberly, S. C. & Spence, D. H. N. (1983) Photosynthetic inorganic carbon use by freshwater plants. J. Ecol. 71: 705—724.

Mathiesen, H. (1971) Summer maxima of algae and eutrophication. Mitt. Int. Ver. theor. angew. Limnol. 19: 161—181.

McCarthy, J. J. (1980) Nitrogen. *In*: Morris, I. (Ed.) The physiological ecology of phytoplankton, pp. 191—233. Blackwell Sci. Publ., Oxford.

McKinley, K. R. & Wetzel, R. G. (1977) Tritium oxide uptake by algae: An independent measure of phytoplankton photosynthesis. Limnol. Oceanogr. 22: 377—380.

Miller, A. G. & Colman, B. (1980) Evidence for HCO_3^- transport by the bluegreen algae (Cyanobacterium) *Coccochloris peniocystis*. Plant Physiol. 65: 397—402.

Moeller, R. E. (1978) Carbon-uptake by the submerged hydrophyte, *Utricularia purpurea*. Aquat. Bot. 5: 209—216.

Nalewajko, C. & Lean, D. R. S. (1980) Phosphorus. *In*: Morris, I. (Ed.), The physiological ecology of phytoplankton pp. 235—258. Blackwell Sci. Publ., Oxford.

Nygaard, G. (1958) On the productivity of the bottom vegetation in Grane Langsø. Verh. Int. Ver. theor. angew. Limnol. 13: 144—155.

Nygaard, G. (1968) On the significance of the carrier carbon dioxide in determinations of the primary production in soft-water lakes by the radiocarbon technique. Mitt. Int. Ver. theor. angew. Limnol. 14: 111—121.

Odum, H. T. (1957) Trophic structure and productivity of silver spring, Florida. Ecol. Monogr. 27: 55—112.

Osmond, C. B., Winter, K. & Ziegler, H. (1982) Functional significance of different pathways of CO_2 fixation in photosynthesis. *In*: Lange, O. L., Nobel, P. S., Osmond, C. B. & Ziegler, H. (Eds.), Physiological plant ecology II. Water relations and Carbon assimilation, pp. 479—547. Springer-Verlag, Berlin.

Peterson, B. J. (1980) Aquatic primary productivity and the $^{14}C\text{-}CO_2$ method: A history of the productivity problem. Ann. Rev. Ecol. Syst. 11: 359—385.

Petterson, K. (1980) Alkaline phosphatase activity and algal surplus phosphorus as phosphorus-deficiency indicators in Lake Erken. Arch. Hydrobiol. 89: 54—87.

Pomeroy, L. R. (1974) The ocean's food web, a changing paradigm. Bio Science 24: 499—504.

Prins, H. B. A., O'Brien, J. & Zanstra, P. E. (1982) (Prins *et al*. 1982a). Bicarbonate utilization in aquatic angiosperms. pH and CO_2 concentrations at the leaf surface. *In*: Symoens, J. J., Hooper, S. S. & Compère, P. (Eds.), Studies on aquatic vascular plants, pp. 112—119. Royal Bot. Soc, Belgium. Brussels.

Prins, H. B. A., Snel, J. F. H., Helder, R. J. & Zanstra, P. E. (1979) Photosynthetic bicarbonate utilization in the aquatic angiosperms *Potamogeton* and *Elodea*. Hydrobiol. Bull. 13: 106—111.

Prins, H. B. A., Snel, J. F., Helder, R. J. & Zanstra, P. E. (1980) Photosynthetic HCO_3^- utilization and OH^- excretion in aquatic angiosperms. Light-induced pH changes at the leaf surface. Plant Physiol. 66: 818—822.

Prins, H. B. A., Snel. J. F. H. and Zanstra P. E. (1982) (Prins *et al.* 1982b) The mechanism of photosynthetic bicarbonate utilization. *In*: Symoens, J. J., Hooper, S. S. & Compère, P. (Eds.), Studies on aquatic vascular plants, pp. 120—126. Royal Bot. Soc. Belgium. Brussels.

Provasoli, L. & Carlucci, A. E. (1974) Vitamins and growth regulators. *In*: Stewart, W. D. P. (Ed.) Algal physiology and biochemistry, pp. 741—787. Blackwell Sci. Publ., Oxford.

Rasmussen, M. B., Henriksen, K. & Jensen, A. (1983) Possible causes of temporal fluctuations in primary production of the microphytobenthos in the Danish Wadden Sea. Mar. Biol. 73: 109—114.

Raven, J. A. (1970) Exogenous inorganic carbon sources in plant photosynthesis. Biol. Rev. 45: 167—221.

Raven, J. A. (1974) Carbon dioxide fixation. *In*: Stewart, W. D. P. (Ed.), Algal physiology and biochemistry, pp. 434—455. Blackwell Sci. Publ., Oxford.

Raven, J., Beardall, J. & Griffiths, H. (1982) Inorganic C-sources for *Lemanea, Cladophora* and *Ranunculus* in a fast-flowing stream: Measurements of gas exchange and of carbon isotope ratio and their ecological implications. Oecologia 53: 68—78.

Rebsdorf, Å. (1972) The carbon dioxide system of freshwater. A set of tables for easy computation of total carbon dioxide and other components of the carbon dioxide system. Booklet: Freshwater Biological Laboratory, Hillerød, Denmark.

Revsbech, N. P. & Jørgensen, B. B. (1983) Photosynthesis of benthic microflora measured with high spatial resolution by the oxygen microprofile method: Capabilities and limitations of the method. Limnol Oceanogr. 28: 749—756.

Riber, H. H., Sørensen, J. P. & Schierup. H.-H. (1984) Primary productivity and biomass of epiphytes on *Phragmites australis* (Cov.) Trin. ex Stendel in a eutrophic Danish lake. Holarctic Ecol. 7: 202—210.

Richardson, K., Beardall, J. & Raven, J. A. (1983) Adaptation of unicellular algae to irradiance: An analysis of strategies. New Phytol. 93: 157—191.

Riemann, B. & Søndergaard, M. (1984) Measurments of diel rates of bacterial secondary production in aquatic environments. Appl. Environ. Microbiol. 47: 632—638.

Riemann, B., Søndergaard, M., Schierup, H.-H., Bosselmann, S., Christensen, G., Hansen, J. & Nielsen, B. (1982) Carbon metabolism during a spring diatom bloom in the eutrophic Lake Mossø. Int. Rev. ges. Hydrobiol. 67: 145—185.

Rodhe, W. (1965) Standard correlations between pelagic photosynthesis and light. Mem. Ist. ital. Idrobiol. 18 (Suppl.): 365—381.

Ruttner, F. (1947) Zur Frage der Karbonatassimilation der Wasserpflanzen. I. Teil. Die beiden Haupttypen der Kohlen stoffaufnahme. Öst. bot. Z. 94: 265—294.

Salisbury, F. B. &. Ross, C. W. (1978) Plant physiology. Wadsworth Publ. Comp., Belmont. 422 pp.

Sand-Jensen, K. (1977) Effects of epiphytes on eelgrass photosynthesis. Aquat. Bot. 3: 55—63.

Sand-Jensen, K. (1978) Metabolic adaptation and vertical zonation of *Littorella uniflora* (L.) Aschers. and *Isoetes lacustris* L. Aquat. Bot. 4: 1—10.

Sand-Jensen, K. (1983) Photosynthetic carbon sources of stream macrophytes. J. exp. Bot. 34. 198—210.

Sand-Jensen, K. & Borum, J., (1984) Epiphyte shading and its effect on photosynthesis and diel metabolism of *Lobelia dortmanna* L. during the spring bloom in a Danish lake. Aquat. Bot. 20: 109—119.

108

Sand-Jensen, K. & Gordon, D. M. (1984) The differential ability of marine and freshwater macrophytes to utilize HCO_3^- and CO_2. Mar. Biol. 80: 247—253.

Sand-Jensen, K., Prahl, C. & Stokholm, H. (1982) Oxygen release from roots of submerged aquatic macrophytes. Oikos 38: 349—354.

Sand-Jensen, K. & Søndergaard, M. (1978) Growth and production of isoetides in oligotrophic Lake Kalgaard, Denmark. Verh. Int. Ver. theor. angew. Limnol. 20: 659—666.

Sand-Jensen, K. & Søndergaard, M. (1979) Distribution and quantitative development of aquatic macrophytes in relation to sediment characteristics in oligotrophic Lake Kalgaard, Denmark. Freshwat. Biol. 9: 1—11.

Sand-Jensen, K. & Søndergaard, M. (1981) Phytoplankton and epiphyte development and their shading effect on submerged macrophytes in lakes of different nutrient status. Int. Rev. ges. Hydrobiol. 66: 529—552.

Schindler, D. W. (1971) Carbon, nitrogen and phosphorus and the eutrophication of freshwater lakes. J. Phycol. 7: 321—329.

Schindler, D. W. & Fee, E. J. (1973) Diurnal variation of dissolved inorganic carbon and its use in estimating primary production and CO_2 invasion in lake 227. J. Fish. Res. Board Can. 30: 1501—1510.

Schindler, D. W., Schmidt, R. V. & Reid, R. A. (1972) Acidification and bubbling as an alternative to filtration in determining phytoplankton production by the ^{14}C method. J. Fish. Res. Board Can. 29: 1627—1631.

Schmitt, M. R. & Adams, M. S. (1981) Dependence of rates of apparent photosynthesis on the phosphorus concentrations in *Myriophyllum spicatum*. Aquat. Bot. 11: 379—387.

Schulthorpe, C. D. (1967) The Biology of Aquatic Vascular Plants. Edward Arnold Publ., London. 610 pp.

Sellner, K. G. (1981) Primary productivity and the flux of dissolved organic matter in several marine environments. Mar. Biol. 65: 101—112.

Shapiro, J. (1973) Blue-green algae: why they become dominant. Science 179: 382—384.

Smith, F. A. & Walker, N. A. (1980) Photosynthesis by aquatic plants: Effects of unstirred layers in relation to assimilation of CO_2 and HCO_3^- and to the carbon isotopic discrimination. New Phytol. 86: 245—259.

Spence, D. H. N. (1967) Factors controlling the distribution of freshwater macrophytes with particular reference to the lochs of Scotland. J. Ecol. 55: 147—170.

Spence, D. H. N. (1975) Light and plant response in fresh water. *In*: Evans, G. G., Bainbridge, R. and Rackham, O. (Eds.), Light as an Ecological Factor: II, pp. 93—133. Blackwell Sci. Publ., Oxford.

Spence, D. H. N. & Chrystal, J. (1970) Photosynthesis and zonation of freshwater macrophytes. II. Adaptability of species of deep and shallow water. New Phytol. 69: 217—227.

Stadlemann, P., Moore, J. E. & Pickett, E. (1974) Primary production in relation to temperature structure, biomass concentration and light conditions at an inshore and off shore station in L. Ontario. J. Fish. Res. Bd. Can. 31: 1215—1232.

Steemann Nielsen, E. (1947) Photosynthesis of aquatic plants with special reference to the carbon-sources. Dansk Bot. Ark. 12: 1—71.

Steemann Nielsen, E. (1952) The use of radioactive carbon (^{14}C) for measuring organic production in the sea. J. Cons. perm. Int. Expl. Mer. 18: 117—140.

Steemann Nielsen, E. (1958) Planteplanktonets årlige produktion af organisk stof i Furesøen (in Danish with an English summary). Folia Limnol. Scan. 10: 104—109.

Steemann Nielsen, E., Hansen, V. K. & Jørgensen, E. G. (1962) The adaptation to different light intensities in *Chlorella vulgaris* and the time dependence on transfer to a new light intensity. Physiol Plant. 15: 505—517.

Steemann Nielsen, E. & Jensen, P. K. (1958) Concentration of CO_2 and rate of photosynthesis in *Chlorella pyrenoidosa*. Physiol. Plant. 11: 170—180.

Stone, S. & Ganf, G. (1981) The influence of previous light history on the respiration of four species of freshwater phytoplankton. Arch. Hydrobiol. 91: 435—462.

Storch, T. A. & Saunders, G. W. (1975) Estimating daily rates of extracellular dissolved organic carbon release by phytoplankton populations. Verh. intern. Ver. theor. angew. Limnol. 19: 952—958.

Stumm, W. & Morgan, J. J. (1981) Aquatic chemistry. An introduction emphasizing chemical equilibria in natural waters. Wiley-Interscience, New York, 2nd ed. 780 pp.

Søndergaard, M. (1979) Light and dark respiration and the effect of the lacunal system on refixation of CO_2 in submerged aquatic plants. Aquat. Bot. 6: 269—283.

Søndergaard, M. (1981) Kinetics of extracellular release of ^{14}C-labelled organic carbon by submerged macrophytes. Oikos 36: 331—347.

Søndergaard, M. & Lægaard, S. (1977) Vesicular-arbuscular mycorrhiza in some aquatic vascular plants. Nature 268: 232—233.

Søndergaard, M. & Sand-Jensen K. (1979a) Physico-chemical environment, phytoplankton biomass and production in oligotrophic, softwater Lake Kalgaard, Denmark. Hydrobiol. 63: 241—253.

Søndergaard, M. & Sand-Jensen, K. (1979b) Carbon uptake by leaves and roots of *Littorella uniflora* (L.) Aschers. Aquat. Bot. 6: 1—12.

Søndergaard, M. & Wetzel, R. G. (1980) Photorespiration and internal recycling of CO_2 in the submersed angiosperm *Scirpus subterminalis*. Can. J. Bot. 58: 591—598.

Talling, J. F. (1965) The photosynthetic activity of phytoplankton in East African lakes. Int. Rev. ges. Hydrobiol. 50: 1—32.

Talling, J. F. (1970) Generalized and specialized features of phytoplankton as a form of photosynthetic cover. *In*: Prediction and measurement of photosynthetic productivity. Proc. IBP/PP technical meeting, Třeboň 1979. p. 431—445. Pudoc, Wageningen.

Talling, J. F. (1976) The depletion of CO_2 from lake water by phytoplankton, J. Ecol. 64: 79—121.

Theodórsson, P. & Bjarnson, J. Ö. (1977) The acid-bubbling method for primary productivity measurements, modified and tested. Limnol Oceanogr. 20: 1018—1019.

Thomas, J. P. (1971) Release of dissolved organic matter from natural populations of marine phytoplankton. Mar. Biol. 11: 311—323.

Thyssen, N. (1980) Factors affecting the oxygen balance of a small, non-poluted stream dominated by submerged macrophytes. With special reference to photosynthetic rate and atmospheric reairation. Ph.D. thesis, Botanical Institute, Univ. Aarhus, Denmark.

Titus, J. E., Adams, M. S., Gustafson, T. D., Stone, W. H. & Westlake, D. F. (1979) Evaluation of differential infrared gas analysis for measuring gas exchange by submerged aquatic plants. Photosynthetica. 13: 294—301.

Titus, J. E. & Stone, W. H. (1982) Photosynthetic response of two submersed macrophytes to dissolved inorganic carbon concentration and pH. Limnol. Oceanogr. 27: 151—160.

Van, T. K., Haller, W. T. & Bowes, G. (1976) Comparison of the photosynthetic characteristics of three submerged aquatic plants. Plant Physiol. 58: 761—768.

Venrick, E. L., Buss, J. R. & Heinbokel, J. F. (1977) Possible consequences of con-

110

taining microplankton for physiological rate measurements. J. exp. mar. Biol. Ecol. 26: 55—76.

Verduin, J. (1975) Photosynthetic rates in lake Superior. Verh. intern. Ver. theor. angew. Limnol. 19: 689—693.

Vincent, W. F. (1980) Mechanisms of rapid photosynthetic adaptation in natural phytoplankton communities. II. Changes in photochemical capacity as measured by DCMU — induced chlorophyll fluorescence. J. Phycol. 16: 568—577.

Vollenweider, R. A. (1965) Calculation models of photosynthesis-depth curves and some implications regarding day rate estimate in primary production estimates. Mem. Ist. ital. Idrobiol. 10 (Suppl.): 425—457.

Vollenweider, R. A. (1968) Scientific fundamentals of the eutrophication of lakes and flowing waters, with particular reference to nitrogen and phosphorus as factors in eutrophication. OECD Paris. DAS/CSI/68.27., 192 pp.

Vollenweider, R. A. & Nauwerck, A. (1961) Some observations on the ^{14}C method for measuring primary production. Verh. intern. Ver. theor. angew. Limnol. 14: 134—139.

Walker, N. A., Beilby, M. J. & Smith, F. A. (1979) Amine uniport at the plasmalemma of charophyte cells: I. Current-voltage curves, saturation kinetics, and effects of unstirred layers. J. Mem. Biol. 49: 21—55.

Wallentinus, I. (1978) Productivity studies on baltic macroalgae. Bot. Mar. 21: 365—380.

Walsh, P. & Legendre. L. 1983. Photosynthesis of natural phytoplankton under high frequency light fluctuations simulating those induced by sea surface waves. Limnol. Oceanogr. 28: 688—697.

Westlake, D. F. (1963) Comparisons of plant productivity. Biol. Rev. 38: 385—425.

Westlake, D. F. (1965) Theoretical aspects of the comparability of productivity data. Mem. Ist. ital. Idrobiol. 18 (Suppl.): 313—322.

Westlake, D. F. (1967) Some effects of low-velocity current on the metabolism of aquatic macrophytes. J. exp. Bot. 18: 187—205.

Westlake, D. F. (1975) Macrophytes. In: Whitton, B. A. (Ed.), River ecology. Studies in Ecology, Vol. 2, pp. 107—128. Blackwell Sci. Publ., Oxford.

Westlake, D. F. (1978) Rapid exchange of oxygen between plant and water. Verh. intern Ver. theor. angew. Limnol. 20: 2363—2367.

Westlake, D. F. (Coordinator) (1980) Primary production. In: Le Cren, E. D. and Lowe-McConnell, R. H. (Eds.), The functioning of freshwater ecosystems. pp. 141—246. IBP 22. Cambridge Univ. Press, Cambridge, U.K.

Westlake, D. F. (1982) The primary productivity of water plants. In: Symoens, J. J., Hooper, S. S. & Compère, P. (Eds.), Studies on aquatic vascular plants, pp. 165—180. Royal Bot. Soc. Belgium, Brussels.

Wetzel, R. G. (1965) Nutritional aspects of algal productivity in marl lakes with particular reference to enrichment bioassays and their interpretation. Mem. Ist. ital. Idrobiol. 18 (Suppl.): 137—157.

Wetzel, R. G. (1966) Productivity and nutrient relationships in marl lakes of northern Indiana. Verh. intern. Ver. theor. angew. Limnol. 16: 321—332.

Wetzel, R. G. (1972) The role of carbon in hard-water marl lakes. In: Likens, G. E. (Ed.), Nutrients and eutrophication: The limiting-nutrient controversy. Spec. Symp. Amer. Soc. Limnol. Oceanogr. 1: 84—97.

Wetzel, R. G. (1979) The role of the littoral zone and detritus in lake metabolism. In: Rodhe, W., Likens, G. E. & Serruya C. (Eds.), Lake metabolism and management. Ergbn. Limnol. 13: 145—161.

Wetzel, R. G. (1983) Limnology. Saunders Corp., Philadelphia. 767 pp.

Wetzel, R. G. & Allen, H. L. (1970) Functions and interactions of dissolved organic matter and the littoral zone in lake metabolism and eutrophication. *In*: Kajak, Z. & Hillbricht-Ilkowska, A. (Eds.), Productivity problems of freshwaters, pp. 333—347. PWN Pol. Sci. Publ. Warsaw.

Wiginton, J. R. & McMillan, C. (1979) Chlorophyll composition under controlled light conditions as related to the distribution of seagrasses in Texas and the U.S. Virgin Islands. Aquat. Bot. 6: 171—184.

Williams, P. J. LeB. (1981a) Incorporation of microheterotrophic processes into the classical paradigm of the planktonic food web. Kieler Meeresf. Sonderh. 5: 1—28.

Williams, P. J. LeB. (1981b) Microbial contribution to overall marine plankton metabolism: direct measurements of respiration. Oceanol. Acta 4: 359—364.

Williams, P. J. LeB. (1982) Microbial contribution to overall plankton community respiration-Studies in enclosures. *In*: Grice, G. D. & Reeve, M. R. (Eds.), Marine mesocosms. Biological and chemical, research in experimental ecosystems. Springer-Verlag, New York.

Wium-Andersen, S., Anthoni, U., Christophersen, C. & Houen, G. (1982) Allelopathic effects on phytoplankton by substances isolated from aquatic macrophytes (Charales). Oikos 39: 187—190.

Zelitch, I. (1971) Photosynthesis, photorespiration and plant productivity. Academic Press, New York. 347 pp.

STRUCTURAL ASPECTS OF
AQUATIC PLANT COMMUNITIES

C. DEN HARTOG & G. VAN DER VELDE

INTRODUCTION

The study of water plants has always lagged behind the study of terrestrial plants. This is clearly reflected by the number of books dealing with them. In fact the number of books especially dedicated to the various aspects of morphology, physiology and ecology of these interesting plants is extremely small (Glück 1905, 1906, 1911, 1924, Arber 1920, Gessner 1955, 1959, Sculthorpe 1967, Hutchinson 1975). Of course there is a number of manuals and there are local or regional floras on aquatic plants, but these give usually pure systematic information, supplemented by notes on occurrence and distribution (Fassett 1957, Mason 1957, Aston 1973, Cook 1974, Beal 1977, Casper & Krausch 1980—1981). Further there are monographs of most aquatic families and genera, but several of these are old and by no means up to date. There are, for example, no recent monographs on such important genera as *Nymphaea*, *Myriophyllum* and *Potamogeton*, while some monographs only deal with a part of the area, e.g. the treatments by Taylor (1964) of *Utricularia* in Africa and by Lowden (1986) of *Najas* in tropical America.

Seen in the light of the weak systematic foundation of the aquatic flora, it is not surprising that a comprehensive work dealing with aquatic vegetation has not yet been written. It does not mean that such studies on vegetation have not been carried out. Scattered in the literature there is a considerable amount of information about the floristic composition of aquatic communities from many areas all over the world. In several areas a systematic classification of the aquatic plant communities, based on species combinations has been attempted.

However, these attempts have resulted in rather superficial descriptions of the various types of vegetation that could be visually recognized. The criterion for their recognition was floristic similarity or dissimilarity; additional criteria were seldom applied. However, similarity as well as dissimilarity of plant communities may be the result of

113

J. J. Symoens (ed.), Vegetation of inland waters. ISBN 90—6193—196—7.
© *1988, Kluwer Academic Publishers, Dordrecht. Printed in the Netherlands.*

a wide scala of causes, and the mere establishment of the difference between them, is not considered acceptable nowadays. Further, although it is recognized that in most communities macrophytes form the bulk of the biomass, there are a large number of other participants in the community, such as microphytes, animals and microbes. Their role is not negligible; on the contrary, many of them are essential in the functioning of the whole system. So, if one wants not only to describe but also to understand the aquatic plant communities the other accompanying categories have to be considered as well.

A community is characterized by its species composition and by additional features derived of this, such as life- and growth-form spectra, diversity indices, etc. It can be considered to be a structural frame which as a unit can be related to the abiotic and biotic surroundings.

An ecosystem is a functional unit consisting of a biotic community, the abiotic environment, and the mutual relations between community and environment; it is more or less selfsupporting, and driven, directly or indirectly, by solar energy (Odum 1971). An ecosystem is characterized by functional parameters, such as primary productivity, turnover rates, energy balance, food webs, geochemical cycles and type of nutrient pool. Generally the concept of functioning can be regarded as a comprehension of all activities and performances of the joint organisms of the community. More specific, non-energetic properties, such as substrate extension, substrate stabilization, shelter, nursery, spawning place, etc. have to be included within the term function. The biotic community can be considered to be the structural biotic component of the ecosystem. Although community and ecosystem are differently defined every-day practice has learnt that this separation is impractible, and that both terms are applied in an interchangeable manner.

The complexity of the relations within an ecosystem has resulted in a wide scala of research approaches, all with the general goal to come to a better understanding of the structure and the functioning of the system. Widely used is the 'blackbox' approach, in which input and output of the system (or a part of it) are determined. From the results one can hypothetisize about the acting mechanisms. In this holistic way the ecosystem is in fact described as a complex of various processes, taking place in an orderly manner, without much attention for the organisms that perform these processes. According to the interest of the investigator a plant can be considered to be a photosynthetic apparatus of which the potential capacity can be expressed as the quantity of active chlorophyll, it can be considered to be biomass containing quantities of carbon, nitrogen and phosphorus in a certain ratio, or it can be considered as a large surface, where exchange processes with the ambient water take place. This approach is of great value to estimate the order of magnitude of the processes that can take place within the system. However, the approach is fully unsuitable for establishing the interrelations between the various biotic components.

The 'assemblage-approach' is another way to study the ecosystem. It is based on the study of separate biotic community components and their mutual relationships. The vision underlying this approach is that it must be possible, by continuous integration of the research results, to arrive at a reasonable structural and functional model of the system. In this approach one usually works with the dominant species and applies autecological and ecophysiological methods. Both the black-box and the assemblage approaches complement each other, and should be applied together, in order to come to a balanced ecosystem model description.

Study of structural and functional components of a system, however, will rarely lead to satisfactory results, because this approach is too static. Most communities are liable to change. The element of change can be related neither to structure nor to function(ing), as changes in structure inevitably cause changes in functioning and function. Therefore, dynamics of the community has to be regarded as a separate aspect in the research or ecosystems. Further, it has to be realized that knowledge of structure, functioning and dynamics of a system is by no means sufficient to explain the often extremely complex relations between the various components, because this knowledge is based on present-day observations. These relations, however, have developed over extremely long periods and are the result of evolution, selection and adaptation processes of each separate component. Consequently, each ecosystem has a firm root in the past and, therefore, the historical aspect must not be omitted from ecosystem research.

Finally, it is a matter of experience that under more or less similar ecological circumstances rather similar communities are found. This may be ascribed to a considerable extent to the above-mentioned complexity of the relations between the various community components. Consequently, by comparison of a great number of rather similar communities the overall traits and features can be worked out to an idealized abstract model of the community, from which all local features are eliminated; in other words, one can elaborate a classification of communities. In such a classification as many parameters as possible should be involved.

Structure consists of at least three major components, which are not completely independent of each other. These components are:

1. Floristic and faunistic composition of the community, in a qualitative as well as a quantitative sense.

2. The arrangement of the organisms in space and time.
3. The relations between the organisms within the community and their relations with the surrounding biotic and abiotic environment.

It appears in practice, that there are very few descriptions in which these three components have adequately been dealt with, as will be shown in the following sections. Barkman (1979) has proposed to restrict the term structure for the spatial arrangement of the organisms, and he introduced the term texture for floristic (and faunistic) composition. Such an unnecessary refinement of the terminology is undesirable and superfluous, and needs to be rejected.

WATER PLANTS AND THEIR LIFE — AND GROWTH FORMS

In the literature the concept 'water plant' has usually been applied rather arbitrarily in the sense of plants occurring in and near the water. Definitions, in the case these have been formulated, vary considerably and often apart from genuine aquatic plants include also plant types which depend on water for only one or a few stages of their life cycle. Here water plants are defined as plants which are able to achieve their generative cycle when all vegetative parts are submerged or are supported by the water (floating leaves), or which occur normally submerged but are induced to reproduce sexually when their vegetative parts are dying due to emersion (den Hartog & Segal 1964).

The first part of this definition comprises the bulk of the aquatic plants, the second part of the definition is added in order to comprise the Podostemaceae, the Hydrostachyaceae and a number of aquatic musci (e.g. *Fontinalis*), which are unable to survive any desiccation. Nevertheless four groups of plants which often are regarded as water plants, have been excluded by this definition. These groups are:

1. The pseudohydrophytes, i.e. plants which frequently occur completely submerged and maintain themselves in this condition for years by vegetative reproduction, but are unable to reproduce sexually under these circumstances. Examples of this category are numerous. In standing waters the waterferns of the genera *Marsilea* and *Pilularia* are generally present in the vegetative stage, but when the water recedes sporocarps are produced. The aquatic forms of *Juncus bulbosus* and *Littorella uniflora* do not flower, but their much smaller, compacter terrestrial forms flower and fruit profusely. In running water species such as *Oenanthe fluviatilis*, *Myosotis palustris* and *Sagittaria sagittifolia* (with linear leaves) may occur in great quantities but sexual reproduction occurs only when due

to drought these plants become emerged. The well-known genus *Cryptocoryne* from tropical Asia contains also many species which only flower when their habitat dries up.

2. The helophytes, i.e. plants which root in the bottom and the basal parts of which are often continually submerged but whose leaves and inflorescences emerge far above the water surface. To this category belong many genera and species characteristic of the vegetation that lines the open water, e.g. *Typha*, many *Sparganium* species, *Phragmites*, *Scirpus* subgen. *Schoenoplectus*, *Acorus*, *Butomus*, *Alisma*, *Pontederia*, many species of *Sagittaria*, etc.

3. The pleustohelophytes, i.e. plants which float freely on the surface of the water with submerged root systems but with all other vegetative parts and inflorescences rising above the water. To this category belong two well-known waterweeds, viz. *Eichhornia crassipes* and *Pistia stratiotes*.

4. The reptohelophytes ('rhizopleustohelophytes' *sensu* Hogeweg & Brenkert 1969b), i.e. plants which root on the banks or in floating vegetable debris but which extend with more or less long runners on or just below the water surface, producing emergent leaves or leaf-bearing shoots and inflorescences. Examples of this category are *Ipomoea aquatica*, *Neptunia oleracea*, *Aeschynomene indica*, *Ludwigia adscendens* and *Sium erectum*.

The strict definition of the concept water plant has the advantage that water plants can be distinguished by a set of characters, instead of by their occurrence in or close to the water. It is, however, true that many plant species which do not fit the above definition are dependent on the aquatic environment, be it often only temporarily. Many species depend on a temporary submersion of the substratum for the germination of their seeds. There are also many species which during their early development show structures which can be interpreted as adaptations to aquatic life, e.g. the linear juvenile leaves of *Alisma plantago-aquatica*.

In the well-known, more general life-form systems of Raunkiaer (1934) and Iversen (1936) in which the plants are classified on account of their adaptation to the unfavourable season and the water factor respectively, the water plants are given a separate status. Raunkiaer recognized the hydrophytes as a special category in his system. They were regarded as plants which have their vegetative parts submerged or floating at the water surface but not projecting into the air and which survive the unfavourable period in the form of submerged buds, these being either attached to a rhizome or lying completely free on the bottom of the water. Unfortunately his concept excludes the short-lived water plants, and includes the species which are able to produce water forms although these can not complete their generative cycle when submerged. Further it seems that Raunkiaer had not realized himself that the unfavourable season for water plants is not necessarily the winter, but that the summer can be unfavourable as well. The adaptations for surviving a drought are often quite different from those for surviving the cold season, even in the same species. Thus *Potamogeton natans* is in the winter a typical hydrophyte but it survives a drought as a hemicryptophyte. When the hydrophytes are compared with the other categories of Raunkiaer's system, it appears that all hydrophytes match certainly one of these categories, so that one can distinguish between hydrochamaephytes, hydrogeophytes, hydrohemicryptophytes and hydrotherophytes. This was recognized already by Braun-Blanquet (1964), who maintained, however, the concept 'hydrophyte'. In our opinion the concept 'hydrophyte' in the sense of Raunkiaer can be rejected as all aquatic plants can be incorporated as special subgroups in the other categories of his system.

Iversen regarded water plants ('limnophytes') as plants which have their vegetative parts submerged or floating at the water surface, but not projecting in the air, and which for the larger part are able to develop vegetatively and generatively reduced land forms. This definition is rather similar to the first section of the definition adopted in this paper. However, it differs, because Iversen (1936) recognized also amphiphytes, a category of plants which are able to achieve their generative cycle in the submerged as well as in the emerged condition. These amphiphytes are defined as plants with emerged aerial leaves and hydromorphic leaves, or which are able to produce water forms. As examples he cited *Eleocharis acicularis, Littorella uniflora* and *Pilularia globulifera*, which reproduce preferentially when emerged, and species of *Ranunculus* subgen. *Batrachium* and *Callitriche*.

The group of species included in the second part of the definition of water plant viz. those which occur normally submerged but reproduce when dying after emersion, certainly would represent a separate category in Iversen's system. He himself did not mention this category in his work, which is mainly based on the Danish flora.

Water plants show a wide variation in the use of the aquatic medium for their vital functions. The species of *Ceratophyllum*, for example, achieve their complete life cycle in the submerged condition. In *Ranunculus circinatus* the flowers are emerged, but the fruits develop submerged; other crowfeet have also emerged flowers but moreover floating leaves which have a part of their metabolic activity in direct contact with the air.

Many *Utricularia* species are fully adapted to water life, except for flowering and fruiting; the inflorescences are often provided with special 'floats' (e.g. in *U. inflata* and *U. stellata*). A subdivision of the aquatic plants, based on this degree of dependence on the aerial medium for their development has been developed by Poplawskaja (1948). She distinguished three categories:

1. The hydatophytes, which achieve their complete life cycle when completely submerged and have no adaptation to aerial life. The numbers of angiosperm species in this category is amazingly small.
2. The 'submerged' aerohydatophytes, which are completely submerged except for the inflorescences.
3. The 'floating' aerohydatophytes, which have submerged and floating vegetative parts, and aerial inflorescences. Strangely enough she used the term 'hydrophyte' for those plants which have only their basal parts in the water, but which mainly project into air.

Hejný (1957, 1960) has also proposed a classification of plants with relation to the factor water; he did not base his categories on morphological characters but on life cycle characters. Three of the 10 categories distinguished by him comprise the water plants, viz.

1. The euhydatophytes of which the vegetative parts are completely adapted to water life, and of which the inflorescences are submerged or project above the water surface. This category coincides with the hydatophytes and submerged aerohydatophytes of Poplawskaja.
2. The hydatoaerophytes, which are bound to the water, but whose vegetative parts are in contact with the air, and with inflorescences rising above the water surface. This category coincides with Poplawskaja's 'floating' aerohydatophytes.
3. The tenagophytes, which is a heterogeneous group of amphibious species. Some of the perennial species mentioned by Hejný are able to achieve their generative cycle when submerged, and can be regarded as water plants. Most species listed by him, however, belong to the ephemeral summer annuals, which are dependent on the dry spell, for their generative reproduction, but are able to survive submersion as seeds and spores.

Although Hejný's system has not received much support, the basic idea that the demands of the various developmental stages of the plants need to be considered in the description and explanation of vegetation structure, is sound and needs more attention.

A further subdivision of water plants has been proposed by Luther (1949) who classified them according to their mode of attachment. He recognized three large groups:

1. The haptophytes, i.e. plants which do not penetrate into the substratum with their basal parts, but are attached to rock, stones and

other solid substrata (e.g. Hydrostachyaceae, Podostemaceae, many algae).

2. The rhizophytes, i.e. plants which penetrate with their basal parts into the bottom, or which have their basal parts covered by the substratum.
3. The pleustophytes, i.e. plants which are not attached at all, and float freely in the water. According to their position in the water column one can distinguish: a. the acropleustophytes, which float on the surface, b. the mesopleustophytes, which float freely between the bottom and the surface, and c. the benthopleustophytes, which are lying loosely on the bottom.[1]

In the systems discussed so far the plants have been grouped according to their adaptation to a certain vital factor: they are life-form systems. There is also a system based on the habit of the plants and originally proposed by Du Rietz (1921, 1930). The habit does not express adaptation to a single factor but is the result of the totality of interactions between the plant and its environment. The original system of Du Rietz has been elaborated for the aquatic plants by den Hartog & Segal (1964), and has been refined since by Segal (1965, 1968), Hogeweg & Brenkert (1969b), and Hutchinson (1975) for fresh water plants, and by den Hartog (1967, 1977, 1981) for the marine and poikilohaline plants. The basic growth forms are well-defined and easy to be recognized. The 'eco-morphological life-form system' of Mäkarinta (1978) is considered here a variation on the same theme. The growth form system can be fruitfully combined with the system of Luther (1949). At present 24 basic forms have been recognized for the pleustophytes and rhizophytes, a number liable to further extension. A further subdivision of the haptophytes has not yet been undertaken.

The following basic growth forms have been distinguished:

1. Lemnids, i.e. small pleustophytes, floating freely on the water surface, with reduced fronds of which the upper side is adjusted to air metabolism, and the under side to life in the water, e.g. *Spirodela, Wolffia, Lemna* subgen. *Lemna, Pseudowolffia, Azolla, Ricciocarpus natans*.
2. Ricciellids ('Wolffiellids'), i.e. small submerged, lanceolate, furcate or lingulate pleustophytes, which often do not produce roots, and which have no adaptations to air metabolism; they only come to the surface for anthesis or sporulation, e.g. *Riccia* subgen. *Ricciella, Lemna* subgen. *Staurogeton, Wolffiella, Wolffiopsis*.

[1] Luther (1949) distinguished the planophytes which he subdivided into macroscopic pleustophytes and microscopic planktophytes. As this paper is only concerned with macroscopic water plants the term planophytes has been disregarded.

3. Ceratophyllids, i.e. submerged pleustophytes with finely divided leaves and without floating leaves, in the summer near the surface of the water but in the autumn sinking to the bottom, hibernating by turions, e.g. *Ceratophyllum, Utricularia, Aldrovanda.*

   Hogeweg and Brenkert (1969a) divide this category into the Ceratophyllids *sensu stricto*, which are completely submerged during their whole cycle and the Utriculariids of which the inflorescences, and infructescences are emerged.

4. Hydrocharids, i.e. stoloniferous pleustophytes which float freely on the water surface with rosettes of specialized floating leaves, in temperate areas hibernating by specialized turions. e.g. *Hydrocharis morsus-ranae* and *Limnobium laevigatum.*

5. Salviniids ('Magno-lemnids' *pro parte*), i.e. pleustophytes which float freely on the water surface with a horizontal stemlike axis along which the floating leaves are arranged, e.g. *Salvinia, Phyllanthus fluitans.*

6. Stratiotids, i.e. stoloniferous rhizophytes, with a rosette of stiff radical leaves of which the upper parts rise above the water surface, loosely anchored by very long roots in soft sapropelium or other organic bottom sediments, in the autumn sinking to the bottom, e.g. *Stratiotes aloides.*

   The stratiotids have been regarded earlier as pleustophytes, but it has appeared that the long roots indeed are anchored in the bottom. This growth form is in fact transitional between rhizo- and pleustophytes.

7. Isoetids, i.e. rhizophytes with a short stem, a rosette of stiff radical leaves with large air lacunae, with or without stolons, e.g. *Isoetes lacustris, Littorella uniflora, Lobelia dortmanna, Lilaeopsis australasica.*

8. Charids, i.e. submerged erect rhizophytes with verticillate branches, e.g. Characeae.

9. Vallisneriids, i.e. submerged rhizophytes with a short stem and a rosette or bundle of long, flabby, linear, radical leaves, with or without stolons, e.g. *Vallisneria, Blyxa* sect *Blyxa.*

10. Magnopotamids, i.e. rhizophytes of considerable size with oblong to lanceolate submerged leaves, caulescent stems arising from a sympodial rhizome, e.g. *Potamogeton perfoliatus, P. lucens.*

11. Parvopotamids, i.e. submerged rhizophytes with sympodial rhizomes, caulescent upright shoots, and linear to oblong leaves; leaf arrangement usually distichous, sometimes more or less decussate and rarely verticillate; there is a tendency towards rhizome reduction, e.g. *Potamogeton* subgen. *Coleogeton, Potamogeton* sect. *Graminifolii, Groenlandia, Zannichellia, Ruppia, Althenia, Lepilaena, Najas.*

12. Elodeids, i.e. submerged rhizophytes without a rhizome, with caulescent upright shoots and whorls of linear or oblong leaves, e.g. *Elodea, Egeria, Maidenia, Blyxa* sect. *Caulescentes, Lagarosiphon, Nechamandra, Hydrilla, Hydrotriche, Mayaca, Callitriche* sect. *Pseudocallitriche.*

Because of the tendency towards rhizome reduction the elodeids and parvopotamids are often difficult to distinguish. Parvopotamids generally have longer leaves. The difference between magnopotamids and parvopotamids is mainly based on size and possibly can only be maintained regionally. *Groenlandia densa* is in western Europe generally a small plant, while in southern Europe it can reach a considerable size in appropriate environments.

13. Myriophyllids, i.e. caulescent rhizophytes with finely dissected, submerged leaves and without specialized floating leaves; generative parts always emerged, e.g. *Myriophyllum* (most species), *Ranunculus circinatus, Hottonia palustris, Cabomba* (most species), *Limnophila indica.*

14. Batrachiids, i.e. caulescent rhizophytes with specialized floating leaves and finely dissected submerged leaves; there is a tendency to develop terrestrial forms; e.g. several species of *Ranunculus* subgen. *Batrachium.* The difference with the myriophyllids is totally based on presence or absence of floating leaves.

15. Peplids, i.e. caulescent rhizophytes with oblong and spatulate leaves, the upper ones forming floating rosettes, adapted to air metabolism, e.g. *Peplis portula, Callitriche* (most species).

16. Trapids, i.e. caulescent rhizophytes with strong, branched stems which produce rosettes of floating leaves, e.g. *Trapa natans.*

17. Otteliids, i.e. rhizophytes with a short stem and a rosette of lanceolate, rhombic or suborbicular, submerged leaves, e.g. *Ottelia alismoides, Barclaya motleyi, Aponogeton ulvaceus.* This category is not very homogeneous and certainly can be differentiated further, e.g. according to the rooting systems and leaf shapes.

18. Nymphaeids, i.e. rhizophytes with a little or not branched stem and longly petiolated floating leaves, sometimes also with linear, lanceolate, rhombic or orbicular submerged leaves.

This group is extremely heterogeneous and certainly has to be further divided. One can distinguish between magnonymphaeids with large floating leaves, usually in rosettes arising from large rootstocks (e.g. *Nymphaea, Nuphar, Victoria, Euryale*) or with stalked rosettes (*Brasenia, Nymphoides*), and parvonymphaeids with small floating leaves. Amongst the latter category is a considerable variety of forms. *Potamogeton natans, P. gramineus* and *P. nodosus* have sympodial rhizomes, while in *P. octandrus* the rhizome is strongly reduced. Some species have no rhizomes at all, but have a short stem with a radical rosette of floating leaves, e.g. *Luronium natans, Caldesia oligococca, C. parnassifolia* and *Ottelia ovalifolia*. In other species the leaf rosettes arise from a tuber, e.g. in *Aponogeton natans, A. hexatepalus* and *Ondinea purpurea*. There are also species with linear floating leaves, e.g. various species of *Sparganium* and *Cycnogeton*, and there are even some ephemeral species, e.g. *Hydrocotyle lemnoides*. Juvenile nymphaeids and plants that occur in streams often fail to produce floating leaves, and in that case can not be distinguished from otteliids.

19. Parvozosterids, i.e. submerged rhizophytes with monopodial rhizomes and fine linear leaves, e.g. *Halodule, Zostera* subgen. *Zosterella*.

20. Magnozosterids, i.e. submerged rhizophytes with monopodial rhizomes, wide linear leaves and sheaths which either decay completely or remain membranous, but never persist as a bunch of fibres, e.g. *Zostera* subgen. *Zostera, Cymodocea, Thalassia*.

   In regions where more species occur the division in parvozosterids and magnozosterids is generally very clear. If, however, one species occurs in a region this species may show the whole scala of sizes of parvo- and magnozosterids, e.g. *Zostera capricorni* in S.E. Australia and *Z. capensis* in S. Africa.

21. Syringodiids, i.e. submerged rhizophytes with monopodial rhizomes and long subulate leaves, e.g. *Syringodium*.

22. Enhalids, i.e. submerged rhizophytes with monopodial rhizomes and leathery, linear or coarse strap-shaped leaves; the remains of their sheaths form thick paintbrush-like fibre bundles at the base of the plants, e.g. *Enhalus, Posidonia* and *Phyllospadix*.

23. Halophilids, i.e. submerged rhizophytes with monopodial rhizomes and elliptic, ovate, lanceolate or linear, very delicate leaves; whole plants devoid of air channels, e.g. *Halophila*.

24. Amphibolids, i.e. submerged rhizophytes with sympodial rhizomes and lignified upright stems; leaves linear, distichously arranged along the stem, e.g. *Amphibolis, Thalassodendron, Heterozostera*.

The growth forms 1—18 are characteristic of fresh water. The growth forms 19—24 represent the sea-grasses. The sea and the fresh water have not one growth form in common. From the poikilohaline brackish waters, and the continental salt waters only two growth forms are known, viz. the parvopotamids and the charids. Both growth forms occur also in fresh water, but are absent from the sea.

The various growth forms are of great importance for the vegetation structure. Each growth form seems to be linked with certain ecological conditions, and the number of growth forms coexisting in one community is always restricted. Hogeweg & Brenkert (1969a) calculated the correlations between the occurrence of various growth forms and found that some of them occur often together, while others are never found together. Therefore, it seems likely that the aquatic plant communities can be largely characterized by their growth-form spectra. The growth-form spectra can give good indications about environmental circumstances such as permanence of the water body, water quality, water movements, and nutritional status, and have been applied for the classification of aquatic plant communities by den Hartog & Segal (1964) and den Hartog (1983a).

Up to now the growth forms have been considered as morphological types; the functional meaning has hardly been taken into account. Recent research has shown that the various growth forms appear not only to be linked with certain environmental conditions, but seem to be also the most efficient solutions for plant life under these conditions. In this connection most attention has been paid to the isoetid growth form. Isoetids are characteristic for poorly buffered water systems in which the CO_2-content is extremely reduced. The isoetid growth-form seems to be an excellent solution to these circumstances. The thick stiff leaves are compact, and cause a reduction of the surface-volume ratio. Further all isoetids have a well-developed system of internal air lacunae, so that CO_2 produced during the photorespiration can be reused again. Further they are able to take up CO_2 with the roots from the interstitial water, where the CO_2 level may be 10—100 times higher than in the overlying water layer, and to release O_2 via the roots into the substratum. In that way they themselves can contribute to the oxidation of organic matter in the bottom, which provides them with CO_2. Furthermore, most of the isoetids apply a special mechanism for photosynthesis, similar to the Crassulacean Acid Metabolism (CAM), and that may be interpreted as an adaptation to environments where CO_2 availability is precarious. Further they show a clear relationship between the development of the underground biomass and the nutrient content of the environment. Under oligotrophic conditions the root system is well-developed. So the isoetids fit functionally very well within the environment where they occur, and this growth form assures the most efficient use of the scarce but essential resources. In richer environments these plants would not stand a chance in the competition with other more demanding species (Søndergaard 1979, Søndergaard & Sand-Jensen 1979, Keeley 1982a, b, Sand-Jensen et al. 1982, Roelofs et al. 1984, Madsen 1985, den Hartog, 1986).

On some nymphaeids also such information is known. Their floating leaves enable the plants to take full advantage of the irradiation but assure them also of a supply of inorganic carbon especially from the air. In this way nymphaeids when fullgrown are more or less independent of the water quality (van der Velde et al. 1986). Absorption rates of phosphorus differ with the absorbing organ and are decreasing in the following sequence: roots, submerged leaves and floating leaves (Twilley *et al.* 1977). The thin submerged leaves are apparently better adapted for nutrient absorption than the thick, leathery floating leaves. Submerged leaves are especially developed in spring for obvious reasons. Thus the morphological differentiation of *Nuphar* leaves has a clear physiological base which explains for a great deal the ecological success of this nymphaeid species. Furthermore, the lacunae in nymphaeids like *Nuphar* fulfill important functions for survival and growth of these plants in anaerobic sediments. The rhizome of *Nuphar* accounts roughly for 80% of the plant's biomass during summer growth (Twilley *et al.* 1977). From the apices of these rhizomes bundles of petioled submerged and floating leaves arise. The gas spaces of *Nuphar* occupy about 60% of the petiole volume and 40% of the roots-rhizome volume. Dacey (1981) and Dacey & Klug (1979, 1982) found that there are internal winds in the *Nuphar* plant. Air enters the lacunae of the youngest floating leaves against a small gradient in total pressure. The pressurization mechanism is purely physical and driven by heat from irradiation and has nothing to do with metabolism. Due to irradiation floating leaves become heatened by which the air in the lacunae builts up a higher pressure. This elevated pressure drives a bulk flow of gas down via the petioles of these young leaves to the rhizome and from the rhizome up via the petioles of older emergent leaves up to the atmosphere, so creating a circulating system via pressure gradients.

The ecological meaning of this system is twofold. At first CH_4 (methane), which is only produced in highly reduced sediments and diffuses into the roots and rhizomes, is removed via the older leaves. Second the rhizome buried in the anaerobic sediment is provided with oxygen. The gas mixture from the rhizome towards the older leaves is rich in CH_4 and CO_2; the CO_2 in the petioles is totally used for photosynthesis during daylight.

Other growth forms have been less intensively studied for their functional meaning. It seems, however, very probable that all growth forms are structural compromises which enable the plants to deal efficiently with the adverse and the advantageous properties of the environment in which they occur. Finely dissected leaves seem to be of advantage in running water for mechanical reasons, but they also have the advantage that due to the enlargement of the leaf surface the exchange processes between the leaves and the surroundings become enhanced without increasing the photosynthetically active surface. Which growth form will turn out the most suitable in a certain type of water depends on a number of external factors and the plant properties themselves. External factors that can be regarded as highly determinant, are the availability and the form of inorganic carbon in the environment, the availability and spatial distribution of the various nutrients in the environment, and the light regime. Plant properties of decisive significance are the form(s) of inorganic carbon that they can use for the photosynthesis, the way how and the locus where they take up their nutrients, and whether they can store surpluses (luxury consumption), the internal transport facilities in the plants, the photosynthetic mechanisms (C_3, C_4, CAM) utilized, and the way they deal with anaerobiosis.

Due to the manifold nature of internal and external factors no generalisations can yet be made. It is considered an important desideratum to arrive at an understanding of the functional meaning of the various growth forms.

FLORISTIC AND FAUNISTIC COMPOSITION

Knowledge of the floristic composition of aquatic plant communities is generally less complete as may have been expected. The investigators usually identify the flowering plants and the water ferns down to the species, but the precision is much less where other plant groups such as Bryophyta, Charophyta and other macro-algae are involved. Micro-algae and fungi are rarely recorded, even not their presence or absence. This incomplete knowledge of the total plant community can be ascribed to various causes: (1) insufficient taxonomic knowledge, and lack of adequate literature in many laboratories, (2) technical problems, as many species can only be collected by special sampling techniques, and need to undergo special treatments before one can identify them (e.g. diatoms, Chrysophyta); identification *in situ* is impossible. A number of algae need culturing under special conditions, to induce sporulation or the formation of gametes before they can be identified. (3) Underestimation of the ecological importance of the non-macro-phytic component of the community.

The fauna is rarely considered in the study of aquatic plant communities. This can at least partly be ascribed to the same causes as mentioned for the non-macrophytes, as many groups can only be obtained adequately with specialized sampling techniques. An additional cause is, however, the lack of integration of botanical and zoological research, where it concerns the study of communities and ecosystems. The functional role of animals in macrophyte systems must not be underestimated.

Therefore, detailed studies of floristic and faunistic composition of aquatic plant communities are still very urgent.

ARRANGEMENT OF THE ORGANISMS IN
SPACE AND TIME

The arrangement of the organisms in space and time, within the aquatic plant communities has only received little attention probably because of the lack of a standardized procedure for this.

The spatial arrangement of the components of a community shows a number of characteristic patterns which can be roughly divided into horizontal, vertical and three-dimensional ones. Horizontal patterns due to a heterogeneous bottom configuration have been described by several authors. Kornaś *et al.* (1960) found in the Polish Baltic Sea a mozaic of *Fucus vesiculosus* and *Zostera marina*, the first growing on scattered stones and the other rooting in the sediment between these stones. Another type of patterns is induced by hydrodynamic conditions. Under quiet conditions, plants may develop patches circular in outline, and if weak currents are involved the patches may be elliptical. This situation is often observed when a new substrate becomes available for plant growth. When the various patches intermingle and/or hinder their mutual expansion an irregular mozaic may come into existence. Under marine circumstances the patchy configuration may remain more permanent, and is known as the 'leopard-skin' pattern (Chassé 1962, den Hartog 1973). In streams plants are usually arranged in longitudinal patches parallel to the current direction. In the intertidal belt plants may be arranged in a stripey zigzag pattern, due to the fact that ebb and flood currents generally are not evenly strong, and moreover do not follow the same route. Where due to topographical factors ebb and flood water have to follow the same route, seagrasses can be arranged in longitudinal stripes, perpendicular to the current direction (tiger-skin pattern, den Hartog, in prep.). These patterns are not linked with properties characteristic of the constituent species, e.g. *Zostera noltii* has been found arranged according to the leopard-skin pattern, the stripey zigzag pattern and the tiger-skin pattern. Even animal groupings may show the same types of patterns, e.g. mussel beds and coral colonies.

Vertical patterns include such phenomena as zonation and stratification. Zonation usually is found on a gradient, and it depends largely on the differences of the various aquatic plant species in their ability to deal with low light intensities; however, hydrodynamic factors such as waves, currents and water level fluctuations may be involved as well, particularly in shallow water bodies. Further hampering of water circulation may cause also the development of a special vegetation, particularly in very shallow water. As an example may be mentioned that the nymphaeid belt along lakes at its lower fringe is bordered by a zone of Characeae. The lower limit of the nymphaeids very probably is

determined by light limitation to the germlings or to the buds in spring
as well as the growth of the petioles which can not be longer than 3
metres (Funke 1951). The vegetation of Characeae usually continues
under the nymphaeids, due to the light interception by the floating
leaves, and the extensive utilization of the bottom by the nymphaeids,
leaving ample space for the superficially 'rooting' Characeae. The land
bound fringe of the nymphaeid belt is usually bordered by helophytes.

Stratification is the situation, that two or more vegetation layers
occur in the same column of water. It is not a rare feature to find
a layer of rhizophytes overlayered by a layer of pleustophytes. Such
a stratification is not permanent because light interception by the
pleustophytes will cause a decline of the rhizophytes. A dense cover of
lemnids does not agree with a well-developed rhizophyte layer, particularly as the lemnid cover overlayers the rhizophytes for several months.
Only *Potamogeton pectinatus* and *Elodea nuttallii* seem to be able to
stand such conditions for a considerable time.

It may also occur that surface floating lemnids overlayer a submerged layer of free floating *Ceratophyllum* or *Utricularia*, and also
these are able to tolerate this situation for a considerable time.

Rhizophyte growths can also show stratification, e.g. the Characeae
subgrowth of the nymphaeids represents a shade-induced extension of
the Characeae belt. It has also been observed that parvopotamids and
elodeids were rooting in a layer of the alga *Vaucheria dichotoma*. The
fact that the floating layers can move around under the influence of
currents and wind and that they generally suffocate the rhizophytes
below them, shows that they have to be considered as separate vegetation units. The same is in fact true for the two cases recorded here of
overlayering of rhizophytes by other rhizophytes. In both cases the
separate layers occur also as independent vegetation units, and there is
no specific interaction between the two vegetation layers.

The three-dimensional pattern or architecture, is defined as the way
in which the community fills up the available space (den Hartog 1976,
1978, 1982). It may be considered as the structural characteristic most
fundamental for the functioning of the community, as it appears that
the various aquatic growth forms utilize the space in very different
manners. The space utilization of nymphaeid communities is very
extensive. From the rhizomes in the substratum groups of long-stalked
floating leaves and peduncles develop; sometimes there are also some
salad-like or lanceolate ground leaves. Although the floating leaves may
cover the surface for 100%, the water column itself contains only the
petioles and peduncles, and possibly some accompanying macrophyte
species. There is ample space to allow large-sized fish to move easily
within the vegetation without obstruction.

In contrast, communities dominated by parvopotamids and elodeids fill up all available space from the water surface down to the bottom, or produce a horizontal canopy in the upper part of the water layer. A fully developed ceratophyllid vegetation may also fill up the total volume of the water body. In these communities water circulation is hampered and oxygen metabolism is precarious, they offer no adequate possibilities for the existence of free-swimming fish (den Hartog 1976, 1978). In an attempt to quantify the complexity of the architecture of aquatic biotic communities the various structural elements that can potentially be recognized within a sea-grass community (den Hartog, 1979, 1983b) and in nymphaeid communities (van der Velde, 1978, 1980, 1981) have been elaborated.

These descriptions are based on situations in which the vegetation had reached its optimum development. However, one has to take into account that the component plant species, including the dominants, have an annual activity rhythm that varies with the species. Architecture depends to a high degree on this activity rhythm. One can distinguish a resting phase in which architecture can be very reduced, because many species including some dominants may occur during this phase only in the form of underground rhizomes with roots, or as turions or seeds. During the extension phase the biomass increases and the space is filled up in a manner that is characteristic for each species within the existing trophic conditions; in this phase the architecture of the community develops. During the consolidation phase the optimum differentiation is reached; in this phase flowering, fruiting and storage of organic and inorganic products take place. This phase is followed by a phase of decline, in which the architecture becomes reduced (see also Westlake 1963).

Other factors that influence the architecture are the morphological differentiation of the dominant species, and the rate of leaf-development and leaf-shedding. These determine the possibilities for settlement for whole categories of organisms. A waterlily is considerably more differentiated than, for example, an isoetid or an elodeid, and has larger and firmer organs and consequently its accompanying epiphytic flora and fauna are much more diversified. Species with a high turnover of the leaves (low leaf age) can at best be colonized by rapidly growing colonizers, but on plants with a low leafing rate the epiphytic flora and fauna may be very diversified. Although the dominant plant species determine to a considerable extent the nature of the accompanying flora and fauna, there are also a number of species whose occurrence seems to be linked with physico-chemical characteristics of the environment, and whose bond with the dominant plant species is weak or non-existent. Jacobs & Huisman (1982) point out that the nature of the microhabitats within a community is also an important criterion. They found partly different faunal assemblages in sandy bottoms and muddy bottoms, although both were occupied by well developed stands of *Zostera marina*. For these reasons, architecture has to be considered in close connection with seasonality, and the growth form of the dominant species whose demographic characteristics are highly decisive for the presence or absence of all kinds of organisms.

The number of structural elements in a community depends on (a) the morphological differentiation of the dominants, their activity rhythm and other demographic features, (b) the trophic condition of the environment and (c) the physical nature of the environment, e.g. the type of substrate (stone, mud, sand), whether standing water is involved or running water, and the degree of permanency of the aqueous phase of the habitat. In principle ca. 25 structural elements can be distinguished, and a further refinement is even possible in some cases. The attention paid to the various elements differs greatly.

The structural elements will be described here in a summarized way according to their spatial arrangement (see Figure 1 and Table I).

1. The dominant macrophytes. The dominant macrophytes form the frame-element of the community, and due to their morphological differentiation a number of subelements can be recognized. In fact this differentiation determines already to a considerable degree the number of microhabitats, and consequently the potential biological diversity. Although the frame element may determine the character of the community, it does not mean that the role of other elements in the community is insignificant. The subelements that can be recognized are:

   (a) partly or totally emerged leaves. The upper parts of the leaves of *Stratiotes aloides* rise above the water level during summer, and due to crowding *Nymphaea alba* may produce aerial leaves.

   (b) floating leaves. The upper side of the leaves is adapted to aerial life, while the underside has a water metabolism. Upper and undersides of the leaves form fundamentally different microhabitats for other organisms.

   (c) submerged leaves. Dependent on shape and position these leaves may offer various microhabitats.

   (d) upright stems, petioles and peduncles.

   (e) generative structures, such as inflorescences, flowers and fruits.

   (f) Underground parts such as roots and rhizomes. The meaning of these structures for other organisms has hardly been studied. As particularly rhizomes and corms function as storage organs, it may be expected that there are organisms that make use of them.

Plant compartments			Environmental compartments	
	B	flower	L	air
	C	aerial leaf		
	D	upperside floating leaf	M	water surface
	E	underside floating leaf		
	F	rolled leaf	N	water
	G	peduncle		
	H	petiole		
	I	submerged leaf		
	J	rhizome	O	bottom surface
A	K	root	P	bottom
plant tissue				

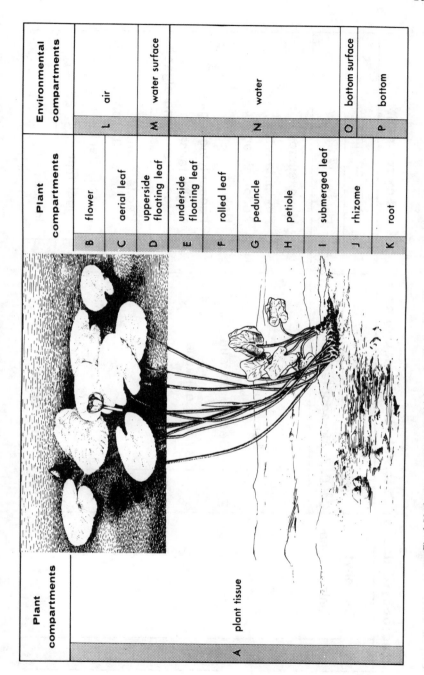

Fig. 1. Various compartments which can be distinguished in a macrophyte-dominated system.

Table 1. Structural elements in macrophyte-dominated systems arranged according to their spatial distribution (see also Fig. 1). The microbial world is present in each compartment; birds and mammals can occur in many compartments and are difficult to include in the scheme.

Macrophyte tissue		External surface macrophyte		Environment	
A	Endophytes Mining fauna	B	Flower visitors	L	Flying animals
		C–D	Air-adapted fauna	M	Epineuston
		E–I	Epiphytes Sessile epifauna Vagile epifauna	N	Hyponeuston Phytoplankton Zooplankton Nekton Loose–lying algae Vagile fauna on loose–lying algae
				O	Vagile fauna on the bottom Sessile benthos Algal film
		J–K	Rhizofauna	P	Infauna

This differentiation applies in extenso to a few species, such as *Nymphaea alba, Nymphoides peltata* and other nymphaeids. In elodeids there are only upright stems, with caulescent small leaves, while the root system (ca. 2% of the total biomass) as well as the generative organs are insignificant from a quantitative point of view. Lemnids have only floating fronds, however, without the above-described differentiation; most species have roots, but the reproductive organs are small and insignificant. Sea-grasses have underground parts and above-ground leaves; their generative parts have a very limited function for some other organisms. Thus the morphological differentiation of most aquatic plants is rather restricted, and consequently their role as a habitat for other organisms.

2. Other macrophytes. When a community consists of a number of species the differentiation of dependent flora components may increase. This is particularly the case as more than one growth form is involved. Van Vierssen (1982) showed that the macrofauna differentiation in brackish and freshwater communities was not linked to the number of plant species but to the number of the growth-forms.

Elements within the macrophyte plant tissue.

3. Endophytes. Within the tissue of the aquatic macrophytes various endophytes may occur. They belong to the bacteria, fungi and algae. Their role is not yet clear; some of them have a function in decomposition and become active when the plants senesce. Lammens and van der Velde (1978) record the occurrence of the fungi *Septoria villarsiae* and *Puccinia scirpi* from the upper surfaces of the leaves of *Nymphoides peltata*. A few of these may be a latent danger to the host, because they may become active under certain unclarified circumstances in an earlier developmental stage of the host, e.g. *Colletotrichum nymphaeae* in *Nuphar lutea* and *Nymphaea alba* (van der Aa 1978). The slime mould *Labyrinthula macrocystis*, a normal endophyte in sea-grass species and several marine algae, has been considered the pathogen in the well-known 'wasting disease', that destroyed most of the beds of *Zostera marina* in the northern Atlantic between 1930 and 1935 (Milne & Milne 1951, Martin 1954, Rasmussen 1977, den Hartog 1987).

It has to be stated here that several endophytes play an extremely important role in the tissue or intercellular spaces of several aquatic animals. Most spectacular is in fact the activity of the dinophycean alga *Symbiodinium microadriaticum* (probably a collective name for a group of related species) which is totally responsible for the high photosynthetic capacity of the tissue of hermatypic corals; without these algae, generally indicated as 'zooxanthellae', the existence of coral reefs would be impossible. In the fresh and marine environment 'zoochlorellae' occur as well, e.g. in the hydra species *Chlorohydra viridissima*, and some Turbellaria such as *Convoluta roscoffensis* and *Dalyellia viridis*. Some marine animals are able to use the chloroplasts or chromatophores of their food items for their own benefit; *Convoluta convoluta* uses the chromatophores of diatoms and some saccoglossan snails are able to utilize the chloroplasts of the algae they consumed. A general study of the endophytes in the aquatic environment is an important desideratum for further study.

4. Mining fauna. The mining fauna consists of animals which live within the tissues of the host plant, where they tunnel in the leaf-blades and the petioles, and so can do great harm. To this category belong insects (particularly larvae of Coleoptera, Lepidoptera and Diptera), worms and some molluscs. The primary miners, phyto-phagous insect larvae, are typical herbivores. In general these animals leave the epidermis intact to prevent water from entering. When the epidermis becomes weak, other secondary miners, which are now able to mine or to use old tunnels filled with water, establish. These animals may be partly herbivorous, but most of them use the tunnels to catch small plankton organisms and seston. The tunnel serves to protect them from predation. Mining insect larvae are not restricted to large plants such as waterlilies (Brock & van der Velde 1983, van der Velde & Hiddink, 1987), but have been recorded also from aquatic weeds like *Myriophyllum* species (Gaevskaya 1969) and *Hydrilla verticillata* (Anonymous 1986). This fauna category needs closer attention, as it may be important for biological water weed management.

Elements occurring on the surfaces of macrophytes.

5. Flower visiting insects. In most aquatic plants flowering takes place above the water level. The importance of flowers for flower visiting insects depends mainly on the size, colour, scent, quantity of products as pollen and nectar, and the shelter they can offer (van der Velde *et al.* 1978, van der Velde & Brock 1980, van der Velde & van der Heyden 1981, Giesen & van der Velde 1983). In some flowering species higher temperatures than those of the surround-ings play also a role as attraction for insects. This is, for example, the case in waterlilies (Prance & Arias 1975, van der Velde & Brock 1980). Consequently the role as a microhabitat varies from species to species.

6. Air-adapted fauna on emerged macrophyte parts. The upper sides

of the floating leaves and emerged other plant parts form a kind of small islands and offer living space to a variety of specialized terrestrial and semiaquatic animals, mainly insects and spiders (van der Velde & Brock 1980, Brock & van der Velde 1983, van der Velde *et al.* 1985, van der Velde 1986, Bergers *et al.* 1986). This fauna component is often neglected in studies on macrophyte-dominated systems. This fauna plays a role in the pollination of the entomophilous flowers of macrophytes and seems to be more effective for cross-pollination than the true flower-visiting insects because of their regular occurrence in the flowers; these insects can also play a role in the distribution of spores of certain fungi on emerged parts (van der Velde & Brock 1980). This air-adapted fauna is also very important for biocontrol, as among them are many phytophagous insect species (Gaevskaya 1969) which are sometimes extremely successful in destroying aquatic weeds like *Salvinia molesta* (Thomas & Room 1986) and *Eichhornia crassipes* (DeLoach & Cordo 1983) which have emerged plant parts, while a large number of species seems promising for that goal e.g. the weevil *Litodactylus leucogaster*, that feeds on the emerged flower spikes of *Myriophyllum spicatum* (Buckingham & Bennett 1981). Therefore, much more effort is necessary to study their biology and their relations with other species (DeLoach & Cordo 1982).

Emerging plant parts play an important role as a route for emerging insects as well as for adult insects which go (partly) under water for oviposition.

7. Epiphytes. Due to the differentiation of the host the growth of epiphytes may be differentiated as well. Among the epiphytes there are many species that are able to settle on aquatic plants, but which never become mature, because of their size, or because their development takes longer than the life-time of the leaves of the host. A restricted number of species appears to be well adapted to the activity rhythm of the host, and these may be found rather frequently. Several of the epiphytic species — in fact the most — can occur on all kinds of substrates, and thus are by no means characteristic. The number of epiphytic species that appear to some degree host-bound is very restricted. The distribution of epiphytes on the aquatic plants is not even, but depends on age of the leaves, their surface structure, the position of the leaves with respect to light influx, and other properties of the leaves (e.g. pH differences between upper and under side of polarized leaves, excretion of allelopathic substances), but also on environmental factors such as currents, wave-action, grazing, etc. On many fresh-water plants the epiphytes are mainly diatom species; on sea-grasses diatoms are the first colonizers, but red and brown algae, and to a lesser degree green algae, also play an important

part. Some of the species occupy the flat sides of the leaves, while other ones colonize the edges and form a fringe.

There is a certain succession in the settlement of the various species, but at the time the leaves or stems start to become senescent the pioneers are still present. A true succession series, where the original community becomes replaced by a new one, has not yet been recorded from aquatic plants, but has been observed on the submerged parts of 2—3 year old reed stalks. In beds of the mediterranean sea-grass *Posidonia oceanica* there is a spatial differentiation in the epiphyte vegetation. On the leaves the photophilous association *Ascocyclo-Giraudietum* occurs, while the leaf-sheats and exposed parts of the rhizomes are occupied by the *Udoteo-Peyssonnellietum*, an association of shaded sublittoral rock in sheltered sites. In beds of *Zostera marina* only communities of photophilous epiphytes occur (den Hartog 1983b). Among the photophilous sea-grass epiphytes several taxa are highly characteristic; these are only found under oligotrophic conditions, where the metabolites of the seagrass may be an important nutrient source. Under more eutrophic conditions ubiquitous marine weeds take over ('fouling').

A certain differentiation has also been recorded from the leaves of nymphaeids. Van der Velde (1980) and Delbecque (1983) found differences in the composition of the epiphytic diatom growths of the marginal parts and the central part of the underside of floating leaves of *Nuphar lutea*, whereas there was also an obvious depth-linked differentiation in the diatom growth on the petioles.

Therefore, it is important to investigate the epiphytic growth of the various plant parts separately, because they might turn out to represent quite different microhabitats due to differences in the physical texture or metabolic activity of the host and/or to differences in the light climate or shelter.

8. Sessile epifauna. Just as described for the epiphytes the differentiation of the sessile epifauna depends entirely on the differentiation of the host. This category comprises a large number of animal groups, e.g. Protozoa (*Vorticella*), sponges, Ectoprocta, hydrozoa, case-building insects, while also many sessile eggs and egg masses of insects, snails etc. can be considered to belong to this structural element. As most of these animals use the plant only as a substratum there are not many specific bonds, the main condition for their survival is, that they fit within the activity rhythm of the host.

9. Vagile epifauna. For this group the differentiation of the host is also of high importance. However, because these animals can move around they are less dependent on the activity rhythm of the host.

The activity of the vagile epifauna is quite varied; in contrast to the sessile animals which have to catch their food from the ambient water, the vagile animals roam around. Most epifauna species are herbivores feeding on epiphyton; there are a very limited number of species, which are feeding on the living macrophyte tissue, because water forms a barrier for most phytophagous insect species (Gaevskaya 1969). However, there is a number of species which could be used for biocontrol (see e.g. Oliver 1983, 1984) and therefore also very important for the development of macro-phyte stands. Furthermore, this structural element contains many different groups feeding on prey, decaying plant material and detritus, which is caught by the various macrophyte organs. For this element the structure of the macrophytes is of key importance as revealed by a number of recent publications (Rooke 1984; Tokeshi & Pinder 1985). Some growth-forms of macrophytes are very rich in epifauna species, e.g. stratiotids (Higler 1975), nymphaeids (van der Velde 1980) and magnopotamids (Soszka 1975), because of offering large surfaces and firm organs.

10. Sessile fauna on other macrophytes.
11. Vagile fauna on other macrophytes. The elements 10 and 11 are only of importance as separate ones if the macrophytes represent another growth-form than the dominant.
12. Rhizofauna. On and around the roots of macrophytes specialized larvae of beetles (e.g. *Donacia, Macroplea*), of flies [e.g. Ephydridae, *Chrysogaster* (Syrphidae)] and of mosquitoes (e.g. *Mansonia*) can be found. These insect species have in common that they have pointed stigmata by which they take oxygen from the intercellular spaces in the roots (Varley 1937). The larvae of the beetles are phytophagous feeding on the macrophyte root tissue, but those of the flies are detritivores and those of the mosquitoes filter-feeders.

An other group of animals belonging to the rhizofauna are the nematodes (Prejs 1977, Esser *et al.* 1985). Prejs (1977) found that the species composition of nematodes around and on the roots differed quantitatively and qua species with those of sediment samples between the macrophytes due to the release of oxygen from the roots towards the sediment, while there is also a number of species using the root cells as food. These species also penetrate the roots and rhizomes.

Elements of the macrophyte environment.

13. Flying fauna. Certain bird and insect species fly above the water surface in search of prey, such as terns, swallows and dragonflies, which feed on fish and/or insects. Some of these species have a preference for hunting above macrophyte stands such as the Black

Tern (*Chlidonias niger*). Floating leaves can therefore be important as shelter for fish against predation from the air.

Another function of macrophyte stands for the flying fauna is as orientation beacon for swarming insects such as Trichoptera and Chironomidae, which use for that purpose emerged vertical and horizontal parts of the macrophytes.

14. Neuston. Neustonic organisms use the surface tension of the water in various ways; there are neustonic organisms resting on the water surface (epineuston) and hanging under water at the surface (hyponeuston). Macrophyte stands lower the surface tension by means of the excretion of organic substances (Wetzel 1975) and a clear influence of macrophytes on the neustonic community can be expected but data about this are almost absent. Dense vegetation hinders the large neustonic insects, which are so characteristic for open water (e.g. large Gerrids); in general in macrophyte stands the smaller neustonic insects and stages (Vepsäläinen & Järvinen 1974) occur, which use also floating leaves and other emerged parts as a refuge at disturbance. Some of them lay their eggs in the macrophyte tissue.

15. Phytoplankton. The role of phytoplankton in macrophyte-dominated communities is by no means negligible. Particularly in communities in which the aboveground biomass is only seasonal, phytoplankton and bottom algae are almost totally responsible for the functional performances in the period when macrophytes are absent. In nymphaeid communities the period of phytoplankton dominance ranges from October to the end of April. Roijackers (1985) found, that on an annual basis the production of the phytoplankton in a stand of *Nymphoides peltata* was 2.4 times as high as the annual production of this water plant itself. Therefore, it is of the utmost importance that more studies are carried out on the phytoplankton of macrophyte-dominated communities.

16. Zooplankton. There are very few studies on the zooplankton in macrophyte-dominated communities. The intensity of isolation from the open water by means of macrophytes is the main cause for differences in the species and quantitative composition of the zooplankton in macrophyte stands and the open water. The intensity of isolation is dependent on the density of the vegetation as well the structure of the dominant macrophytes (Gliwicz & Rybak 1976). Zooplankton typical for the open water avoids macrophyte stands because of repellents present there (Pennak 1973, Dorgelo & Koning 1980, Dorgelo & Heykoop 1985). There are some indications that various zooplankton species show a horizontal migration, i.e. they move during the night into the

open water, but retract in the daytime to the macrophytes (Timms & Moss 1984). Further studies on this category are desirable.

17. Nekton. Free-swimming animals which are not confined to a particular part of the community but nevertheless permanent members of it; some of them rest on the macrophytes. This category comprises such groups as fish, prawns and many water insects which swim freely between the water plants, and extend their feeding activity to practically all compartments. The ratio between space occupied by water plants and open space, and the scale of this pattern is probably very important. For water insects and prawns the pattern of open spaces and plant growth may be small-scaled, but for schools of fish this pattern needs to be large-scaled. Macrophyte stands offer shelter against predation for juveniles of many fish species of which the adults occur in the open water (Werner *et al.* 1983a, b; Lammens 1986). Furthermore, there is a number of fish species characteristic for macrophyte stands because of their kind of food such as snails, insects and macrophytes themselves (Lammens 1986).

18. Loose-lying algae. Loose-lying algae and aegagropilae may form impressive masses between the aquatic plants. They occur entangled between the macrophyte shoots, which prevent the algae from being washed away. The species participating in the algal mat may have a very different origin. In the marine environment part of the loose-lying algae are in fact detached epilithic algae, which are able to continue growth within the shelter of seagrasses. Another part of the algae caught between aquatic plants, consists of specimens that germinated on small objects as sand grains or small shells, but grew so large that they became uplifted due to hydrostatic pressure. Further there are a number of species that seem to be obligately free-floating, i.e. they have never been found attached to a substratum. In fresh water free-floating *Cladophora* and filamentous Conjugatae form often dense masses known as 'blanket weed'. These loose-lying algae have not received much attention in spite of the fact that they can constitute an important part of the above-ground biomass, and have a high photosynthetic activity. In the case of eutrophication 'blanket weed' may suffocate the rooting water plants.

19. Vagile fauna on loose-lying algae. The loose-lying algae have seldom been studied, and about the fauna of this structural element even less is known. In seagrass beds the vagile fauna of the looselying algae shows a high diversity, particularly where the amphipods are concerned (Nelson 1979, den Hartog & Jacobs 1980, Jacobs & Huisman 1982).

20. Algal film on the bottom substrate. On sandy and coarse-grained bottoms diatoms and various other small algae may form very dense growths. Particularly in situations where the macrophyte cover is not closed and in wintertime these microalgae may contribute the larger part of the primary production, and only for this reason they should not be omitted. In closed macrophyte growths the role of these algae is much more limited. This structural element has hardly been studied as an integral part of an aquatic macrophyte system.

21. Sessile fauna on the bottom. This structural element is only of significance if the substratum is coarse-grained or stony or when empty shells or living mussels are present.

22. Vagile fauna on the bottom. On the bottom dead and senescent plant and animal material is deposited. Although the initial decomposition of this material starts already at the surface or in the water column the main decomposition processes take place at the bottom. This detritus is a source of food for many animal species, which actively fragment the material (Brock 1984). Although none of these detritivores are exclusively restricted to plant-dominated systems, their functional role in the material cycles of the aquatic ecosystem is of key importance.

23. Infauna. The infauna comprises the animals living in the sediment. They can be divided into macrofauna, e.g. anthozoa, bivalves, many polychaetes, oligochaetes, chironomids, sipunculids, and crustaceans, and meiofauna mainly consisting of nematodes. In macrophyte dominated systems a differentiation may occur, because the fauna in the immediate surroundings of the root systems of the macrophytes may differ from that of the sediment not affected by the underground parts of the macrophytes. Prejs (1977) found in both types of sediment qualitatively and quantitatively different nematod assemblages.

24. Temporary residents. This category comprises a heterogeneous group of species with an action radius extending far beyond the limits of the macrophyte communities, or which live for the main part of their life elsewhere. The large grazers and predators which visit the aquatic plant communities at irregular intervals, such as swans, geese, ducks, turtles and some mammals, exercise an enormous influence on the structure of the community. Further herbivores and predators which need an environment built up of a number of quite different habitats, belong to this category, e.g. muskrats and herons. Further, in the intertidal belt there is daily a considerable migration; with the incoming flood water fish and crustaceans invade the area for feeding, and when the tide recedes crows, gulls and waders arrive. Further, adult insects and spiders

from terrestrial environments could be included in this category, but as these often form the bulk of some of the other structural elements, this is not practical. In communities with a seasonal above-ground biomass most of the fauna may belong to this category.

25. The microbial world. Bacteria and other micro-organisms rarely figure in community descriptions. Their activity, however, is of paramount importance to the material cycles in the communities. For this reason the microbial assemblages need more attention in ecological work.

Table II shows the potential differentiation into structural elements of some aquatic communities; although the various fauna compartments have been omitted, the variation in the degree of complexity is very obvious. Terrestrial communities of herbaceous plants consist of less structural elements than the aquatic communities, but usually the

Table II. Compartmentation of some monodominant aquatic plant communities.

		Lemna gibba	*Lemna trisulca*	*Ceratophyllum demersum*	*Stratiotes aloides*	*Zostera marina*	*Ruppia cirrhosa*	*Potamogeton pectinatus*	*Nuphar lutea*
Macrophytes	emerged leafblades				+				
	floating leafblades	+							+
	flowers + fruits				+			+	+
	petioles + peduncles						+		+
	submerged leafblades			+	+	+	+	+	+
	rhizomatic parts				+	+	+	+	+
Epiphytes on	floating leafblades	+							+
	petioles + peduncles								+
	submerged leafblades			+	+	+	+	+	+
	rhizomatic parts				?	−	−	−	?
Microalgal film on substrate						+	+	?	?
Loose-lying algae						+	+	?	?
Endophytes		+	+	?	+	+	+	+	+
Other macrophytes		+	+	+	+	−	−	+	+

frame-element is composed of a considerably greater variety of plant species of various growth-forms, resulting in a great variety of the accompanying fauna with many specific relationships and intricate relations. An exact comparison can not be made because of insufficient data.

RELATIONS BETWEEN THE ORGANISMS WITHIN THE COMMUNITY

The structure of the community is determined to an important degree by the relations that exist between the participating organisms. Some of these relations have already been mentioned. It has been shown that many categories of species are fully dependent on the activity rhythm of the dominant macrophytes which determine the frame of the community. Conversely the dominants are kept in check by large polyphagous grazers, mainly vertebrates such as herbivorous birds and various mammals (including domestic animals) and fish and by certain phytophagous insects. Apart from these generalists a large number of organisms are specialized, and restrict their activity to a certain plant family, a certain species or even a certain organ of a plant. The activities can be divided into various types, e.g. mere consumption, parasitism, symbiosis and disease. The relations between the organisms do not only concern the dominants, but they play a part on all levels within the community. Thus, grazing invertebrates, such as snails, small crustaceans and insect larvae keep the epiphytic algae under control to the benefit of the macrophytes, as a too luxuriant development of epiphytes could have an adverse effect on the exchange processes with the surrounding water, and as a result of increased shading photosynthesis could be hampered as well. One may expect that in a well-balanced aquatic plant community the development and activity of the various structural elements are well adjusted. Disturbance of one or more of the structural elements may lead to a disturbance of the whole community. Thus application of insecticides may reduce the invertebrate fauna, causing unrestricted growth of algae. Eutrophication of the water-layer also may result in an increase of epiphytic algae and/or phytoplankton, and to a final destruction of the whole community (Phillips *et al.* 1978).

It appears in practice that there are large differences in the degree of structural organisation within the communities. In various vegetation types dominated by lemnids, by elodeids or by parvopotamids the disappearance of certain plant species does have hardly any influence or has no influence at all on the companion species, because of the absence of any specific relationships between the participating species. The growth form of the plants appears to be more important to the coexisting fauna than the actual species. In such vegetations the internal coherence is low, and one may wonder whether they deserve to be regarded as communities; the term 'assemblage' seems more appropriate. This kind of communities are not very sensitive to external disturbances. In other vegetation types, e.g. those dominated by nymphaeids, by *Stratiotes aloides* and by seagrasses, there are clear mutual relationships between a part of the coexisting species, and a reduction of the dominants has severe consequences for the community as a whole. These communities are more sensitive to external influences.

The relations between the various participants of the community are usually presented as a linear food-chain or a three-dimensional food web. However, the trophic relationships are much more complex, because such chains and webs play a role at several scales in space and time. There is a complete trophic network at the level of the micro-organisms. On a submerged leaf of a macrophyte one can recognize producents, and consuments of the first order and of the second order. As the surface-volume ratio of small organisms is much more favourable than that of larger organisms, the processes in these mini-foodwebs are more intensive; and because of the short lifetime of the organisms the turnover is considerably more rapid. One can conclude that within the community there is a wide variety of trophic cycles which proceed simultaneously with different speed and different intensity. Further it has to be stressed that beside food value of the organisms all kinds of other factors are involved in the structural relationships which make a species assemblage a community, e.g. behaviour of the prey, taste of the prey, behaviour of the predator, etc.

It occurs often that in aquarium experiments fish and crustaceans take almost all kinds of food items that are offered, while from research of the contents of the intestinal tract of specimens collected in the natural environment one would conclude that they are 'specialized on' or 'prefer' a certain food item. However, it may be that some preys are easier observed by predators because of behaviour, colour or other characteristics. As the stimuli and attractions are different for all taxa, this may lead to food partitioning in the natural environment. The various feeding mechanisms, existing in aquatic organisms contribute also to the development of the community structure.

In spite of the existence of numerous positive and negative relations between the members of the community, each species has, however, to fulfill quite independently all the requirements necessary for its survival within the circumstances existing in the environment. Therefore, diversity can only be high as within the various guilds the coexisting species exploit or endure the environmental conditions in sufficiently different manners, to avoid strong competition.

RELATIONSHIPS OF THE ORGANISMS WITH THE SURROUNDING ENVIRONMENT

As the communities are commonly considered as spatial units, they can occupy areas, or volumes of water, and it is possible to map their occurrence. Consequently they are exposed to all kinds of influences exercised by the surroundings, and some of these may be quite essential. The communities themselves exercise also a certain influence on the surroundings.

In running water the aquatic plants are continually flushed by other water; but also in shallow standing waters a circulation occurs due to temperature differences and wind action. Biotic influences are also manifold. The flowers of several aquatic plants are visited by terrestrial insects, such as the honeybee, several bumblebees and (hover)flies and these are responsible for the pollination, and thus contribute to the reproductive success. It has also been shown that the occurrence of a number of varied environments in a limited area contributes to the richness in species, because there are many species that need different environments for the various stages of their life, e.g. aquatic insects like midges and dragonflies, frogs, and also fish species. Aquatic plant communities may be grazed by migrating waterfowl, that were borne thousands of kilometres away.

The aquatic plant communities exercise their influence mainly by the production of organic material. This may be washed ashore creating suitable circumstances for rapidly developing communities of Chenopodiaceae and species favouring debris. Conversely the organic material may sink to the bottom and collect in the deepest places, where it will decompose. The nature of the decomposition processes and their speed depends greatly on the hydrochemical conditions, such as oxygen content, alkalinity, pH and nutrient availability. In alkaline waters a fine gyttja is produced, while in acid waters hardly any decomposition takes place, which leads to the formation of peat.

TEMPORAL PATTERNS

The temporal pattern or seasonality of aquatic plant communities depends mainly on climatic factors as well as the physico-chemical characters of the water in which they occur. The main acting climatic factors are temperature (alternation of cold and warm season), precipitation (alternation of wet and dry seasons, resulting in level fluctuations, changes in current velocity, etc.), and wind strength. In temperate areas photoperiodicity plays an important role in triggering vegetation processes.

The temporal sequence in the activity of plant growth has already been described on p. 129, as the development of community architecture. Although the sequence in this development is generally the same in all communities there is quite a variation in timing. Under permanent submerged conditions the aquatic plants can be perennial in such a way that they overcome the unfavourable period as a rootstock or a corm, as is the case in many nymphaeids, or they may remain present above the substrate as well, e.g. in sublittoral communities of *Zostera marina* (Jacobs 1979). The seasonal vegetations which survive unfavourable periods as seeds or turions are restricted to the summer period, e.g. the growth of *Potamogeton trichoides* (van Wijk & Trompenaars, 1985). Other communities start their development in autumn, are wintergreen and break down at the beginning of the summer, e.g. communities of *Ranunculus aquatilis, Hottonia palustris* and *Callitriche* species. The animal species dependent on macrophyte growth in the seasonally developed vegetation need special adaptations to survive the season in which the macrophytes are not available: most of them are temporary inhabitants with well-timed life-cycles. Research is still needed to determine the exact cause of the end of the extension phase of the aquatic plant growth. Possibly a critical temperature may stop further extension, but other possibilities are nutrient limitation or self-shading. The decline is usually caused by gradually decreasing temperatures, but may also be brought about by complete desiccation, as is usually the case in vegetations of the *Callitricho—Batrachion*. In areas with a marked dry and wet season, the vegetation may be abruptly destroyed at the beginning of the wet season, due to rapidly rising water levels and strongly increasing current velocities. Although most environments occupied by aquatic plants are permanently submerged, there are some environments that are temporarily exposed to the air. According to the period in which exposure takes place, the duration of the exposure and the frequency of exposure the aquatic communities may respond very differently. The communities of *Zostera marina* and *Z. noltii* in the intertidal belt of the seashore are strikingly reduced in complexity when compared with sublittoral *Zostera* stands (Jacobs & Huisman 1982). The stands of *Littorella uniflora* in the shallow-water zone of soft-water pools seem well-adapted to protracted exposure during the summer and show no sign of a reduced diversity. In the contrary, within such *Littorella* stands a community of ephemeral terrestrial annuals can develop, with *Cicendia filiformis, Juncus mutabilis* and *Radiola linoides*. Communities dominated by *Ranunculus aquatilis* and *Callitriche* species

146

occur in harder and more eutrophic shallow standing waters which become emerged for a short period at the end of the spring or early summer. In permanent standing waters these species form only an insignificant component in the aquatic communities. When emersion lasts too long, these species become overgrown by rapidly developing terrestrial species, e.g. *Glyceria fluitans*, and such a change in the vegetation is usually not reversible.

In the south-western part of Australia associations of aquatic plants occur, which have their complete above-ground development within the three-month period of the rainy season. These plants develop from corms or seeds between the terrestrial plants which are submerged and inactive during that period. Examples are *Aponogeton hexatepalus, Crassula helmsii* and *Hydrocotyle lemnoides.*

Due to the variations in the climatic circumstances which occur from year to year, the vegetations of aquatic plants may also show year-to-year variation, which can be noticed as quantitative differences in the development of the participating plant species, or differences in the timing of the life performances. Usually no qualitative differences occur, but when extreme conditions prevail also qualitative differences may be observed. Some normally submerged communities may rather exceptionally become exposed, e.g. during very protracted dry summers. In that case special communities may develop from the seed bank contained in the substratum of the aquatic community. In stands of *Nymphoides peltata* exposed to emergence dense growths of *Oenanthe aquatica* and *Rorippa amphibia* may develop (see also Westhoff & den Held 1969, Hejný & Husák 1978). These species maintain themselves after reinstatement of the normal, flooded situation, but being biannuals they disappear after two years.

Seasonality is generally related to the year cycle of the macrophytes. It has to be kept in mind, however, that many of the invertebrates and algae, normally occurring as members of the community, have considerably shorter life spans, and thus show much larger fluctuations in numbers and development as a response to weather variations.

SIZE AND AREA

The size of the water bodies is usually decisive for the type of vegetation which can develop. Lemnid vegetation, for example, develops in small ponds and ditches; in larger bodies of water such vegetation is restricted in its occurrence and may be found in sheltered places along the banks. Lemnids seem to possess almost all faculties to be able to cover the surface of lakes, by being independent of the bottom, living in the interface between air and water, and by depending on contact pollination which can be easily fulfilled by small water movements. However, wind and wave action prevent them to cover large areas of water; they are driven together to pluri-layered masses and succumb. The larger salviniids (e.g. *Salvinia auriculata* and *S. molesta*) and pleustohelophytes (e.g. *Pistia stratiotes* and *Eichhornia crassipes*), however are able to expand over large water-surfaces, choking lakes and rivers, especially outside their original distribution areas.

In water bodies of small size small species come to development, e.g. representatives of the parvopotamids, elodeids, ceratophyllids, etc. while in larger water areas the magnopotamids, magnonymphaeids and vallisneriids develop well-structured communities. This size-area relation has been recorded by Segal (1965).

In this connection it should be mentioned that the area occupied by a certain aquatic plant community has no relation to its diversity. Large, homogeneous stands of a waterplant community are often considerably poorer in accompanying species than small stands bordering on other communities. In the latter case a continual exchange between the adjacent communities is assured and favourable circumstances are present for species which need different habitats for their various life stages. Generally an environment exhibiting small-scaled heterogeneity presents the highest diversity.

RECOMMENDATIONS FOR COMMUNITY STUDIES

In this chapter an overall survey has been given of the various components of structure. It is obvious, that up to now the structure of aquatic plant communities has been studied only inadequately and superficially. Most descriptions of aquatic plant communities are merely based on the macrophyte element, the other structural elements being largely neglected. For the understanding of the functioning of the aquatic plant communities knowledge of the structure is of paramount importance. For this purpose thorough analyses of the other structural elements of the communities are urgently needed. Further desiderata are autecological and ecophysiological studies of the main dominant species, and studies on the interactions between the various community components. Laboratory studies are unavoidable, since it is a well-known fact that various processes which take place in the communities are triggered off by 'situations', which can be easily missed during field research. The pattern of the relationships between the various participants of a community can only be studied in the form of well-planned case studies, which have to be projected against the background of the whole community.

Up to now community research has not taken place in a coordinated

148

manner. Although there are many data available in the literature, they are generally insufficient to construct a coherent image of the communities under study. For example, in one area an author produces a list of diatoms occurring on a certain water plant species in April; in another area another author studies the productivity of epiphytes on the same plant in summer, and in again another area somebody investigates the effect of grazing by snails on epiphytes on the same plant. Usually these investigators do not know about each other's activities. In order to attain real progress in the study of structure and functioning of aquatic plant communities it is absolutely necessary that multidisciplinary research is conducted with a high intensity in a restricted number of adequately chosen sites. The field research must be backed by experimental research in the laboratory. The totality of these activities must be carefully planned. Such a programme is a task for research teams rather than for individual scientists.

REFERENCES

Anonymous (1986) *Hydrellia* on *Hydrilla*. Aquaphyte 6(1): 12.
Arber, A. (1920) Waterplants, a Study of Aquatic Angiosperms. Cambridge, I—XVI + 436 pp.
Aston, H. (1973) Aquatic Plants of Australia. Melbourne University Press, Melbourne, 358 pp.
Barkman, J. J. (1979) The investigation of vegetation texture and structure. *In*: Werger, M. J. A. (ed.) The Steady of Vegetation. Dr. W. Junk Publ., The Hague, pp. 125—160.
Beal, E. O. (1977) A Manual of Marsh and Aquatic Vascular Plants of North Carolina with habitat data. North Carolina Agricultural Experiment Station Technical Bulletin No. 247: 1—298.
Bergers, P. J. M., van der Velde, G. & Bartels, H. A. (1986) Spatial and temporal distribution of adult Nematocera on the upper surfaces of the floating leaves of nymphaeids in a Dutch pond. General results. Proc. 3rd Eur. Congr. Entomol. (Amsterdam): 71—74.
Braun-Blanquet, J.-J. (1964) Pflanzensoziologie. ed. 3. Springer, Vienna, 865 pp.
Brock, Th. C. M. (1984) Aspects of the decomposition of *Nymphoides peltata* (Gmel.) O. Kuntze (Menyanthaceae). Aquat. Bot. 19: 131—156.
Brock, Th. C. M. (1985) The ecological role of the white, yellow and fringed waterlily: a synthesis. Thesis, Nijmegen, pp. 167—190.
Brock, Th. C. M. & van der Velde, G. (1983) An autecological study on *Hydromyza livens* (Fabricius) (Diptera, Scatomyzidae), a fly associated with nymphaeid vegetation dominated by *Nuphar*. Tijdschr. Ent. 126: 59—90.
Buckingham, G. & Bennett C. A. (1981) Laboratory biology and behavior of *Litodactylus leucogaster*, a Ceutorhynchine Weevil that feeds on Watermilfoils. Ann. entomol. Soc. Am. 74: 451—458.
Casper, S. J. & Krausch, H.-D. (1980) Pteridophyta und Anthophyta, 1 Teil: Lycopodiaceae bis Orchidaceae. Süsswasserflora von Mitteleuropa, Bd. 23: 1—403.
Casper, S. J. & Krausch, H.-D. (1981) Pteridophyta und Anthophyta, 2 Teil: Saururaceae bis Asteraceae. Süsswasserflora von Mitteleuropa, Bd. 24: 413—943.

Chassé, C. (1962) Remarque sur la morphologie et la bionomie des herbiers de Monocotylédones marines tropicales de la province de Tuléar (République Malgache). Rec. Trav. Stn. mar. Endoume Suppl. no. 1, Trav. Stat. mar. Tuléar: 237—248.

Cook, C. D. K. (1974) Waterplants of the World. A manual for the identification of the genera of freshwater macrophytes. Junk, The Hague, I—VIII + 568 pp.

Dacey, J. W. H. (1981) Pressured ventilation in the yellow waterlily. Ecology 62: 1137—1147.

Dacey, J. W. H. & Klug, M. J. (1979) Methane efflux from lake sediments through waterlilies. Science 203: 1253—1255.

Dacey, J. W. H. & Klug, M. J. (1982) Ventilation by floating leaves in *Nuphar*. Am. J. Bot. 69: 999—1003.

Delbecque, E. J. P. (1983) A comparison of the periphyton of *Nuphar lutea* and *Nymphaea alba*. The distribution of diatoms on the undersides of floating leaves. Chapter 6: 41—45. *In*: Wetzel, R. G. (ed.), Periphyton of Freshwater Ecosystems. Dev. Hydrobiol. 17: 1—346.

DeLoach, C. J. & Cordo, H. A. (1982) Natural enemies of *Neochetina bruchi* and *N. eichhorniae*, two weevils from Waterhyacinth in Argentina. Ann. entomol. Soc. Am. 75: 115—118.

DeLoach, C. J. & Cordo, H. A. (1983) Control of Waterhyacinth by *Neochetina bruchi* (Coleoptera: Curculionidae: Bagoini) in Argentina. Environ. Entomol. 12: 19—23.

den Hartog, C. (1967) The structural aspect in the ecology of seagrass communities. Helgoländer Wiss. Meeresunters. 15: 648—659.

den Hartog, C. (1973) The dynamic aspect in the ecology of seagrass communities. Thalassia Jugoslavica 7: 101—112.

den Hartog, C. (1976) Aquatische Oecologie en Waterplanten. Gedachten over structuur en dynamiek van waterplantenbegroeiingen. Inaugural address. Catholic University Nijmegen, 18 pp.

den Hartog, C. (1977) Structure, function and classification in seagrass communities. *In*: McRoy, C. P. & Helfferich, C. (eds.), Seagrass Ecosystems, a scientific perspective. Marine Science 4: 89—121.

den Hartog, C. (1978) Structural and functional aspects of macrophyte dominated aquatic systems. Proc. EWRS. 5th Symp. on Aquatic Weeds (Amsterdam): 35—41.

den Hartog, C. (1979) Seagrasses and seagrass ecosystems, an appraisal of the research approach. Aquat. Bot. 7: 105—117.

den Hartog, C. (1981) Aquatic plant communities of poikilosaline waters. Hydrobiologia 81: 15—22.

den Hartog, C. (1982) Architecture of macrophyte dominated aquatic communities. *In*: Symoens, J. J., Hooper, S. S. & Compère, P. (eds.), Studies on Aquatic Vascular Plants. Roy. Bot. Soc. Belgium, Brussels, pp. 222—234.

den Hartog, C. (1983a) Synecological classification of aquatic plant communities. Colloq. phytosoc. X. Végétations aquatiques (Lille, 1981), pp. 171—182.

den Hartog, C. (1983b) Structural uniformity and diversity in *Zostera*- dominated communities in Western Europe. Mar. Techn. Soc. J. 17(2): 6—14.

den Hartog, C. (1986) The effects of acid and ammonium deposition on aquatic vegetations in The Netherlands. Proc. 1st. Internat. Symp. Watermilfoil (*Myriophyllum spicatum*) and related Haloragaceae species, 1985, Vancouver, B.C., pp. 51—58.

den Hartog, C. (1987) 'Wasting disease' and other dynamic phenomena in *Zostera* beds. Aquat. Bot. 27: 3—14.

den Hartog, C. & Jacobs, R. P. W. M. (1980) Effects of the 'Amoco Cadiz' oil spill on an eelgrass community at Roscoff (France) with special reference to the mobile benthic fauna. Helgoländer Meeresunters. 33: 182—191.

150

den Hartog, C. & Segal S. (1964) A new classification of the waterplant communities. Act. Bot. Neerl. 13: 367—393.

Dorgelo, J. & Heykoop, M. (1985) Avoidance of macrophytes by *Daphnia longispina*. Verh. Internat. Verein. Limnol. 22: 3369—3372.

Dorgelo, J. & Koning W. (1980) Avoidance of macrophytes and additional notes on avoidance of the shore by *Acanthodiaptomus denticornis* (Wierzejski, 1887) from Lake Pavin (Auvergne, France). Hydrobiol. Bull. (Amsterdam) 14: 196—208.

Du Rietz, E. G. (1921) Zur methodologischen Grundlage der modernen Pflanzensoziologie. Thesis, Uppsala, 272 pp.

Du Rietz, E. G. (1930) Vegetationsforschung auf soziationsanalytischer Grundlage. Abderhalden. Handb. biol. Arbeitsmeth. 11(5): 293—480.

Esser, R. P., Buckingham, G. R., Bennett, C. A. & Harkcom, K. J. (1985) A survey of phytoparasitic and free living Nematodes associated with aquatic macrophytes in Florida. Soil and Crop Soc. Florida 44: 150—155.

Fassett, N. C. (1957) A Manual of Aquatic Plants. ed. 2. University of Wisconsin Press, Madison, 405 pp.

Funke, G. L. (1951) Waterplanten. Gorinchem, Noorduijn's Wetenschappelijke Reeks no. 38: 1—250.

Gaevskaya, N. S. (1969) The role of higher aquatic plants in the nutrition of the animals of fresh-water basins. Vols. I, II, & III. National Lending Library for Science and Technology, Boston Spa, England, 629 pp.

Gessner, F. (1955) Hydrobotanik. Die physiologischen Grundlagen der Pflanzenverbreitung im Wasser. I. Energiehaushalt. Deutscher Verl. der Wissenschaften, Berlin, 517 pp.

Gessner, F. (1959) Hydrobotanik. Die physiologischen Grundlagen der Pflanzenverbreitung im Wasser. II. Stoffhaushalt. Deutscher Verl. der Wissenschaften, Berlin, 701 pp.

Giesen, Th. G. & van der Velde, G. (1983) Ultraviolet reflectance and absorption patterns in flowers of *Nymphaea alba* L., *Nymphaea candida* Presl and *Nuphar lutea* (L.) Sm. (Nymphaeaceae). Aquat. Bot. 16: 369—376.

Gliwicz, Z. M. & Rybak, J. I. (1976) Zooplankton. Chapter V. *In*: E. Pieczynska (ed.), Selected problems of Lake Littoral Ecology. University of Warsaw, Institute of Zoology, Department of Hydrobiology, 238 pp.

Glück, H. (1905) Biologische und morphologische Untersuchungen über Wasser- und Sumpfgewächse. I. Die Lebensgeschichte der europäischen Alismaceen. Fischer, Jena, 312 pp.

Glück, H. (1906) Biologische und morphologische Untersuchungen über Wasser- und Sumpfgewächse. II. Untersuchungen über die Mitteleuropäischen *Utricularia*-Arten, über die Turionenbildung bei Wasserpflanzen, sowie über *Ceratophyllum*. Fischer, Jena, 256 pp.

Glück, H. (1911) Biologische und morphologische Untersuchungen über Wasser- und Sumpfgewächse. III. Die Uferflora. Fischer, Jena, 644 pp.

Glück, H. (1924) Biologische und morphologische Untersuchungen über Wasser- und Sumpfgewächse. IV. Submerse und Schwimmblattflora. Fischer, Jena, 746 pp.

Hejný, S. (1957) Ein Beitrag zur ökologischen Gliederung der tschechoslowakischen Niederungsgewässer. Preslia 29: 349—368.

Hejný, S. (1960) Ökologische Characteristik der Wasser- und Sumpfpflanzen in den slowakischen Tiefebenen (Donau- und Theissgebiet). Bratislava, 487 pp.

Hejný, S. & Husák S. (1978) Higher plant communities: *In*: Dykyjová, D. & Květ, J. (eds.), Pond Littoral Ecosystems, Structure and Functioning. Springer, Berlin. Ecological Studies 28: 23—64.

Higler, L. W. G. (1975) Analysis of the macrofauna-community on *Stratiotes* vegetations. Verh. internat. Verein. theor. angew. Limnol. 19: 2773—2777.

Hogeweg, P. & Brenkert-van Riet A. L. (1969a) Structure of aquatic vegetation: a comparison of aquatic vegetation in India, The Netherlands and Czechoslovakia. Trop. Ecol. 10: 139—162.

Hogeweg, P. & Brenkert-van Riet A. L. (1969b) Affinities between the growth forms in aquatic vegetation. Trop. Ecol. 10: 183—194.

Hutchinson, G. E. (1975) A Treatise of Limnology. III. Limnological Botany. Wiley-Interscience, New York, London, Sydney, Toronto, 660 pp.

Iversen, J. (1936) Biologische Pflanzentypen als Hilfsmittel in der Vegetationsforschung. Thesis Copenhagen, 224 pp.

Jacobs, R. P. W. M. (1979) Distribution and aspects of the production and biomass of eelgrass, Zostera marina L. at Roscoff, France. Aquat. Bot. 7: 151—172.

Jacobs, R. P. W. M. & Huisman W. H. T. (1982) Macrobenthos of some Zostera beds in the vicinity of Roscoff (France) with special reference to the relationships with community structure and environmental factors. Proc. Kon. Ned. Ak. Wetensch. ser. C. 85: 335—356.

Keeley, J. E. (1982a) Distribution of diurnal acid metabolism in the genus Isoetes. Am. J. Bot. 69: 254—257.

Keeley, J. E. (1982b) Distribution of diurnal acid metabolism in submerged aquatic plants outside the genus Isoetes. Photosynthetica 16: 546—553.

Kornaś, J., Pancer, E. & Brzyski B. (1960) Studies on seabottom vegetation in the Bay of Gdańsk off Rewa. Fragm. Florist. Geobot. 6: 3—92.

Lammens, E. H. R. R. (1986) Interactions between fishes and the structure of fish communities in Dutch shallow, eutrophic lakes. Thesis Wageningen, 100 pp.

Lammens, E. H. R. R. & van der Velde G. (1978) Observations on the decomposition of Nymphoides peltata (Gmel.) O. Kuntze (Menyanthaceae) with special regard to the leaves. Aquat. Bot. 4: 331—346.

Lowden, R. M. (1986) Taxonomy of the genus Najas in the Neotropics. Aquat. Bot. 24: 147—184.

Luther, H. (1949) Vorschlag zu einer ökologischen Grundeinteilung der Hydrophyten. Act. Bot. Fenn. 44: 1—15.

Madsen, T. V. (1985) A community of submerged aquatic CAM-plants in Lake Kalgaard, Denmark. Aquat. Bot. 23: 97—108.

Mäkarinta, U. (1978) Ein neues ökomorphologisches Lebensformen-System der aquatischen Makrophyten. Phytocoenologia 4: 446—470.

Martin, A. C. (1954) A clue to the eelgrass mystery. Trans. 19th North Am. Wildlife Conf. Washington DC, pp. 441—449.

Mason, H. L. (1957) A Flora of the Marshes of California. Univ. of California Press, Berkeley, VIII + 878 p.

Milne, J. L. & Milne, M. J. (1951) The eelgrass catastrophe. Sci. Am. 184(1): 52—55.

Nelson, W. G. (1979) An analysis of structural pattern in an eelgrass (Zostera marina L.) amphipod community. J. Exp. Mar. Biol. Ecol. 39: 231—264.

Odum, E. P. (1971) Fundamentals of Ecology. W. B. Saunders. Philadelphia, London, Toronto, 3rd ed., 574 pp.

Oliver, D. R. (1983) Chironomidae (Diptera) associated with Myriophyllum spicatum in Okanagan Valley Lakes, British Columbia. Can. Ent. 115: 1545—1546.

Oliver, D. R. (1984) Description of a new species of Cricotopus van der Wulp (Diptera: Chironomidae) associated with Myriophyllum spicatum. Can. Ent. 116: 1287—1292.

Pennak, R. W. (1973) Some evidence for the aquatic macrophytes as repellents for a limnetic species of Daphnia. Int. Rev. ges. Hydrobiol. 58: 569—576.

Phillips, G. L., Eminson, D. & Moss, B. (1978) A mechanism to account for macrophyte decline in progressively eutrophicated freshwaters. Aquat. Bot. 4: 103—126.

Poplavskaja, G. I. (1948) Ekologija rastenij. Sov. nauka, Moscow, 295 pp. (in Russian).

152

Prance, G. T. & Arias, J. R. (1975) A study of the floral biology of *Victoria amazonica* (Poepp.) Sowerby (Nymphaeaceae). Acta amazonica 5: 109—139.

Prejs, K. (1977) The Nematodes of the root region of aquatic macrophytes, with special consideration of nematode groupings penetrating the tissues of roots and rhizomes. Ekol. pol. 25: 5—20.

Rasmussen, E. (1977) The wasting disease of eelgrass (*Zostera marina*) and its effects on environmental factors and fauna. *In*: McRoy, C. P. & Helfferich, C. (eds.), Seagrass Ecosystems, a Scientific Perspective. Marcel Dekker, New York, Basel. Marine Science 4: 1—51.

Raunkiaer, C. (1934) The life forms of plants and statistical plant geography. Oxford, 632 pp.

Roelofs, J. G. M., Schuurkes, J. A. A. R. & Smits, A. J. M. (1984) Impact of acidification and eutrophication on macrophyte communities in soft waters. II. Experimental studies. Aquat. Bot. 18: 389—411.

Roijackers, R. M. M. (1985) Phytoplankton studies in a nymphaeid-dominated system. Thesis, Nijmegen, 172 pp.

Rooke, J. B. (1984) The invertebrate fauna of four macrophytes in a lotic system. Freshwat. Biol. 14: 507—513.

Sand-Jensen, K., Prahl, C. & Stokholm, M. (1982) Oxygen release from roots of submerged aquatic macrophytes. Oikos 38: 349—359.

Sculthorpe, C. D. (1967) The Biology of Aquatic Vascular Plants. Edw. Arnold. London, xviii + 610 pp.

Segal, S. (1965) Een vegetatie-onderzoek van hogere waterplanten in Nederland. Wetensch. Meded. K. N. N. V. 57: 1—80.

Segal, S. (1968) Ein Einteilungsversuch der Wasserpflanzengesellschaften. *In*: R. Tüxen (ed.), Pflanzensoziologische Systematik. Junk, The Hague, pp. 191—218.

Søndergaard, M. (1979) Light and dark respiration and the effect of the lacunal system on refixation of CO_2 in submerged aquatic plants. Aquat. Bot. 6:269—283.

Søndergaard, M. & Sand-Jensen, K. (1979) Carbon uptake by leaves and roots of *Littorella uniflora* (L.) Aschers. Aquat. Bot. 6: 1—12.

Soszka, J. (1975) Ecological relations between invertebrates and submerged macrophytes in the lake littoral. Ekol. pol. 23: 393—415.

Taylor, P. (1964) The genus *Utricularia* L. (Lentibulariaceae) in Africa (south of the Sahara) and Madagascar. Kew Bull. 18: 1—245.

Thomas, P. A. & Room, P. M. (1986) Taxonomy and control of *Salvinia molesta*. Nature 320, No. 6063: 581—584.

Timms, R. M. & Moss, B. (1984) Prevention of growth of potentially dense phytoplankton populations by zooplankton grazing in the presence of zooplanktivorous fish, in a shallow wetland ecosystem. Limnol. Oceanogr. 29: 472—486.

Tokeshi, M. & Pinder, L. C. V. (1985) Microhabitats of stream invertebrates on two submersed macrophytes with contrasting leaf morphology. Holarct. Ecol. 8: 313—319.

Twilley, R. R., Brinson, M. M. & Davis, G. J. (1977) Phosphorus absorption, translocation, and secretion in *Nuphar luteum*. Limnol. Oceanogr. 22: 1022—1032.

van der Aa, H. A. (1978) A leaf spot disease of *Nymphaea alba* in The Netherlands. Neth. J. Pl. Path. 84: 109—115.

van der Velde, G. (1978) Structure and function of a nymphaeid-dominated system. Proc. EWRS 5th Symp. on Aquatic Weeds (Amsterdam): 127—133.

van der Velde, G. (1980) Remarks on structural, functional and seasonal aspects of nymphaeid-dominated systems. Thesis Nijmegen, pp. 11—58.

van der Velde, G. (1981) A project on nymphaeid-dominated systems. Hydrobiol. Bull. (Amsterdam) 15: 185—189.

van der Velde, G. (1986) Temporal and spatial distribution of flies (Diptera Brachycera and Cyclorrhapha) in nymphaeid zones in a Dutch pond. Proc. 3rd Eur. Congr. Entomol. (Amsterdam): 151—154.

van der Velde, G. & Brock, Th. C. M. (1980) The life history and habits of *Notiphila brunnipes* R.-D. (Diptera, Ephydridae), an autecological study on a fly associated with nymphaeid vegetations. Tijdschr. Ent. 123: 105—127.

van der Velde, G., Brock, Th. C. M., Heine, H. & Peeters, P. M. P. M. (1978) Flowers of Dutch Nymphaeaceae as a habitat for insects. Act. Bot. Neerl. 27: 429—430.

van der Velde, G., Custers, C. P. C. & de Lyon, M. J. H. (1986) The distribution of four nymphaeid species in the Netherlands in relation to selected abiotic factors. Proc. EWRS/AAB 7th symp. on Aquatic Weeds, 1986 (Loughborough): 363—368.

van der Velde, G. & Hiddink, R. (1987) Chironomidae mining in *Nuphar lutea* (L.) Sm. Ent. scand. Suppl. 29: 253—264.

van der Velde, G., Meuffels, H. J. G., Heine, M. & Peeters, P. M. P. M. (1985) Dolichopodidae (Diptera) of a nymphaeid-dominated system in the Netherlands: species composition, diversity, spatial and temporal distribution. Aquatic Insects 7: 189—207.

van der Velde, G. & van der Heijden, L. A. (1981) The floral biology and seed production of *Nymphoides peltata* (Gmel.) O. Kuntze (Menyanthaceae). Aquat. Bot. 10: 261—293.

van Vierssen W. (1982) The ecology of communities dominated by *Zannichellia* taxa in western Europe. II. Distribution, synecology and productivity aspects in relation to environmental factors. Aquat. Bot. 13: 385—483.

van Wijk, R. J. & Trompenaars, H. J. A. J. (1985) On the germination of turions and the life cycle of *Potamogeton trichoides* Cham. et Schld. Aquat. Bot. 22: 165—172.

Varley, G. C. (1937) Aquatic insect larvae which obtain oxygen from the roots of plants. Proc. R. ent. Soc. Lond. (A) 12(4—6): 55—60.

Vepsäläinen, K. & Järvinen, O. (1974) Habitat utilization of *Gerris argentatus* (Het. Gerridae). Ent. scand. 5: 189—195.

Werner, E. E., Gilliam, J. F., Hall, D. J. & Mittelbach, G. G. (1983a) An experimental test of effects of predation risk on habitat use in fish. Ecology 64: 1540—1548.

Werner, E. E., Mittelbach, G. G., Hall, D. J. & Gilliam, J. F. (1983b) Experimental tests of optimal habitat use in fish: the role of relative habitat profitability. Ecology 64: 1525—1539.

Westhoff, V. & den Held, A. J. (1969) Plantengemeenschappen in Nederland. Thieme, Zutphen, 324 pp.

Westlake, D. F. (1963) Comparisons of plant productivity. Biol. Rev. 38: 385—425.

Wetzel, R. G. (1975) Limnology. W. B. Saunders, Philadelphia. 743 pp.

THE PHYTOSOCIOLOGICAL APPROACH TO THE DESCRIPTION AND CLASSIFICATION OF AQUATIC MACROPHYTIC VEGETATION

E. P. H. BEST

INTRODUCTION

The aim of this paper is to present methods for describing aquatic vegetation in a manner which emphasizes plant—plant co-occurrence.

The general concepts and notions are derived largely from studies on terrestrial vegetations. However, many of these notions are in modified form applicable to aquatic vegetations. Therefore pertinent factors unique to aquatic ecosystems are discussed in relation to the terrestrially derived ones.

The merits of several classification systems which emphasize various aspects of aquatic vegetation are discussed, representing Danish, Scandinavian and French—Swiss schools of thought.

Finally, several examples of vegetation analysis on aquatic plant communities are presented.

1. THE VEGETATION AND ITS CHARACTERIZATION

Most definitions and schemes for describing aquatic plant communities have been adapted from schemes originally applied to terrestrial plant assemblages. Therefore pertinent terrestrial concepts will be discussed first.

1.1. *A vegetation*

A vegetation can be defined as a characteristic collection of individual plants in connection with their growth site. It exhibits a structure which changes in space and time, due to selection by dispersal potential, physico-chemical requirements and biological interactions.

The structure of a vegetation has a physical as well as a temporal component. The physical element, i.e. the vegetation layers, can be

155

J. J. Symoens (ed.), Vegetation of inland waters. ISBN 90—6193—196—7.
© *1988, Kluwer Academic Publishers, Dordrecht. Printed in the Netherlands.*

considered as the fundamental elements of the vegetation and may be called synusiae.

A vegetation consists usually of at least one plant community. A plant community or phytocoenosis is a characteristic collection of plants, which is in equilibrium with respect to the number and combination of species, and with respect to the abundance of individuals. It occupies a more or less homogeneous site (biotope).

1.2. *Classification*

A classification system is a syntaxonomical scheme, proposed intuitively for a limited area (often determined by climate) by investigators with a thorough floristic knowledge of that area.

A vegetation can be classified according to its: (a) site (geographic location and physico-chemical factors), (b) plant morphology and life form (physiognomy), (c) community structure, (d) floristic composition and (e) the final stage of a succession series to which that particular vegetation belongs. The first two viewpoints are usually chosen for surveys of unknown areas.

Classification according to site is merely descriptive and it is difficult to extrapolate the results to unknown areas.

Classification according to physiognomy requires no detailed knowledge of species and it is possible to extrapolate these results to other areas. Vegetation units based on physiognomic criteria are usually called formations (Lam & Westhoff 1948).

In contrast, classification according to floristic composition or plant sociology requires detailed knowledge of species. This method describes vegetations partly or completely in a quantitative and qualitative way. Such a description is called a 'relevé'.

Classification according to final succession stage can not be applied to aquatic vegetations because these vegetations are usually reckoned to the early stages of succession series.

Ordination is a type of classification which involves the grouping of data using numerical methods. Phytosociology ordination is often applied to relevés of less well-known areas where less floristic knowledge is available. There are several methods commonly used in numerical taxonomy (Sneath & Sokal 1973, Anderberg 1973, Orlóci, 1978) based on the most frequently used resemblance functions in phytosociology (Van der Maarel 1979). Ordination does not have to be limited to plant relevé's: it can also be used to relate plant groups to environmental or other factors. Results of intuitive classification schemes and ordination can be comparable. For instance the intuitive and the numerically developed classifications, using an information function and discriminant analysis, were applied to a set of relevés from eutrophic, stagnant water in a badly known area in Italy (Figure 1; Feoli & Gerdol 1982). Comparison revealed that the different numerical methods simulate different intuitive schemes, but the results of the numerical classifications were always judged superior. The increasing interest in environmental impact and the availability of computers, however, have greatly stimulated the use of ordination techniques in vegetation research during the last decade.

1.3. *Methods and notions in phytosociology*

The methods used to investigate a vegetation depend on the choice of the classification system. In western Europe three basic methods have been developed, the Danish, Scandinavian and French—Swiss schools. The latter has most followers globally. According to the French—Swiss school the degree of similarity between the various relevés is estimated from the floristic composition and the entire vegetation structure, and not from the predominant, constant or frequent species. Furthermore geographical aspects are considered.

The aim of the French-Swiss method is to determine to which extent certain species occur more frequently in one community than in certain other communities with which the studied community is compared or even all other communities of the flora area investigated. These flora-areas should be more or less homogeneous units in floristic and climatological respect, e.g. the mediterranean region. Originally much value was set on the occurrence of 'faithful species', i.e. species which are more numerous in one community than in any other within the area investigated. Because these species do not usually occur in high frequencies and are often rare, later on more use was made of 'differential species', i.e. species which have a higher frequency in a certain community than in other comparable communities and thus are relatively faithful in one plant community. In fact fidelity indicates that the vegetation unit discriminated according to this criterion has a characteristic relation with its environment. The various degrees in which species are bound to communities can be considered as reflections of their respective ecological requirements (tolerance areas, ecological amplitudes; Braun-Blanquet 1913).

The fundamental vegetation unit, the association, is defined as a plant community with a more or less constant floristic composition, characterized by faithful species and constant accompanying species. Associations with many faithful and (or) differentiating species have sharp boundaries, which become more vague when the constant species increase i.e. the vegetation becomes more homogeneous.

A relevé starts with delimiting a suitable experimental plot, which should be homogeneous and should at least have the size of the

158

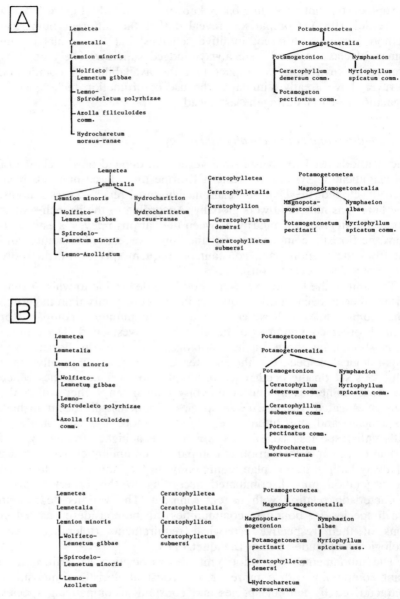

Fig. 1. Evaluation of intuitive and numerically developed syntaxonomic schemes for aquatic plant communities. After Feoli & Gerdol (1982). The schemes were applied to a set of relevés from eutrophic stagnant water in a badly known area in Italy. (A) Intuitive classification schemes: upper, according to Oberdorfer (1977); lower, according to den Hartog & Segal (1964). (B) Proposed classification schemes, developed by applying numerical methods: upper, in two classes; lower, in three classes.

smallest area representative for its normal distribution (minimum area). Sometimes an experimental transect is chosen when a plot is not convenient. Subsequently, the species are listed per vertical structural unit (for terrestrial vegetation: layers of trees, shrubs, herbage, mosses). The contributions of the individuals of each species to the vegetation cover are estimated, according to two criteria: the abundance (number of individuals) and the cover percentage (% of total surface area covered by the vertical projection of the perimeter of the individuals on the bottom). Both criteria are combined in a scale with 6 parts (+, 1, 2, 3, 4, 5; Braun-Blanquet 1928). Also the sociability is quantified according to a scale in five parts. Thus the spatial arrangement (solitary, groups etc.) of the individuals of a species is evaluated. Finally the vitality is described, i.e. the degree in which the individuals achieve their life cycle and the phenological stage at the time of the relevé.

For a proper description of the associations several (for instance 10) relevés of the same plant community should be made, combined in a table. On the basis of presence (the percentage of relevés in which the various species of the table occur) and by comparing abundance, cover %, sociability and vitality the relative fidelity of the species in the tables can be determined. In this way faithful species are established. Associations which have certain species in common and are synecologically related are combined to a higher unit, the alliance. Alliances are combined to orders and orders to classes. Classes can be grouped in different ways, for instance according to sociological progression (i.e. from low to high organization level of the plants; Braun-Blanquet 1928) or landscape (freshwater and swamps, salt water and beaches; Ellenberg 1986). Associations are marked by the suffix '-etum' following the genus name of one or two constant, faithful or differentiating taxa, e.g. *Riccietum fluitantis* and *Wolffio-Lemnetum gibbae*. Alliances, orders and classes in analogy are usually marked by the suffixes '-ion', '-etalia' and '-etea', respectively; subassociations with '-etosum'.

To investigate a vegetation phytosociologically one starts making relevés as described. For the interpretation of the relevés there are two possibilities:

(1) If a reliable classification system for that flora area exists it can be applied to the new relevés. In that case, it is useful to consult an alphabetical list of the faithful and differentiating species with records of the association, alliance and class to which they belong (e.g. for The Netherlands: Westhoff & den Held 1969 and den Held 1979). In this way it becomes apparent which already defined sociological units are present.

(2) If there is no classification system for the flora area concerned, either one should be developed in the way described, or other

ordination techniques can be applied depending on the aim of the study.

2. PLANT CONSTITUENTS OF AQUATIC COMMUNITIES

2.1. *The concept of aquatic plant*

Several definitions of aquatic plant exist. The majority of these definitions is based on structural and life form characteristics, as discussed by den Hartog & van der Velde (1987).

Den Hartog & Segal (1964) defined aquatic plants as plants which are able to achieve their generative cycle with all vegetative parts submerged or supported by the water (floating leaves), or normally occurring submerged but being induced to reproduce sexually when their vegetative parts are dying due to emersion. This definition excludes three groups of plants which in literature are often considered as aquatic plants. The first group comprises plants, which frequently occur fully submerged, multiplying vegetatively but not able to complete their generative cycle, e.g. *Sagittaria sagittifolia* f. *vallisneriifolia* (pseudohydrophytes). A second group consists of bottom roting plants with their basal parts almost continually submerged, but with leaves and inflorescences above the water surface, e.g. *Typha, Phragmites* (helophytes). The third group contains plants which, except for their submerged root systems, drift on the water surface e.g. *Eichhornia crassipes, Calla palustris* (pleustohelophytes).

I suggest that we consider as aquatic plants all macroscopic, herbaceous plants which are part of the vegetation of littoral zones or marshes. This definition is broader than that of den Hartog & Segal (1964). It comprises the three plant groups which they excluded but also various species which are able to grow under other, mostly dryer, conditions. However, for the investigation of littoral vegetation all species are included in the relevé, without the omission of those species which are present but not fitting the definition of 'aquatic plant'. This point of view has been taken in the past, although not literally stated, by Hoogers (1966) and de Lange (1972).

2.2. *The aquatic ecosystem*

Several aspects of aquatic ecosystems are important to consider when performing aquatic vegetation analysis (Wetzel 1987; see also Haslam 1978).

The water column of a water body can be filled completely by aquatic vegetation. Terrestrial vegetation is ultimately limited by the need to be supported above the soil. Aquatic ecosystems present generally a less fluctuating environment than terrestrial ecosystems because of the moderating effects of the water column. This is apparent from the often wide geographical distribution and the relatively high number of cosmopolitan species among aquatic plants (van Steenis & Ruttner 1932), which are even higher in pelagic algae (Behre 1966). Terrestrial plants absorb their water and nutrients from the substratum and gaseous compounds from the air. Aquatic plants are more ambidexterous, using water and sediment for metabolism and support.

In aquatic ecosystems, light, temperature, inorganic carbon and nutrient availability are important factors for green plants (see Wetzel 1987). Many submerged aquatics are also sensitive to the action of water currents and turbulence, as waves caused by surface winds (Figure 2).

3. CLASSIFICATION SYSTEMS FOR AQUATIC VEGETATIONS AND THEIR CRITERIA

3.1. *General classification systems developed for all vegetations*

For a long time phytosociological research centered on terrestrial plants and aquatic plants were classified in a relatively low number of groups.

Most aquatic plant communities are relatively poor in species but are seldom monospecific. Mass development of a single species is common. Usually the earliest occupying species becomes predominant suggesting a rather limited competition. Non-dominant species with similar growth forms vary from site to site and result mostly in transitions between two or more communities. Probably for these reasons no special classification system pertaining only to aquatic vegetations in general was developed until recently (Tüxen 1975).

Du Rietz (1921, 1930) classified plants according to growth form. For Europe 11 basic aquatic plant types were distinguished: isoetids, myriophyllids, nymphaeids, ceratophllids, hydrocharids, stratiotids, etc.

In some cases using Du Rietz's classification system is difficult because the boundary between 'in the bottom rooting' and 'buoyant' is not clear. *Ceratophyllum demersum* and *Utricularia vulgaris*, for example, are able to form rhizoids with which they anchor themselves to the bottom. *Hydrocharis* and *Stratiotes* can also attach themselves temporarily to the sapropelium layer with their roots. *Polygonum amphibium* has a nymphaeid water form, but it can also occur in terrestrial form ashore, making classification as aquatic plant doubtful.

162

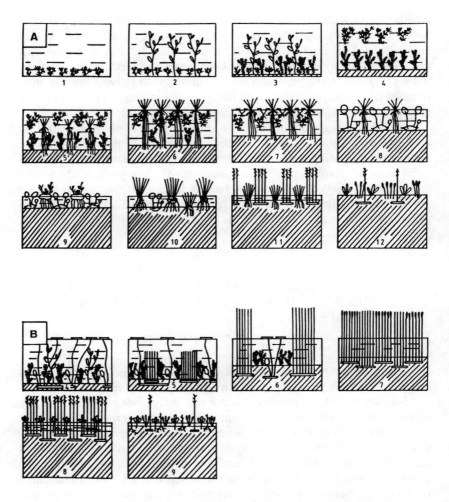

Fig. 2. Comparison of succession series of aquatic vegetation in a eutrophic freshwater lake under conditions differing in degree of exposure to wind action. After Segal (1965). (A) Wind-sheltered: 1. Characeans. 2. Appearance of magnopotamids. 3. Appearance of parvopotamids. 4. Appearance of ceratophyllids. 5. Appearance of submerged stratiotids. 6. Emerged stratiotids, last stage elodeids. 7. Appearance of hydrocharids. 8. Appearance of pleustohelophytes (*Calla*). 9. Formation of floating tufts of trembling bog (Cariceto-Cicutetum). 10. Large floating tufts of trembling bog (*Carex paniculata*). 11. Appearance of helophytes. 12. Formation of a trembling bog. (B) Wind-exposed: The stage 1, 2 and 3 are identical to those of the A-series. 4. Appearance of magnonymphaeids. 5. Appearance of submerged helophytes, last stage of hydrophytes. 7. Optimum stage of helophytes. 8. Appearance of trembling bog species (*Menyanthes*). 9. Trembling bog. The succession depends on the thickness of the sapropel layer (diagonally striped).

The original system of Du Rietz has been elaborated for the aquatic plants by den Hartog & Segal (1964) and repeatedly refined thereafter. Combining the growth form system with Luther's (1949) system based on the mode of attachment, den Hartog & van der Velde (1987) distinguish 24 basic growth forms. Regardless of some difficulties, the description of aquatic plants according to growth form is still useful for some physiognomic classification systems of the vegetations.

Rübel (1933) grouped the aquatic plant communities of Switzerland into one formation class 'Submersiherbosa' (a formation, defined by Grisebach 1872, is a vegetation unit with a structural-physiognomic character). The Submersiherbosa comprise the vegetations formed by the aquatic plants as defined by den Hartog & Segal (1964), but also some haptophytes and emerged plants. This class was subdivided into three orders according to their attachment to the substrate:

Order: *Potametalia* (rhizophytes, with the alliances *Potamion eurosibiricum, Littorellion, Characion* and *Nanocyperion flavescentis*);
Order: *Hydrocharitetalia* (pleustophytes, with the alliance *Hydrocharition*);
Order: *Encyonematetalia* (haptophytes, with the alliance *Encyonemation*).

Braun-Blanquet & Tüxen (1943) grouped in their system for central Europe the submerged, floating(-leaved) and some emerged plants according to sociological progression into two classes, notably:

Class: *Potametea* (order *Potametalia*, with the alliances *Potamion euro-sibiricum*; order *Zosteretalia*, with the alliances *Ruppion maritimae* and *Zosterion*);
Class: *Littorelletea* (order *Littorelletalia*, with the alliances *Littorellion* and *Helodo-Sparganion*).

The class *Potametea*, originally consisting of only one order *Potametalia*, was split by them into two orders with salt content of the water as criterion.

Lohmeyer *et al.* (1962) developed a classification system for central Europe more recently, which is also based on the concept of sociological progression. This system split the aquatic plant communities, covering submerged, floating(-leaved) and some emerged plants, up into 5 classes, notably:

Class 1: *Lemnetea* (order *Lemnetalia*, alliance *Lemnion minoris*);
Class 2: *Zosteretea* (order *Zosteretalia marinae*, alliance *Zosterion marinae*);

Class 17: *Ruppietea maritimae* (order *Ruppietalia*, alliance *Ruppion maritimae*);
Class 18: *Potametea* (order *Potametalia*, alliances *Eu-Potamion, Nymphaeion* and *Ranunculion fluitantis*);
Class 19: *Littorelletea* (order *Littorelletalia*, alliances *Sphagno-Utricularion, Hypericion elodis* and *Littorellion*).

Westhoff & den Held (1969) developed a classification system for The Netherlands based on the formation concept. They used the terms characteristic taxon and differentiating taxon, instead of species, because in this system distinction can concern species as well as genera and subspecies units. The submerged and floating(-leaved) plants are classified in one formation, the amphiphytes. The emerged plants, therophytes, geophytes, telmatophytes, helophytes, hemicryptophytes and chamaephytes, are usually found in four other formations (Table I). The aquatic plants as defined by den Hartog & Segal (1964) are classified in all 6 classes of one formation, whereas those according to my definition are classified in 14 classes and 19 orders in one formation which I believe is applicable to all aquatic ecosystems.

Ellenberg (1986) based a classification system for the vegetation of central Europe on the concept of sociological progression, which includes a sequence of the eight landscape units as well as the classes. It incorporates the systems of Oberdorfer (1979) and Runge (1980). The vegetation units are distinguished by faithful and differentiating species (Table II). The first 2 vegetation types occur in aquatic ecosystems which differ in salt content of the water, and the third vegetation type emphasizes the stress tolerance of species. The aquatic plants as defined by den Hartog & Segal (1964) are classified in 6 classes of the first two vegetation types. The aquatic plants as defined by me are classified within the first 3 of the total of 8 vegetation types, in 11 classes and 14 orders.

Although initially developed for the vegetation of temperate regions, such phytosociological systems were successfully extended to the vegetation units of tropical areas. In this respect, Lebrun (1947) and Léonard (1950, 1952) did pioneering work in Zaire and were followed by Germain (1952), Mullenders (1954) and Schmitz (1963, 1971). Some classes of aquatic vegetation units recognized by these authors have a wide distribution, but are represented in Africa by tropical orders and alliances: e.g. the *Potametea*, with the order *Nymphaeetalia loti* (alliances: *Pistion, Nymphaeion loti*, etc.). Other classes are exclusively tropical, like the *Saxopodostemetea* (order *Leiothylacetalia*) grouping the epilithic associations of Podostemaceae of fast-flowing rivers and waterfalls.

Table I. Survey of the plant communities of The Netherlands (Westhoff & den Held 1969). Only the formations with aquatic plants and their respective classes and orders are indicated.

```
-------------------------------------------------------------------------
Formation I.    Vegetation composed by aquatic plants and amphibious species
                (amphiphytes)
    Class 1:    Lemnetea (order Lemnetalia, alliance Lemnion minoris)
    Class 2:    Zosteretea (order Zosteretalia, alliance Zosterion)
    Class 3:    Ruppietea (order Ruppietalia, alliance Ruppion maritimae)
    Class 4:    Charetea (order Charetalia)
    Class 5:    Potametea (order Magnopotametalia, alliances Magnopotamion,
                    Nymphaeion)
                (order Parvopotametalia, alliances Parvopotamion,
                    Hydrocharition, Callitricho-Batrachion)
                (order Luronio-Potametalia, alliance Potamion
                    graminei)
    Class 6:    Littorelletea (order Littorelletalia, alliance Littorellion
                uniflorae)

Formation III. Vegetation principally composed by therophytes; one- to
                multi-species primary or secundary pioneer vegetations, mostly
                short-lived; on open, more or less N-rich, humus-poor bottoms.
    Class 10:   Isoeto-Nanojuncetea (order Nanocyperetalia, alliance
                Nanocyperion flavescentis)
    Class 11:   Bidentetea tripartiti (order Bidentetalia tripartiti, alliance
                Bidention)

Formation IV.  Vegetation principally composed by geophytes; species-poor,
                virtually non-layered, more or less open pioneer vegetations of
                perennial terrestrial plants; on moving substrata liable to
                accumulation and erosion.
    Class 14:   Spartinetea (order Spartinetalia, alliance Spartinion)

Formation VI.  Pioneer vegetations principally composed by telmatophytes and
                other helophytes; structurally showing a coarse mosaic pattern,
                in which polycorms of large gramineae and Cyperaceae predominate;
                in more or less eutrophic swamps, with the groundwater at or
                above the bottom surface the whole year round.
    Class 19:   Phragmitetea (order Nasturtio-Glycerietalia, alliances
                    Glycerio-Sparganion, Apion nodiflori, Cicutium
                    virosae)
                (order Phragmitetalia, alliance Phragmition)
                (order Magnocaricetalia, alliance Magnocaricion

Formation VIII. Vegetation, principally composed by hemi-cryptophytes,
                sometimes largely of chamaephytes; closed, seldom open
                grass vegetations; on wet bottoms influenced by groundwater
                levels or tide.
    Class 24:   Asteretea tripolii (order Glauco-Puccinellietalia, alliance
                Halo-Scirpion)
    Class 25:   Molinio-Arrhenatheretea (order Molinietalia, alliance Calthion
                palustris)
    Class 27:   Parvocaricetea (order Caricetalia-nigrae, alliance
                    Caricion-nigrae)
                (order Tofieldietalia, alliance Caricion
                    davalliane)
    Class 28:   Scheuchzerietea (order Scheuchzerietalia palustris, alliance
                    Rhynchosporion albae)
-------------------------------------------------------------------------
```

166

Table II. Summary of the survey of vegetation types and species of central Europe (Ellenberg 1986). Only the vegetation units with aquatic plants and their respective classes and orders are indicated.

```
-----------------------------------------------------------------------
Group I. Vegetations of freshwaters and swamps
    Class I.1. Lemnetea (order Lemnetalia, alliance Lemnion minoris)
                      (order Utricularietalia, alliance Sphagno-Utricularion)
    Class I.2. Potamogetonetea (order Potamogetonetalia, alliances
               Potamogetonion, Nymphaeion, Ranunculion fluitantis)
    Class I.3. Littorelletea (order Littorelletalia, alliances Deschampsion
               litoralis, Isoetion, Eleocharition acicularis, Hydrocotylo-
               Baldellion)
    Class I.4. Phragmitetea (order Phragmitetalia, alliances Phragmition,
               Sparganio-Glycerion fluitantis, Magnocaricion, Bolboschoenion
               maritimi
    Class I.6. Scheuchzerio-Caricetea nigrae (order Scheuchzerietalia,
               alliances Rhynchosporion albae, Caricion nigrae, Caricion
               lasicarpae)
                                    (order Tofieldietalia, alliances
               Caricion davallianae, Caricion maritimae)
    Class I.8. Charetea fragilis (order Nitelletalia flexilis, alliances
               Nitellion flexilis, Nitellion syncarpotenuissimae)
                                    (order Charetalia hispidae, alliance Charion
               asperae)

Group II. Vegetations of saltwaters and beaches
    Class II.1.Zosteretea (order Zosteretalia, alliance Zosterion marinae)
    Class II.2.Ruppietea (order Ruppietalia, alliance Ruppion maritimae)
    Class II.3.Spartinetea (order Spartinetalia, alliance Spartinion)

Group III. Herbaceous vegetations of often disturbed habitats
    Class III.1.Isoeto-Nanojuncetea (order Cyperetalia fusci, alliance
               Nanocyperion)
    Class III.2.Bidentetea (order Bidentetalia tripartitae, alliances Bidention
               tripartitae, Chenopodion rubri)
-----------------------------------------------------------------------
```

3.2. Classification systems developed for aquatic vegetations

3.2.1. Total aquatic vegetation

Luther (1949) distinguished three groups of aquatic plants according to their mode of attachment to the substratum. The first group, the haptophytes, consists of plants attached to the substratum (usually hard), but not penetrating it with their basal parts. This group includes the benthic algae, all aquatic lichens and most aquatic mosses (*Fontinalis*) and liverworts (*Scapania undulata*). There are no European phanerogamic species among the haptophytes, but this group includes the tropical freshwater Podostemonaceae and the epilithic seagrasses (*Phyllospadix*) along the Pacific coasts of North America and Japan. The second group, the rhizophytes, is composed by plants with basal parts penetrating into the bottom or covered with substratum. Many algae (Charophyta, Vaucheriaceae, *Caulerpa*) and the majority of the aquatic phanerogams fit in this group. The third group, the planophytes, consists of aquatic plants which are not attached to the substratum. The group is subdivided into the benthopleustophytes — phytoplankton and macro-algae resting on the bottom (e.g. *Cladophora aegagropila*) — the mesopleustophytes — plants between substrate and water surface (e.g. *Lemna trisulca, Utricularia vulgaris*) — and acropleustophytes — plants floating on the water surface (e.g. *Lemna minor, Azolla filiculoides*). The separation between the three subgroups of planophytes is only clear during the growth season, since in winter most species occur on the bottom.

Hejný (1960) developed a system for water and swamp plants using relevés of the Danube valley (Czechoslovakia) based on their ecological adaptations to water stress. His system can be applied for the study of environments with fluctuating water levels. The first group, the euhydatophytes, has vegetative parts which are completely adapted to life in the water, inflorescences which are submerged or rise above the water surface. The second group, the hydatoaerophytes, is bound to the water but has vegetative parts in contact with air. The floating leaves have special adaptations and the inflorescences rise above the water surface. The third group, the tenagophytes, consists of amphibious plants characteristic of banks along waters with a strongly fluctuating water level. Some species of the last group are able to complete their generative cycle in submerged state, e.g. *Littorella uniflora*. Others are ephemeral summer annuals of the alliance *Nanocyperion flavescentis* Koch 1926, which tolerate submergence but can only complete their generative cycle when emerged. The latter plants fit the definition given by Best (1987).

Den Hartog & Segal (1964) made a classification system for aquatic plants, ss. their own definition of 'aquatic plant', in Europe (Table III). It incorporates largely the systems of Luther (1949) and Du Rietz (1921, 1930). The criteria for this classification are floristic composition as well as attachment to the substratum, growth form and ecological factors (light, water movement and water chemistry). They decided to use other criteria besides floristic composition to distinguish the vegetation units, because species exclusively bound to one association are rare amongst aquatic plants, and because the vertical structural characteristics of terrestrial vegetations due to the occurrence of layers of trees, shrubs etc. are not comparable to those in vegetated waters. In the latter the stratification of the water itself results in a variety of niches and stratification of the vegetation also depends on anchorage potential. This system comprises largely the pleustophytes and rhizophytes. The haptophytes have not been included since in the country where the authors made their relevés (The Netherlands), the haptophytic freshwater communities were not surveyed and a study on the salt water communities was already published (den Hartog 1959). The plants were divided over 9 classes and 11 orders.

168

Table III. Summary of a classification system for European aquatic plant communities (den Hartog & Segal 1964). Only the groups, classes and orders are indicated.

```
Group I. Communities of pleustic water plants
    Class 1.   Lemnetea (order Lemnetalia, alliances Lemnion minoris, Lemnion
               trisulcae)
    Class 2.   Ceratophylletea (order Ceratophylletea, alliance Ceratophyllion)
    Class 3.   Utricularietea (order Utricularietalia, alliance Utricularion)
    Class 4.   Stratiotetea (order Stratiotetalia, alliance Stratiotion)

Group II. Communities of rhizophytic water plants
    Class 5.   Charetea (order Charetalia, alliance Charion)
    Class 6.   Zosteretea (order zosteretalia, alliance Zosterion)
    Class 7.   Ruppietea (order Ruppietalia, alliance Ruppion maritimae)
    Class 8.   Potametea (order Magnopotametalia, alliances Magnopotamion,
               Nymphaeion albae)
                        (order Parvopotametalia, alliances Parvopotamion,
               Callitricho-Batrachion)
                        (order Luronio-Potametalia, alliance Potamion
               polygonifolii)
    Class 9.   Littorelletea (order Littorelletalia, alliance Littorellion
               uniflorae)

Group III. Communities of haptophytic water plants
           **No complete survey available
```

De Lange (1972) classified ditch vegetation in The Netherlands using floristic composition, growth form and attachment to the substrate as distinctive criteria (Table IV). The term 'growth form' in this case pertains to the synusial classes high-emerged, low-emerged, floating, high-submerged and low-submerged, and involves less detail than the same term used by den Hartog & Segal (1964) after Du Rietz (1921). This system comprises all plants in and on the shores of the ditches. The system uses for its classification principal, associated and differentiating species, emphasizing the first two groups. It segregates the various types of lemnid vegetation at the level of the alliance in contrast to the class level (Westhoff & den Held 1969). Lemnids are linked at the level of the association with the submerged species because it is claimed that there is a great deal of interdependence of the vegetational layers in the aquatic environment also, and is one of the

Table IV. Summary of a classification system for ditch vegetations in The Netherlands (De Lange 1972). The classes, orders and alliances are indicated.

```
Class 1.    Lemno-Potametea (order Lemnopotametalia, alliances Ruppion
            maritimae, Parvopotamion, Lemnion gibbae, Elodeion,
            Hydrocharition)
                     (order Magnopotametalia, alliance Nymphaeion)
Class 2.    Phragmitetea (order Phragmitetalia, alliance Phragmition
            (communis)
                     (order Nasturtio-Glyceretalia, alliance
            Glycerio-Sparganion)
Class 3.    Bidentetea tripartiti (order Bidentetalia tripartiti, alliance
            Bidention)
```

theoretical prerequisites of their unification into phytosociological units as pointed out by Braun-Blanquet (1928). In de Lange's study the variety in sites has been limited to ditches: aquatic plant communities characteristic for large water bodies like lakes, oligotrophic and saline waters are lacking. The plants were divided over 3 classes and 5 orders.

The Russian system of Katanskaya (1956) is widely used in eastern European countries. However the author did not have access to the publication or its translation, therefore the classification system is merely mentioned and not characterized in this paper.

3.2.2. Segments of aquatic vegetation

No general complete system for the classification of aquatic plants exists. However, several systems for specific aquatic plant communities have been generally accepted.

The vegetation of oligotrophic lakes occurring mainly in boreal-atlantic regions has been studied by Iversen (1936), Lillieroth (1950) and Malmer (1960). These communities have been classified in the order *Littorelletalia* (Koch 1926) and contain the following faithful species: *Deschampsia setacea, Elatine hexandra, Juncus bulbosus, Littorella uniflora, Ranunculus polygonifolius, Veronica scutellata* (Oberdorfer 1979).

The vegetation of unpolluted, chalk-rich lakes and gravel pits has been studied extensively and a classification system for the characean meadows was developed (Krause & Lang 1977). In such systems the class *Charetea fragilis*, with faithful species *Chara fragilis* and *Nitella batrachosperma*, can be divided into two orders (Oberdorfer 1979). The first order *Nitelletalia flexilis*, occurring in more or less acid waters, has two alliances: *Nitellion flexilis* and *Nitellion syncarpotenuissimae*, and 6 faithful species (*Chara braunii, Nitella flexilis, N. mucronata, N. opaca, N. syncarpa, N. tenuissima*). The second order, *Charetalia hispidae*, occurs in hard waters. It has one alliance, *Charion asperae*, with the suballiance *Charion vulgaris*, and 8 faithful species (*Chara aspera, Ch. contraria, Ch. hispida, Ch. intermedia, Ch. strigosa, Ch. tomentosa, Nitellopsis obtusa, Tolypella glomerata*).

The bottom-rooting and free-floating vegetation of standing waters in central Europe have been investigated with respect to the trophic status of the water body (Müller & Görs 1960, Weber-Oldecop 1969). These aquatic plant communities contained several associations of the classes *Lemnetea* and *Potametea* (Table V). Vollmar (1947) divided the *Potamion* into *Parvo-* and *Magnopotamion*. The classification of the *Lemnetea* is a point of discussion. This discussion has been summarized recently by Landolt (1986). Some authors are inclined to consider the lemnids as a separate class apart from the vegetation below the

Table V. Bottom rooting and free-floating aquatic plant communities in eastern Lower-Saxony and two floating fern communities from the south of central Europe. H, species of Lemnetea (*Hydrocharis-* and *Stratiotes* communities); L, species of Lemnetea (*Lemna*-communities); N, species of Potametea (Nymphaeion); P, species of Potametea (Potamion, Po, Potamogetonetalia). After tables of Weber-Oldecop (1969) and Müller & Görs (1960).

Class	No.	1	2	3	4	5	6	7	8	9	10	11	12
		Potametea						Lemnetea					
Alkalinity low (≤ 1.5)		0.7		1.5				1.5				–	–
average		2.4		3.3				1.8				–	–
				3.6				3.0				–	–
high (> 4.0)				4.6				4.6				–	–
Trophy (oligo-, meso-, eutrophic)		o	e	m	e	e	e	m	e	e	e	e	e
	Myriophyllum alterniflorum	4											
N	Nuphar lutea	5	5										
P	Potamogeton obtusifolius	5	2	2									
P	P. alpinus	2	1	1	1								
N	Hottonia palustris	4	1	5	3								
H	Stratiotes aloides	3	1	1	3							1	
N	Utricularia australis	4	1	1	1			2					
L	Riccia fluitans			1	3			5					
N	Myriophyllum verticillatum		1	4	3				3				
	Callitriche hamulata		1	2	1					4			
H	Hydrocharis morsus-ranae	4	2	5	5		2	4	5			3	3
L	Lemna trisulca	5	3	5	4	1		4	5		2	3	2
Po	Ranunculus peltatus	5	2	4	2	5							
N	Potamogeton natans	4	5	4	2	2	2						
P	P. crispus		4	2		5	5						
Po	Elodea canadensis	5	4	5	4	4	2	5			2		
P	Callitriche platycarpa f.natans	5	3	3	2	4	3	2	5	5	2	3	2
L	Spirodela polyrrhiza	5	5	5	5	5	5	5	5		5	3	2
L	Lemna minor	5	5	5	5	5	5	5	5	5	5	2	1
Po	Potamogeton friesii	2	2	1		5	5			4			
	Ceratophyllum demersum	5	4	3		5	5		5				
L	Lemna gibba		3	3		5	5		5				
Po	Myriophyllum spicatum	2		2	3			3					
P	Potamogeton pectinatus	1		4	2				2				
P	P. pusillus	1		2	3				1				
H	Utricularia vulgaris	1											
N	Nymphaea alba	1											
N	Polygonum amphibium f.natans	1										1	1
P	Potamogeton lucens	1											1
	Glyceria fluitans			5	5								
	Sium latifolium			2	2								
	Oenanthe aquatica			1	1								
	Butomus umbellatus			2	3								
	Alisma plantago			1	2								
L	Ricciocarpus natans							2					
Po	Montia fontana								2				
	Nasturtium officinale								2				
	Zannichellia palustris								2				
L	Salvinia natans											4	2
L	Azolla filiculoides												5

water surface on which the lemnids usually float (den Hartog & Segal 1964, Westhoff & den Held 1969, Ellenberg, 1986), others relate the lemnids completely (de Lange 1972, referring to the class *Lemno-Potametea*) or partly (Passarge 1978, referring to the orders *Lemno-Utricularietalia* and *Hydrocharietalia*) to the submerged and floating-leaved plants with which they often associate. Landolt (1986) distinguished one class (*Lemnetea*), 2 orders and 6 alliances (Table VI). He indicated that this classification system is incomplete because of the scarcity of sociological relevés on lemnids available.

Table VI. Classification of lemnaceae communities in a single class *Lemnetea minoris* W. Koch et R. Tx. (in lit. 1954) apud R. Tx. 1955, according to Landolt (1986). The orders and alliances are indicated.

```
Order 1.    Lemnetalia minoris (alliance Lemnion minoris, 10 associations)
                               (alliance of L.turionifera, 5 associations)
                               (alliance of L.japonica, 2 associations)
                               (alliance of L.obscura, 3 associations)
                               (alliance Lemnion gibbae, 7 associations)
                               (alliance of L.disperma, 2 associations)

Order 2.    Lemnetalia aequinoctalis (17 associations)
```

The communities of reeds and sedges have been studied in many eastern and western European countries and several authors have developed classification systems pertaining to their respective countries (Pignatti 1953, Tüxen 1955, Balatova-Tulackova 1963a, 1963b, Holub *et al.* 1967). The emerged marsh vegetations are usually dense and solar irradiation is prevented from reaching the water surface. Therefore marshes are poor in submerged plants. The marsh vegetations belong to one class, *Phragmitetea*, and one order, *Phragmitetalia* (Table VII). The most competitive species in the high emerged reed marshes is *Phragmites australis*, which has a wide ecological and sociological amplitude. *Typha angustifolia* and *T. latifolia* are slightly less competitive than *Phragmites*, followed by *Schoenoplectus lacustris*. *Glyceria maxima* outrivals *Phragmites* at heavily eutrophied sites. The *Cladietum marisci* prevails at chalk-rich areas. The marshes of large sedges develop usually on the thickening sapropel layer due to the accumulation of decaying reed plants. These vegetations still have several species in common with the reed marshes and therefore they are reckoned to the same order as the reed vegetations. The order *Phragmitetea* has 4 alliances, *Phragmition, Bolboschoenion, Sparganio-Glycerion, Magnocaricion*, and comprises 8 faithful species, *Acorus calamus, Alisma plantago aquatica, Equisetum fluviatile, Iris pseudacorus, Poa palustris, Sagittaria sagittifolia, Schoenoplectus mucronatus, Sch. tabernaemontani*, according to Ellenberg (1986).

The successful use of phytosociological methods for the description and classification of Zairean aquatic vegetation types has already been

Table VII. Survey of the communities of reeds and large sedges, and their faithful species, based on Tüxen, Preising, Oberdorfer and others (after Ellenberg 1986). A: faithful species of a community (mostly association). C: faithful species of a class.

```
Faithful species of the class and order Phragmitetea (Phragmitetalia)
```

Acorus calamus	Poa palustris
Alisma plantago-aquatica	A Sagittaria sagittifolia
Equisetum fluviatile	Schoenoplectus mucronatus
Iris pseudacorus	Sch. tabernaemontani

```
Faithful species of the 4 alliances
```

1.Phragmition	3.Sparganio-Glycerion	4.Magnocaricion
Butomus umbellatus	Apium nodiflorum	A Carex appropinquata
C Eleocharis palustris	A Berula erecta	C. distans
A Glyceria maxima	Epilobium parviflorum	A C. paniculata
Hippuris vulgaris	E. roseum	A C. pseudocyperus
C Oenanthe aquatica	Glyceria fluitans	A C. riparia
CA Phragmites australis	G. plicata	A C. vesicaria
Ranunculus lingua	Nasturtium officinale	A C. vulpina
Rorippa amphibia	Scrophularia umbrosa	Cicuta virosa
C Rumex aquaticus	Veronica anagallis-aquatica	Cyperus longus
R. hydrolapathum	V. beccabunga	Eleocharis uniglumis
A Schoenoplectus lacustris		Galium palustre
Sium latifolium		Oenanthe fistulosa
C Sparganium emersum		Peucedanum palustre
C Sp. erect.ssp.polyedrum		Scutellaria galericulata
A Typha angustifolia		Teucrium scordium
A T. latifolia		

		4a In suballiance
		Caricion elatae:
2.Bolboschoenion		A Carex elata
	1/4 Communities intermediate	A C. rostrata
	between reed and large sedges	Lysimachia thyrsiflora
A Bolboschoenus maritimus	A Cladium mariscus	Senecio paludosus
Schoenoplectus americanus	CA Phalaris arundinacea	
Sch. triquetrus		4b In suballiance
		Caricion gracilis:
		A Carex gracilis

mentioned (Section 3.1.). The remarkable vicariance in the hydrosere communities of a lakelet in Lower Zaire vs. those of similar habitats in Western Europe was emphasized by Duvigneaud & Symoens (1951), while recent research led to the description of new aquatic and semi-aquatic plant associations, mainly in the Central Basin of the Zaire River: e.g. *Utricularieto exoletae — Nymphaeetum loti, Leersietum hexandrae* and *Ludwigio-Rhynchosporetum corymbosae* (Szafranski & Apema 1983), *Eleocharetum acutangulae* (Szafranski *et al.* 1983), *Lipocarpho-Cyperetum haspan* (Szafranski *et al.* 1986a), *Hydrolea glabra* community (Szafranski *et al.* 1986b).

4. APPLICATIONS OF VEGETATION ANALYSIS OF AQUATIC PLANT COMMUNITIES

Vegetation analysis can be used to describe aquatic plant communities in a manner which provides more information than only lists of species. Most common macrophytes in central European waters were surveyed and factors describing the distribution of species identified (3000 relevés from several countries; Wiegleb 1978a). These factors included co-distribution of species (sociological connection). 47 Species occurred most often in central Europe, mainly in hard waters. These species were grouped sociologically by calculating their contingency-coefficient and Hamming-value. Four large groups were found (Table VIII). The first group contained pleustophytes. The second group consisted mainly of magnonymphaeids and magnopotamids, the third and fourth group both of parvopotamids and some batrachiids. A good correspondence was found generally between these groups and those distinguished by the intuitive classification systems described earlier. However, subdivisions within the four large groups became clearer than in any existing classification system.

Vegetation analysis was used to classify and study succession patterns of *Schoenus* communities for which no detailed classification system existed (Tyler 1979). The syntaxonomy was calculated using various numerical programmes. Clusters distinguished by TABORD analysis at the 0.40 homotoneity level were regarded as associations, provided that they consisted of enough relevés (Figure 3). Clusters at the next (0.60) homotoneity level were regarded as potential subassociations.

Vegetation analysis has been applied to assess anthropogenic stress on aquatic ecosystems. It showed that changes in land-use from permanent pasture to cereal farming in Great Britain caused deterioration of the aquatic (plant and macrofauna) communities (Palmer 1986).

The relationships between water quality and aquatic vegetations are important to determine because of: (1) the great effect, (2) ecological indicators and (3) management. The macrophyte vegetation of standing and running waters of central Europe has been studied in relation to water quality (Wiegleb, 1978b, 1981). Unfortunately, rather limited chemical analyses were performed. The ecological amplitudes for several chemical factors of the aquatic plants in several French ponds were related to the occurrence of these plants in their natural aquatic habitats using factor analysis (Felzines 1982a). Ecological groups were established in this way. Species with the same growth form and with very similar ecological niches never occurred in the same phytosociological unit (Figure 4). Kohler (1975) suggested to use the occurrence of submerged aquatic plants as indicator for the pollution of aquatic ecosystems. Aquatic plant communities rapidly respond to changes in water quality and are thus a sensitive indicator. However, many species are rapidly eliminated leaving only the species which are relatively resistant.

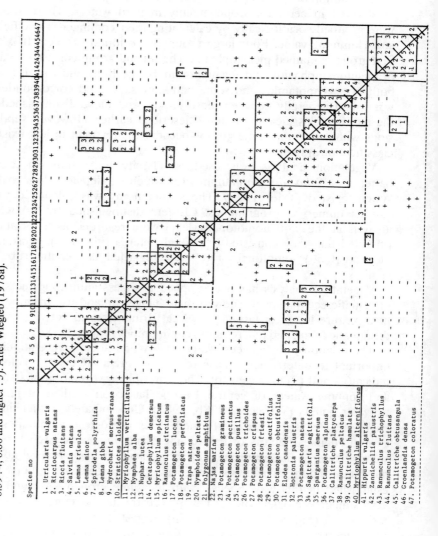

Table VIII. A contingency table of the aquatic macrophytes which occur most frequently in the fresh waters of central Europe. The contingency coefficient was calculated for all species, and the values were in a scale with 6 parts divided (0.10–0.16 : +; 0.17–0.19 : 1; 0.20–0.29 : 2; 0.30–0.39 : 3; 0.40– 0.59 : 4; 0.60 and higher : 5). After Wiegleb (1978a).

Species no

1. Utricularia vulgaris
2. Ricciocarpus natans
3. Riccia fluitans
4. Salvinia natans
5. Lemna trisulca
6. Lemna minor
7. Spirodela polyrrhiza
8. Lemna gibba
9. Hydrocharis morsus-ranae
10. Stratiotes aloides
11. Myriophyllum verticillatum
12. Nymphaea alba
13. Nuphar lutea
14. Ceratophyllum demersum
15. Myriophyllum spicatum
16. Ranunculus circinatus
17. Potamogeton lucens
18. Potamogeton perfoliatus
19. Trapa natans
20. Nymphoides peltata
21. Polygonum amphibium
22. Najas marina
23. Potamogeton gramineus
24. Potamogeton pectinatus
25. Potamogeton pusillus
26. Potamogeton trichoides
27. Potamogeton crispus
28. Potamogeton friesii
29. Potamogeton acutifolius
30. Potamogeton obtusifolius
31. Elodea canadensis
32. Hottonia palustris
33. Potamogeton natans
34. Sagittaria sagittifolia
35. Sparganium emersum
36. Potamogeton alpinus
37. Callitriche platycarpa
38. Ranunculus peltatus
39. Callitriche hamulata
40. Myriophyllum alterniflorum
41. Hippuris vulgaris
42. Zannichellia palustris
43. Ranunculus trichophyllus
44. Ranunculus fluitans
45. Callitriche obtusangula
46. Groenlandia densa
47. Potamogeton coloratus

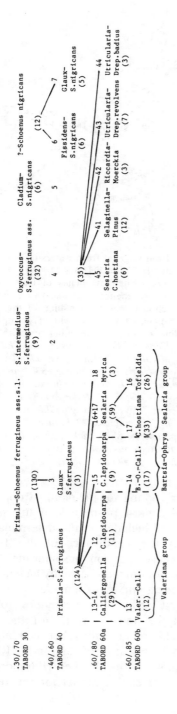

175

Fig. 3. Arrangement of *Schoenus* communities into phytocoena using a numerical method (TABORD). After Tyler (1979). In southern Sweden 189 relevés were made. All relevés and the most frequent (200 of the 289) species were included in the data matrix. The threshold and fusion levels are given at the left side of the figure. The number of the cluster is indicated above the name of the phytocoenon. The number of relevés per cluster is written within brackets under the name of the phytocoenon. The clusters at the 0.40 homotoneity level of TABORD nos. 1 to 6 could be regarded as associations. However, the nos. 2, 3, 5 and 6 were considered as potential nuclei for associations, because they composed only a few relevés. The clusters at the following homotoneity level were potential subassociations.

176

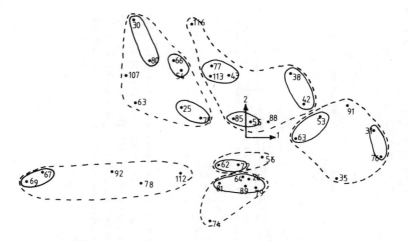

Fig. 4. Arrangement of aquatic plants from stagnant fresh waters in central France in ecological groups using numerical methods. After Felzines (1978a). The ecological profiles of the various associations were treated by means of factor analysis and hierarchical analysis. Only the axes 1 and 2 (of the total of 5 axes) are presented, giving the most information on the multidirectional spatial orientation of the species. The subgroups (encircled by interrupted line) consist of species with very similar ecological niches, the groups (encircled by solid line) of species with similar ecological niches. The species are indicated by numbers: 25 *Ceratophyllum demersum*; 26 *Chara contraria*; 30 *Elatine hexandra*; 31 *Eleocharis acicularis*; 35 *Elodea canadensis*; 38 *Glyceria fluitans*; 42 *Hottonia palustris*; 43 *Hydrocharis morsus-ranae*; 53 *Lemna minor*; 54 *Spirodela polyrhiza*; 55 *Lemna trisulca*; 56 *Littorella uniflora*; 62 *Myriophyllum alterniflorum*; 63 *M. spicatum*; 64 *M. verticillatum*; 66 *Najas minor*; 67 *Nitella flexilis*; 68 *Nuphar lutea*; 69 *Nymphaea alba*; 72 *Pilularia globulifera*; 74 *Polygonum amphibium*; 75 *Potamogeton crispus*; 76 *P. gramineus*; 77 *P. lucens*; 78 *P. natans*; 79 *P. nodosus*; 80 *P. obtusifolius*; 81 *P. pectinatus*; 85 *P. trichoides*; 88 *R. peltatus*; 89 *R. circinatus*; 91 *Riccia fluitans*; 92 *Ricciocarpus natans*; 107 *Trapa natans*; 112 *Utricularia australis*; 115 *Chara braunii*; 116 *Nitella translucens*.

The occurrence of the bottom-rooting aquatic plants strongly depends on water quality and physicochemical properties of the substratum. Thus bottom properties should be included in the determinants of the ecological amplitude. However the significance of this factor only recently became apparent, because data on mechanisms of nutrient availability and uptake from the bottom were virtually lacking. De Lyon and Roelofs (1986a, b) in a recent survey of aquatic plants and their sites in The Netherlands used ordination techniques to calculate ecological response factors of the aquatic plants for the quality of ambient water and bottom. The relationships between aquatic and swamp plants and their environment near Groet and in the River Vecht area, The Netherlands, were shown more detailed (Barendrecht *et al.* 1985, Wassen *et al.* 1986). A correlative model for the response of

177

these plants to environmental factors was developed for The Netherlands. Although the model pertains at present to a rather limited area it can be used in the future to predict effects of changes in management on aquatic and swamp plants.

Analysis of aquatic vegetation was used for the classification of lakes (Jensen 1979). Clustering was performed by calculating the degrees of similarity between the relevés of the helophytes (H), the elodeids, lemnids and isoetids (ELI), and the nymphaeids (N) of the lakes, using the numerical program TABORD and choosing different fusion limits. The three classifications yielded different results. The H-classification distinguished 10 different types of lakes, the ELI-classification 11 types and the N-classification 6 types at fusion limit 0.80. Reciprocal averaging showed similar results for the N-classification (Figure 5). Each plant group or association distinguished by calculating their degree of similarity can be treated again as relevé by ascribing an average abundance-dominance coefficient to it. In this manner units at a higher sociological level (sigmassociation) can be calculated (Felzines 1982b).

Descriptions of vegetation are often difficult to understand and judge for people who did not make the relevés themselves. These people

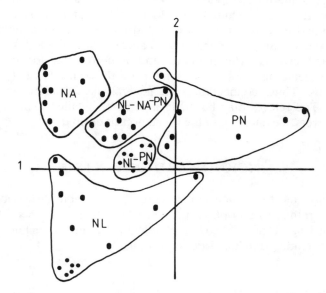

Fig. 5. Classification of lakes in southern Sweden on basis of their macrophyte composition using numerical methods. After Jensen (1979). Reciprocal averaging ordination of 48 lakes on the basis of their nymphaeid vegetation. The types distinguished in the classification are encircled by solid lines. The predominant species were: NA *Nymphaea alba*; NL *Nuphar lutea*; PN *Potamogeton natans*.

would be greatly helped by an easy to handle evaluation system. Such a system was developed for aquatic biotopes based on the composition of the macrophytic vegetation by de Lange & van Zon (1983). The system consists of a structural evaluation number (SEN), based on the percentage cover per vegetation layer, the number of species and the quantity of filamentous algae, and a geographical rareness number (GRN), which is based on the rareness of the species present. De Lange & van Zon (1978) used this system for an evaluation of the effect of different treatments to control excessive plant growth in ditches. Both SEN and GRN values increased after mechanical cleaning from July onwards, after application of herbicides with non-persistent activity from June onwards and after the introduction of grass carp.

5. CONCLUDING REMARKS

Aquatic vegetations can be characterized in various ways and for various purposes. In any case a clear definition of aquatic plant is required. The definition given by me comprises all vegetations which can be found in aquatic ecosystems. Vegetation analysis should not be limited to a mere description of structures because extrapolation to unknown areas is difficult. It should be aimed at understanding the structure and function of vegetations and therefore use taxonomical as well as functional (i.e. concerning characteristics of life-history and physiology) criteria. The methods and notions used for describing aquatic vegetations should be similar to those used for terrestrial vegetations. Thus, connection with general classification systems is possible. The relevés may be made in a more detailed manner, i.e. in vertical layers, if evaluation of the vegetations is desired.

ACKNOWLEDGEMENTS

Thanks are due to W. van der Zweerde for valuable discussions and criticism on the manuscript. H. L. Fredrickson (Dept. of Chemistry and Chemical Eng., Delft Univ. of Technology) is gratefully acknowledged for critical reading and linguistic corrections.

REFERENCES

Anderberg, M. R. (1973) Cluster analysis for applications. Academic Press, New York & London.
Balatova-Tulackova, E. (1963a) Zur Systematik der europaischen Phragmitetea. Preslia 35: 118—122.

Balatova-Tulackova, E. (1963b) Abhangigkeit einiger Magnocaricetalia- und Molinie-talia-Gesellschaften vom Pufferungsvermogen ihrer Boden. Biologia 18: 713—729.

Barendregt, A., Smidt, J. T. de, Wassen, M. J. (1985) Relaties tussen milieufactoren en water- en moerasplanten in de vechtstreek en de omgeving van Groet. Report Interfac. Dept. of Environmental Science, Utrecht, 47 pp. and annexes (In Dutch).

Behre, K. (1966) Zur Algensoziologie des Süsswassers (unter besonderer Berücksichti-gung der Litoralalgen). Arch. Hydrobiol. 62: 125—164.

Braun-Blanquet, J. (1913) Die Vegetationsverhaltnisse der Schneestufe in den rätisch-Lepontischen Alpen. N. Denkschr. Schweiz. Nat. Ges. 48.

Braun-Blanquet, J. (1928) Pflanzensoziologie. Grundzüge der Vegetationskunde. Biol. Studienbucher. Springer Verlag. Berlin.

Braun-Blanquet, J. & Tüxen, R. (1943) Übersicht der hoheren Vegetationseinheiten Mitteleuropas. Comm. S.I.G.M.A. 84, Montpellier, 11 pp.

Duvigneaud, P. & Symoens, J. J. (1951) Contribution à l'étude des associations tourbeuses du Bas-Congo. Le Rhynchosporetum candidae à l'étang de Kibambi. Verhandl. internat. Verein. theor. angew. Limnol. 11: 100—104.

Ellenberg, H. (1986) Vegetation Mitteleuropas mit den Alpen in ökologischer Sicht. Vierte Aufl., Verlag Eugen Ulmer: 900—915.

Felzines, J.-C. (1982a) Un traitement des profils écologiques des macrophytes des eaux douces stagnantes et leurs associations à l'aide de l'analyse factorielle des corres-pondances et de l'analyse hiérarchique. In: Symoens, J. J., Hooper, S. S. & Compère, P. (eds.), Studies on aquatic vascular plants. Brussels, pp. 241—248.

Felzines, J.-C. (1982b) Contribution à l'étude symphytosociologique des groupements végétaux des étangs du centre de la France. In: Symoens, J. J., Hooper, S. S. & Compère, P. (eds.), Studies on aquatic vascular plants. Brussels, pp. 284—289.

Feoli, E. & Gerdol, R (1982) Evaluation of syntaxonomic schemes of aquatic plant communities by cluster analysis. Vegetatio 49: 21—27.

Germain, R. (1952) Les associations végétales de la plaine de la Ruzizi (Congo belge) en relation avec le milieu. Publ. I.N.E.A.C., Sér. scient., no. 52, 321 pp., 83 photos.

Grisebach, A. (1872) Die Vegetation der Erde nach ihrer klimatischen Anordnung. Band 1. Engelmann, Leipzig. 2. Aufl.

Hartog, C. den (1959) The epilithic algal communities occurring along the coast of The Netherlands. Wentia 1: 1—241.

Hartog, C. den & Segal S. (1964) A new classification of the water-plant communities. Acta Bot. Neerl. 13: 367—393.

Hartog, C. den & Velde, G. van der (1988) Structural aspects of aquatic plant commu-nities. In: Symoens, J. J. (ed.), Handbook of Vegetation Science, vol. 15. Vegetations of inland waters. Dr. W. Junk. This volume, pp. 113—153.

Haslam, S. M. (1978) River plants. The macrophytic vegetation of watercourses. Cambridge Univ. Press, Cambridge etc., 396 pp.

Hejný, S. (1960) Ökologische Charakteristik der Wasser- und Sumpfpflanzen in den slowakischen Tiefebenen (Donau- und Theissgebiet). Verlag Slowakischen Akad. Wissensch., Bratislava, 492 pp.

Held, J. J. den (1979) Beknopt overzicht van Nederlandse Plantengemeenschappen. Wetenschappelijke Mededelingen KNNV. 134, 86 pp. (In Dutch).

Holub, J., Hejný, S., Moraver, J. & Neuhausl, R. (1967) Übersicht der hoheren Vegetationseinheiten der Tschechoslowakei. Academia. Prague.

Hoogers, B. J. (1966) De groeicyclus van waterplanten. Jaarb. I. B. S., Wageningen, 1966: 31—40 (In Dutch).

Iversen, J. (1936) Biologische Pflanzentypen als Hilfsmittel in der Vegetationsfors-chung. Thesis. Copenhagen, 224 pp.

Jensen, S. (1979) Classification of lakes in Southern Sweden on the basis of their macrophyte composition by means of multivariate methods. Vegetatio 39: 129—146.

180

Katanskaya, V. M. (1956) [The method of investigation of higher aquatic vegetation]. *In*: Zhizn presnykh vod SSSR 4: 160—182 (In Russian).

Koch, W. (1926) Die Vegetationseinheiten der Linthebene, unter Berucksichtigung der Verhaltnisse in der nordostschweiz. Jb. Naturw. Ges. St. Gallen 61: 144 pp.

Kohler, A (1975) Submerse Markrophyten und ihre Gesellschaften als Indikatoren der Gewasserbelastung. Beitr. naturk. Südw.-Dtl. 34: 149—159.

Krause, W. & Lang, G. (1977) Klasse Charetea fragilis (Fukarek 1961 n. n.) Krausch 1964. *In*: Süddeutsche Pflanzengesellschaften, Part I. Ed: E. Oberdorfer. Pflanzensociologie 10: 78—88, 2nd ed., VEB Gustav Fischer, Jena.

Lam, H. J. & Westhoff, V. (1948) Plantenoecologie en -geografie. *In*: Inleiding E.N.S.I.E. VI: 127—129 (In Dutch).

Landolt, E. (1986) Biosystematic investigations in the family of duckweeds (Lemnaceae) (Vol. 2). The family of Lemnaceae — a monographic study. Vol. 1, 566 pp.

Lange, L. de (1962) An ecological study of ditch vegetation in The Netherlands. Thesis, 112 pp.

Lange, L. de & Zon, J. C. J. van (1978) Evaluation of the botanical response of different methods of aquatic weed control, based on the structure and floristic composition of the macrophytic vegetation. Proc. Symp. Eur. Weed Res. Soc., 6th, 279—286.

Lange, L. de & Zon, J. C. J. van (1981) A system for the evaluation of aquatic biotopes based on the composition of the macrophytic vegetation. Biol. Conserv. 25: 273—284.

Lebrun, J. (1947) La végétation de la plaine alluviale au sud du lac Edouard. *In*: Exploration du Parc National Albert, Mission J. Lebrun (1937—1938), fasc. 1, 2 vols.

Léonard, J. (1950) Les groupements végétaux. *In*: Encyclopédie du Congo belge. Bieleveld, Bruxelles, 1: 345—389.

Léonard, J. (1952) Aperçu préliminaire des groupements végétaux pionniers dans la région de Yangambi (Congo belge). Vegetatio 3 (4—5): 279—297.

Lillieroth, S. (1950) Über Folgen kulturbedingter Wassersenkungen fur Makrophyten- und Planktongemeinschaften in seichten Seen des Südschwedischen oligotrophiegebietes. Acta Limnol. 3: 288 pp.

Lohmeyer, W., Matuskiewicz, A. & W., Merker, H., Moore, J. J., Muller, Th., Oberdorfer, E., Poli, E., Seibert, P., Sukopp, H., Trautmann, W., Tüxen, J., Tüxen, R., & Westhoff, V. (1962) Contribution a l'unification du système phytosociologique pour l'Europe moyenne et nord-occidentale. Melhoramento 15: 137—151.

Luther, H. (1949) Vorschlag zu einer ökologischen Grundeinteilung der Hydrophyten. Act. bot. fenn. 44: 1—15.

Lyon, M. J. H. de & Roelofs, J. G. M. (1986a) Waterplanten in relatie tot waterkwaliteit en bodemgesteldheid. Deel 1. Lab. v. Aquatische Oecologie, Katholieke Univ. Nijmegen, 106 pp. (In Dutch).

Lyon, M. J. H. de & Roelofs, J. G. M. (1986b) Waterplanten in relatie tot waterkwaliteit en bodemgesteldheid. Deel 2. Lab. v. Aquatische Oecologie, Katholieke Univ. Nijmegen, 126 pp. (In Dutch).

Maarel, E. van der (1979) Multivariate methods in Phytosociology with reference to the Netherlands. *In*: M. J. A. Werger (ed.), The study of vegetation. Junk, The Hague.

Malmer, N. (1960) Some ecologic studies on lakes and brooks in the South Swedish Uplands. Botan. Not. (Lund) 113: 87—116.

Mullenders, W. (1954) La végétation de Kaniama (Entre-Lubishi-Lubilash, Congo belge). Publ. I.N.E.A.C., Sér. scient., no. 61, 499 pp., 18 pl.

Müller, Th. & Görs, S. (1960) Pflanzengesellschaften stehender Gewässer in Baden-Wurttemberg. Ebenda 19: 60—100.

181

Oberdorfer, E. (1977) Süddeutsche Pflanzengesellschaften. Pflanzensoziologie 10: 564 pp.

Oberdorfer, E. (1979) Pflanzensociologische Exkursionsflora, 4. Aufl. Verlag Eugen Ulmer, Stuttgart: 997 pp.

Orlóci, L. (1978) Multivariate analysis in vegetation research. 2nd ed., Junk, The Hague.

Palmer, M. (1986) The impact of a change from permanent pasture to cereal farming on the flora and invertebrate fauna of watercourses in the Pevensey levels, Sussex. Proc. EWRS/AAB 7th Symp. on Aquatic Weeds: 233—238.

Passarge, H. (1978) Uber Erlengesellschaften im Unterharz. Hercynia, N. F., 15: 399—419.

Pignatti, S. (1953) Introduzione allo studio fitosociologico della pianure Veneta orientale. Atti Inst. Bot. Univ. Lab. Crittogamico Pavia, Series 5, 9: 92—258 (In Italian).

Raunkiaer, C. (1934) The life forms of plants and statistical plant geography. Oxford, 632 pp.

Rietz, E. G. Du (1921) Zur methodologischen Grundlage der modernen Pflanzensoziologie. Thesis, Uppsala, 272 pp.

Rietz, E. G. Du (1930) Vegetationsforschung auf soziationsanalytischer Grundlage. Aberhalden's Handb. Biol. Arbeitsmethoden 11(5): 293—480.

Rübel, E. (1933) Versuch einer Ubersicht uber die Pflanzengesellschaften der Schweiz. Ber. geobot. Forsch. Inst. Rübel. 1932: 19—30.

Runge, F. (1980) Die Pflanzengesellschaften Mitteleuropas. Eine kleine Übersicht. 6/7. Aufl. 1980: 278 pp. Aschendorff, Munster i.W.

Schmitz, A. (1963) Aperçu sur les groupements végétaux du Katanga. Bull. Soc. r. Bot. Belg., 96(2): 233—447.

Schmitz, A. (1971) La végétation de la plaine de Lubumbashi (Haut-Katanga). Publ. I.N.E.A.C., Sér. scient., no. 113, 388 pp., 32 photos.

Segal, S. (1965) Een vegetatieonderzoek van de hogere waterplanten in Nederland. Wetensch. Meded. KNNV 57, 80 pp.

Sneath, P. H. A. & Sokal, R. R. (1973) Numerical taxonomy. The principles and practice of numerical classification. Freeman, San Francisco.

Steenis, C. G. G. J. van & Ruttner, F. (1932) Die Pteridophyten und Phanerogamen der Deutschen limnologischen Sunda-Expedition. Arch. Hydrobiol., Suppl. 12: 231—387.

Szafranski, F. & Apema, A. K. (1983) Contribution à la connaissance des groupements végétaux aquatiques et semi-aquatiques dans les environs de Kisangani (Haut-Zaïre). I. Bull. Soc. r. Bot. Belg. 116(1): 93—106.

Szafranski, F., Apema, A. K. & Nyakabwa, M. (1983) Contribution à la connaissance des groupements végétaux aquatiques et semi-aquatiques dans les environs de Kisangani (Haut-Zaïre). II. L'association à Eleocharis acutangula. Bull. Soc. r. Bot. Belg. 116(2): 189—194.

Szafranski, F., Apema, A. K. & Nyakabwa, M. (1986a) Contribution à la connaissance des groupements végétaux aquatiques et semi-aquatiques dans les environs de Kisangani (Haut-Zaïre). III. L'association à Cyperus haspan et Lipocarpha chinensis. Bull. Soc. r. Bot. Belg. 119(1): 81—86.

Szafranski, F., Apema, A. K. & Nyakabwa, M. (1986b) Contribution à la connaissance des groupements végétaux aquatiques et semi-aquatiques dans les environs de Kisangani (Haut-Zaïre). IV. Le groupement à Hydrolea glabra. Bull. Soc. r. Bot. Belg. 119(1): 87—91.

Tüxen R. (1955) Das System der nordwestdeutschen Pflanzengesellschaften. Mitt. Flor.-soz. Arbeitsgem., N.F., 5: 155—176.

Tüxen, R. (1975) Prodromus der europäischen Pflanzengesellschaften. Lief. 2: Littorelletea uniflorae. J. Cramer, Vaduz, 149 pp.

Tyler, C. (1979) Classification of Schoenus communities in South and Southeast Sweden. Vegetatio 41: 69—84.

Vollmar, F. (1947) Die Pflanzengesellschaften des Murnauer Moores. Teil I. Ber. Bayerischen Bot. Ges. Erforsch. heim. Flora 27: 13—97.

Wassen, M. J., Barendregt, A., Lippe, E., Smidt, J. T. de, Witmer, M. C. H. (1986) Een model voor de responsie van water- en moerasplanten op de watersamenstelling. In: Proc. Symp. Mathematische Ecosysteemmodellen, 1986, 12 pp. (In Dutch).

Weber-Oldecop, D. W. (1969) Wasserpflanzen-gesellschaften im ostlichen Niedersachsen. Diss. T. H. Hannover, 171 pp.

Westhoff, V. & Held, A. J. den (1969) Plantengemeenschappen in Nederland. N. V. W. J. Thieme & Cie, Zutphen, 324 pp. (In Dutch).

Wetzel, R. G. (1987) Water as an environment for plant life. In: Symoens, J. J. (ed.), Handbook of Vegetation Science, vol. 15. Vegetations of inland waters. Dr. W. Junk. This volume, pp. 1—30.

Wiegleb, G. (1978a) Der soziologische Konnex der 47 haufigsten Makrophyten der Gewasser Mitteleuropas. Vegetatio 38: 165—174.

Wiegleb, G. (1978b) Untersuchungen uber den Zusammenhang zwischen hydrochemischen Umweltfaktoren und Makrophytenvegetation in stehenden Gewassern. Arch. Hydrobiol. 83: 443—484.

Wiegleb, G. (1981) Application of multiple discriminant analysis on the analysis of the correlation between macrophyte vegetation and water quality in running waters of Central Europe. Hydrobiologia 79: 91—100.

ALGAL COMMUNITIES OF CONTINENTAL WATERS

JEAN-JACQUES SYMOENS, ELSALORE KUSEL-FETZMANN
& JEAN-PIERRE DESCY

INTRODUCTION

Recognition, description and classification of natural communities originated from the beginning of the XIXth century. However, it is mainly
from the turn of the century that an ever-increasing literature has been
devoted to these aspects of ecology, and that the concepts and methods
initially developed for the study of forest, scrub, heath or grassland
vegetation, would be applied to the study of algal communities.

The community-unit most used through the XIXth century was the
formation, a term applied by Grisebach (1838, p. 160) to a group of
plants possessing a definite physiognomic character, as a meadow,
forest, etc. The word 'formation' has been used afterward by different
authors in different senses, but there has been much later an increasing
agreement that it should keep its original physiognomical meaning and
thus be applied to a community-type defined by dominance of a given
growth-form in the uppermost stratum (or the uppermost closed stratum)
of the community, or by combination of dominant growth-forms. As
the communities of freshwater algae, especially the microscopic ones,
are mostly difficult to visualize, they were rarely introduced into
physiognomical schemes and therefore they were classified more often
according to their habitat than to their physiognomy. An abundant
vocabulary has been created to define and classify the algal habitats: on
the whole, we shall follow the terminologies defined and used by
Hutchinson (1967) and Round (1981).

Another approach to the study of natural communities gave more
emphasis to their floristic composition and culminated with the use of
the concept of *association*. A. von Humboldt (1805, p. 17) has already
used the word association and Lorenz (1858) recognized vegetation-
units he named by the *-etum* suffix.

An event of some importance in the history of the study of natural
communities was the agreement of two leading figures of the phyto-
geography of the years 1900, Flahault and Schröter, on a definition of

183

J. J. Symoens (ed.), Vegetation of inland waters. ISBN 90—6193—196—7.
© *1988, Kluwer Academic Publishers, Dordrecht. Printed in the Netherlands.*

the association which they proposed to the IIIrd International Botanical Congress: "An association is a plant community of definite floristic composition, presenting a uniform physiognomy, and growing in uniform habitat conditions. The association is the fundamental unit of synecology" (Flahault & Schröter 1910). This definition became the basis for a broad acceptance of the association as a vegetation unit defined — at least partly — by species composition.

Intensive work was then carried out in order to recognize, describe and classify plant associations, although very different conceptions of a plant association soon appeared in different schools, and even notable divergences within the same school. The rapid development of the science of plant associations, the phytosociology, did not leave the algologists indifferent. After a period of intense taxonomic and floristic research, the algologists began thus to recognize algal associations. Allorge (1921–22, 1925), Denis (1924, 1925), Messikommer (1927), and Budde (1928) were pioneers in this algal sociology of which the history has been reviewed by Fetzmann (1956), the bibliography compiled by Schroevers (1973) and, for running waters, by Johansson et al. (1977).

INDIVIDUALITY AND SYNECOLOGICAL RANK OF ALGAL COMMUNITIES

The integration of the algae into a syntaxonomical system is obvious only in two cases: the case where the vegetation consists exclusively of algae ('echte Algengesellschaften' of Panknin 1945), and the case where the size of the algae is comparable to that of the higher plants which they accompany. The former case is that of the vegetation of cold or thermal springs, rocky lake shores, etc.: the algal associations may then be described and classified according to the same principles applied to the description and the taxonomy of the associations of higher plants. The latter case is that of the Charophyta which sometimes accompany vascular plants in associations of the Classes Potametea or Littorelletea; in this case, the algae are themselves macrophytes and thus constituents of macrophytic communities: they represent characteristic species in associations of the class Charetea, and differentiating or accompanying species in other associations.

In the other case, the phytosociologist stands in a dilemma already clearly explained and discussed by Allorge (1925): must the algae, whatever their size, be treated as constituents of the association of the co-occurring higher plants? Or may they be considered separately, their assemblage possibly deserving the status of association?

Due to its usually small size, an algal assemblage is often spatially included in a stand of an association of higher plants: epiphyton on submerged macrophytes, desmid and diatom assemblage in the hollows ('Schlenken') of the peat-bogs, etc. The algal assemblages are thus spatially subordinate to associations of higher plants: apparently they belong to the latter ('zugehörige' or 'zugeordnete Algengesellschaften' of Panknin 1941, 1945). Some phytosociologists consequently did not hesitate to define an association by listing in the same 'relevé' all the plants, unicellular algae as well as vascular plants, occurring together within the limits of the higher association: a striking example of such a 'mixed' association is the *Micrasterieto-Sparganietum* described by Guinochet (1938).

In fact, the opposite solution has been adopted already by Allorge (1925) who, having found the same algal community, his *Micrasterietum*, in different associations of higher plants, consequently emphasized its relative autonomy of occurrence among the latter.

Symoens (1951b, 1983) definitely rejected the global treatment of such co-occurring population groups and Guinochet himself (1973, pp. 8—10) abjured his own previous point of view. A necessary character of a vegetation entity to be described (a stand, or an 'association individual' in the sense of the school of Braun-Blanquet) is its relative homogeneity in floristic composition. This homogeneity makes it possible to recognize its spatial limits. A stand of higher plants may be homogeneous on a relatively large area and nevertheless possess on this same area micro-habitats different enough to harbour very different assemblages of algal populations. A peat-bog of which the higher vegetation appears homogeneous over several ares has in its hollows a desmid and diatom assemblage very different from that which is revealed by the expression of the *Sphagnum*-hummocks. To integrate in the higher plant association all the co-occurring algae would conceal the structure of the community which only appears from a more synusial analysis.

Similarly within the limits of *Potamogeton*-dominated vegetation in a pond, one may recognize an association of epiphytic algae, an association of epipelic algae covering the sediments, and, in the water between the macrophytes, a planktonic association more or less 'contaminated' with epiphytic and epipelic species. Here again, treatment of the whole as a unique association would obscure completely the structural aspects of the biocœnosis.

Especially in the aquatic habitat where most of the algae are not bound to the soil, surviving cells may be transported for considerable distances from their primary habitats by currents. Thus, much more than in terrestrial habitats, where the plants are bound to the soil, the assemblage present in any given water sample may consist of populations derived from different habitats and adapted to respond to substantially different sets of conditions. Such assemblages should not be considered as associations, although they deserve no less attention from the ecologist.

Incidentally, it may be noted that, even at the level of higher vegetation, den Hartog (1983) emphasized the independence of the synusia in the aquatic environment. Several communities may occur in the same place as a mixture, or spread in time. They should be treated as separate associations.

In terrestrial ecology also, some synusia of small-sized plants may be independent of the associations of the higher plants where they occur: this is the case for the epiphytic growth of lichens and bryophytes on the tree trunks in forests. The epiphytic communities are not bound to one type of forest association; they occur often on isolated trees. These synusial communities were thus generally described and treated by phytosociologists as independent associations (e.g. in the monumental work of Barkman 1958). Some authors (Wilmanns 1962, Barkman 1969) used the term *union* instead of *association* for such communities and it could similarly be used for the algal microcenoses we deal with, but we are reluctant to encourage a multiplication of the terms in use for vegetation units.

TAXOCENE OR LIMNETIC ASSOCIATIONS?

The term *taxocene* was proposed by Chodorowski (1959) for a group of species, all members of a supraspecific taxon and occurring together in the same association (e.g. the littoral taxocene of monocotyledons, the epilithic taxocene of diatoms, the benthic taxocene of chironomid larvae).

Many authors described associations based on only one algal taxocene: Cholnoky (1953, 1955), Cholnoky & Schindler (1953), Jørgensen (1948), Coste & Ricard (1982), Gasse *et al.* (1983) described diatom associations; Donat (1926), Laporte (1931), and more recently Coesel (1975) described desmid associations.

Although it is often difficult for one author to identify taxa of very different groups, the treatment of a single group is only justified in the case where the algal vegetation consists only, or nearly only, of one algal group, such as in the case of the cyanophycean vegetation of thermal springs and, to some extent, the epilithic diatom vegetation in streams. Taxocene studies may be of use in some applied researches, but are incomplete from a phytosociological point of view: indeed, in most cases, the algal vegetation comprises algae of various taxonomic groups, none of which should be neglected (Symoens 1951b).

Many of the gaps in our present knowledge of the algal communities are due to the fact that their smallest-celled components were often neglected, especially when plankton communities were described from net-obtained samples. At present, more attention than previously is given to the very small fractions of phytoplankton, referred to as nanoplankton, ultraplankton, and micro-algae by different authors; Stockner & Antia (1986) provided recently a comprehensive review on the smallest-celled plankters (size <3 µm) forming the picoplankton in ocean and lake ecosystems. Small-celled algae occur in a wide variety of habitats, such as oligotrophic lakes, especially in high mountains (Pechlaner 1971), subarctic areas (Rodhe 1955, Nauwerck 1968), the Arctic (Wright 1964, Kalff & Welch 1974, Hobbie 1980), and the Antarctic (Light *et al.* 1981). They occur also in eutrophic situations: massive production of a chroococcoid organism (1 µm diameter, 1.5— 3 µm length) was observed in a sewage oxidation pond, in E. Germany, causing a supersaturation of oxygen (Drews *et al.* 1961); species of small-celled green algae may develop also into large populations in eutrophic situations subsequent to the period of exponential growth of larger phytoplankters or during the winter when larger algae are not increasing rapidly in density (Priddle & Happey-Wood 1983).

Similarly, in the epipelic communities inhabiting sediments within the euphotic zone, small flagellates and coccoid green algae of cell volume <100 µm^3 have been found in considerable numbers (Happey-Wood 1976, 1978, Ong & Happey-Wood 1979) and were found to be the dominant algae in a shallow oligotrophic pool (Happey-Wood 1980); green flagellates were present throughout the whole year and often contributed more than 50% of the algal population on sediments of mountain lakes (Happey-Wood & Priddle 1984).

The small size of these micro-algae is really no good reason to omit them in the study of algal communities since they may account for high percentages of the biomass and production, e.g. 85% of the planktonic biomass in Lake Superior, N. America, and 80% of the primary production in Lake Biwa (Tanaka *et al.* 1974). Even higher percentages for carbon fixation by the nanoplankton (95—98% of the total) had been reported during the spring outburst after the ice melt in Lake Erken, Sweden, by Rodhe *et al.* (1958).

Further, in view of the similar dependence of all organisms, plant and animal, on some factors of the aquatic environment, and of their interrelations, it is advantageous to consider them all in the analysis of the limnetic association. An additional argument for considering the animals in the aquatic microcenoses is their relatively high biomass. For example, planktonic systems maintain a much higher heterotroph biomass, e.g. 21 to 32 g dry matter m^{-2} (Riley 1956, Harvey 1950), than do forests, e.g. 0.25 to 0.31 g m^{-2} (Satchell 1971, Reichle *et al.* 1973).

Margalef (1947, 1949) successfully extended the algal association to the whole limnetic association including protozoa, rotifers, cladocera, copepods, chironomid larvae in the associations where they are living. Similarly, Thunmark (1945) characterized plankton associations by both the dominant plant and animal form, as in the *Anabaena flos-aquae — Holopedium gibberum* association recorded in Lake Allgunnen, Sweden; Symoens (1951a) listed the thecamoebae with the algae of the *Penio silvae-nigrae — Cosmarietum obliqui*, a desmid dominated association of the hollows of *Sphagnum* bogs; Höhn-Ochsner (1978) included protozoa, rotifers and other helminths, crustacea, etc. in the description of microcenoses of fen biotopes.

Although all micro-organisms should be considered in the study of microcenoses, their sizes may differ by several orders of magnitude. Water ecosystems may be compared to fractal objects, the study of which might possibly require a synusial approach at more than one level.

TEMPORAL CHANGES IN THE COMPOSITION OF ALGAL COMMUNITIES

Nature and factors of temporal changes

Inland waters and, with them, the biocenoses they harbour, particularly their algal communities, are subject to continuous temporal changes. On the geological time scale, they are, in contrast to the more stable oceans, short-lived (very few lakes and rivers have existed continuously for more than a million years). Lakes are subject to greater changes and, even without the influence of man, they undergo a double evolution: (a) a raising of the bottom by silting or by accumulation of organic debris; (b) a natural tendency from oligotrophy to eutrophy or, sometimes, chthoniotrophy. Further, due to the short duration of a generation of most algae, especially of the unicellular species, the assemblages of algae undergo much quicker quantitative and qualitative changes than those of higher plants.

There may be relative compositional stability only at either end of the gradient frequency of disturbance.

The equatorial Lake George, W. Uganda, rather extended (250 km²), but very shallow (2.5 m), supports a nearly constant planktonic community, particularly of *Microcystis aeruginosa* and *Thermocyclops hyalinus*, all year round with very little change in densities of the total community and of most major species (with the exception of *Anabaena flos-aquae*) (Burgis 1971a, b, Ganf 1974).

When an extreme ecological factor exerts a permanent action, strongly selective on the composition of the associations, the latter show but little variation in their composition over the whole year ('Dauergesellschaften'). This is the case for the algal communities of the cold springs of mountain streams, for those of thermal springs, to a certain extent also for those of the hollows ('Schlenken') of peat-bogs (Symoens 1957, Leher 1958). In these cases, the permanence of an extreme environmental factor is responsible for this stability of the algal association: constant temperature in cold or in thermal springs, constant acidity and humidity in the bogs.

If frequency of disturbance is very high, there may be a constant presence of species adapted to that disturbance and local extinction of species not so adapted (White 1979). This is the case for the algal crusts on humid rocks (Jaag 1945): the extreme instability of the temperature and humidity is a constant characteristic of such rocks, with a highly selective effect.

Other associations have been described as appearing over the whole year and showing only minor deviations. Nevertheless some associations which seem to prevail throughout the year in some regions may be of shorter duration elsewhere: this would be the case for the *Cocconeis — Ulvella* community recorded as permanent in some British rivers by Butcher (1940, 1946, 1947), but only found as a summer assemblage in the River Verkeån, Southern Sweden, by Klasvik (1974). Caljon (1983) points out the different temporal behaviour of some species in the waters of the Flemish lowland: e.g. *Chlamydomonas umbonata, Scourfieldia cordiformis, Pseudopedinella pyriforme*, seasonally indifferent in mesohaline waters, form a characteristic winter-spring group in the oligohaline waters.

In most aquatic habitats, temporal changes represent a universal characteristic of algal populations. The factors responsible for these changes are many and complex: temperature, light, turbulence, inorganic nutrients, organic materials, grazing, parasitism. Their action is concomitant and they all are likely to affect competition. Reynolds (1984) emphasized the fact that the environmental factors that have been considered important in the past have been those which are relatively easy to measure (temperature, light, nutrients). Until recently, insufficient account has been taken of those factors which are more difficult either to measure (e.g. the effects of organic chelating agents, allelopathy) or to interpret (the interaction between algal requirements for micronutrients and their chemistries in natural waters). Moreover, there has been an evident tendency to relate the specific abundance, rather than the specific growth of the algae, to particular environmental variables, an approach which assumes the widespread existence of stable environmental equilibria, whereas this is unlikely to be the case (Harris 1980, 1983).

A rather old approach to the causality of temporal changes in natural communities distinguishes between agents internal and external to the community. Tansley (1916, 1920) used the terms autogenic and allogenic to distinguish between these causes. Autogenesis is community change driven by the properties of the species present and their effects on site environment (e.g. pond filling, depletion of available nutrients). Allogenesis is community change driven by the variation, sometimes of a catastrophic nature, of the physical environment (e.g. climatic changes, meteorological adversities). These concepts have been critically reviewed (Drury & Nisbet 1973, Horn 1976, White 1979) and a search to ascribe all causes of community changes to either of these types of factors is probably futile. However, the distinction is, to a certain extent, useful and has been successfully applied by Reynolds (1984) to the temporal changes of algal communities considered as responses to critically changing resource-ratio gradients (autogenic changes) and to the variability of the physical environment (allogenic changes).

Meteorological adversities may severely affect the composition of algal communities. In the Neusiedlersee, short-term aperiodic fluctuations within the reed belt are chiefly the result of currents caused by the wind. Northwesterlies, for example, cause large patches of the reed to be left dry near Neusiedl, whilst the quickly-rising turbid water resulting from southeast winds washes away large numbers of algae. A period of calm is necessary before the algal communities can re-establish themselves (Kusel-Fetzmann 1979).

When a lake is subject to severe changes of level (sometimes to complete dryness) and, consequently, of chemical composition (drying phase often implies saline conditions), the algal communities may undergo very severe and rapid changes. This is especially the case of endorheic lakes, i.e. lakes with no outlet. A good example is Lake Chilwa, Malawi, for which Moss (1979) has summarized the changes in the phytoplankton around a period of severe drying. During the drying phase of 1967—1968, when very alkaline and saline, Lake Chilwa supported a planktonic community, poor in species, but very dense, dominated by *Spirulina platensis*, accompanied by *S. major* and *Anabaenopsis circularis*. As the lake filled in late 1968/69, phytoplankton was scarce, but small organisms formed a surface film (neuston), initially consisting mostly of Volvocales and Euglenophyceae, later of *Synura* and Volvocales. When the filling phase came to an end, with the decrease in lake level, the flora changed again: *Anabaena torulosa* and *Anabaenopsis circularis* dominated in 1971—1972, followed, when lake was specially high, by a community much richer in species, dominated by *Oscillatoria*. A rather parallel evolution of the zooplankton has been described (Kalk & Schulten-Senden 1977, Kalk 1979).

Also during the filling phase of man-made lakes, rapid changes in the environment induce great ranges of plankton fluctuations in space and time, as reported for reservoirs in Russia by Rzoska (1966, p. 150), in Africa by McLachlan (1974, p. 374), in South America by van der Heide (1982, p. 283), etc.

Seasonality

The temporal variations which have stimulated the greatest effort from limnologists are the periodic, seasonally-bound, cycles of species dominance in aquatic, and especially planktonic communities.

Lakes show different patterns of seasonal variation, and in some cases, the changes seem rather irregular. A good review of phyto-plankton periodicities and the cycles of particular species in selected European and North American lakes was given by Hutchinson (1967, pp. 407—426). Since then, at least 200 descriptions of such cycles in other lakes and water-bodies have been added to the literature, most of them concerning lakes in Europe and North America, others referring mainly to Africa, Australia, Israel and Japan (Reynolds, 1984).

In large lakes, the cycles are rather regular and consequently more or less predictable. For example, some poorly productive subalpine lakes show a spring dominance of centric diatoms (e.g. *Cyclotella comensis* and *Rhizosolenia eriensis*) and a summer dominance of dinoflagellates (e.g. *Peridinium willei, Ceratium hirundinella*), accompanied by some desmids (*Staurodesmus* . . .), with a single summer or early autumn biomass peak. In other lakes, as in the sea, the phytoplankton regularly shows a spring and an autumn maximum in diatoms. In more produc-tive lakes, water blooms of blue-green algae appear rather regularly after midsummer in temperate regions, at the transition dry-rainy season in tropical regions.

In small lakes (less than 10 ha in surface areas), the phytoplankton assemblages that they might harbour at any given time of year are more difficult to predict since short retention times, reduced areas of open water and increased susceptibility to small-scale environmental fluctua-tions reduce ecological stability compared with larger water bodies. However, this does not prevent the algal population in some small lakes from showing a certain degree of constancy in their annual recurrence (see e.g. Rees *et al.* 1984).

When communities are multispecific, limited knowledge of the deter-minism of their temporal changes can be derived from measurements of the environmental parameters, if the nutrient requirements and uptake kinetics for each species are unknown. There is now an increas-ing tendency to consider that the algae rather than the environment provide the most direct and relevant answers regarding algal behaviour (Sakshaug & Olsen 1986).

With such a view, Reynolds (1982, 1984) adopted an initially sociologically-based approach to the problem of the determination of planktonic periodicity. He recognized a certain number of 'assemblages' of species, in fact ceno-ecological groups, which share closely similar phasing of increase, relative abundance and decrease in lakes of different trophic status (Figure 1). According to the same concept, Rott (1984) described similar trends in Tyrolean low-and mid-altitude lakes.

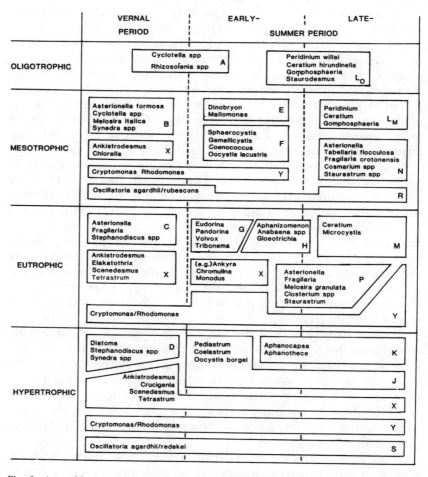

Fig. 1. Assemblages of temperate freshwater phytoplankton (*A—S, X, Y*) and some representative species, one or more of which may grow well, relatively well or become abundant in the types of lake and during the approximate seasons of the year indicated (separated by broken lines). (After Reynolds 1984.)

The annual overturn in monomictic lakes brings hypolimnic nutrients to the euphotic layer and thus contributes to the maintenance of lake productivity. However, it is a catastrophic process in profoundly disturbing the planktonic late-summer community. The community which develops thereafter may then be compared to a pioneer community and the succession of vernal, early-summer and late-summer communities may be compared to a seral succession in the terrestrial vegetation: increase in the maturation of the community organisation, shown by the increase in the number of species, and the shifting from *r*-strategist species (highly productive, fast-growing, investing in reproduction of new individuals) to *K*-strategist species (growing slower, but well adapted to operate close to the resource-determined carrying capacity of the environment) (Reynolds 1980, 1982, Kilham & Kilham 1980, Sommer 1981). Examples of *r*-strategist plankters in temperate lakes are the green-algae *Ankistrodesmus, Elakatothrix, Scenedesmus, Tetrastrum*, and the chrysophycean flagellates *Mallomonas* and *Chromulina*; examples of *K*-strategists are the blue-green *Microcystis* and the dinoflagellates *Peridinium* and *Ceratium* (Reynolds 1984).

The dynamics of the epipelic communities developing on sediments have been well investigated in the lakes of the English Lake District by Round (1957a, 1957b, 1957c, 1960, 1961a, 1961b). The diatoms dominate these epipelic communities which show a general reproductibility from year to year. Their growth commences in early spring and reaches maximum cell numbers in April to May, after which a decline occurs in midsummer prior to a smaller autumnal peak that is over by November. While this pattern is similar to that of planktonic species, the epipelic flora is slower to react. On shallow sediments, the epipelic algal populations tend to exhibit early winter and spring peaks, followed by midsummer maxima. No species extends over the whole year in significant numbers, which was confirmed in a three-years study of the epipelic algae of two small pools in Birmingham, England (Round 1972). For each species and, sometimes even for each variety of a same species, the population increases and declines rapidly.

Fetzmann (1961, 1963) described the epipelic association of *Oscillatoria limosa* and diatoms forming a dark brown carpet on the mud of old arms of the Danube in Austria. Its optimal development occurs from November to March. In March and April, euglenae, especially creeping species (e.g. *Euglena adhaerens*), accompany the dominant species; in May and June, the latter are exposed to the competition of luxuriant flakes of *Conjugatae* and of *Callitriche*. After periods of high light intensity, gas bubbles are formed within the tapetic layer which then may be detached from the bottom, rise to the surface and be transported by water currents.

The seasonal population dynamics of epiphytic algae growing on submersed portions of higher plants are complicated by changes in their substrate as the plants grow, senesce and decompose. In Lake Fureso, Denmark, the epiphytic diatoms growing on the submersed portions of the reed *Phragmites* compete for water silica with the planktonic species, mainly *Stephanodiscus hantzschii*, during the early growth season. Later in summer and fall, the epiphytes apparently utilize the silica from the *Phragmites* which gives them a competitive advantage over planktonic forms. However things are not simple, as it appears that when plankton is abundant, organic substances are excreted by planktonic green algae and diatoms which hold back the development of epiphytic diatoms (Jørgensen 1957).

As for lakes, twelve-month records of the periodicity of freshwater algae from streams have been obtained mostly in Europe and North America, the first studies dating from the beginning of the century. Since the review on the subject by Blum (1956), many new studies have been made for benthic as well as for planktonic stream communities. The results are very diverse and few generalizations can be drawn. Some streams exhibit considerable constancy, whilst others show, even more than lakes, sharp differences in temporal patterns, due to great year-to-year differences in hydrological conditions, and the frequent occurrence of catastrophic events (flash flooding and washout). Broadly, the maximum development of benthic algae occurs mostly in warm months, the minimum in cold months, but many algae behave in different ways. Regarding potamoplankton, many streams are characterized by two separate periods of maximum abundance; mostly, the principal pulse, if there is one, is to be expected at some time during the warm season. Large rivers, like lakes and ponds, may occasionally develop blooms, mostly due to the sudden reproduction of a single species (often a blue-green alga or a diatom). Sörensen (1948) however records a river bloom which involved several members of Volvocales.

In the aphotic ground waters below the active streams of rivers of the Morava Basin, Czechoslovakia, Poulíčkova (1987) observed a varied algal community of allochthonous origin, depending on the colonization of the surface stream; this community also is subject to seasonal variations, showing, at the horizon of 15 cm, a shift compared with the situation on the bottom surface.

Panknin (1941, 1945) has discussed the classification of seasonal communities which assume temporary dominance in a given site, and concluded that they should not be regarded as seasonal associations but alternatively as representing seasonal aspects of the entire association, a view shared by Behre (1966). The diatoms which form a peak in spring, and the green and blue-green algae which develop massively in summer, must well be present the whole year, because they develop according to the same rhythm the next growing season. Consequently, these authors consider that these seasonal aspects may be described as are the phenophases of an association of higher plants (even if they are definitely not homologous).

Another view may be taken. When the succession of the algal vegetation shows a regular shift, like succession from r-dominated to K-dominated assemblages, the seasonally-bound assemblages may be considered as distinct associations. Indeed their occurrence is short only if compared with the absolute duration of the associations of higher plants, but if we take as a time unit the duration of a life cycle (one week or less for the phytoplankton, 5—7 years for a grassland, one or several centuries for a forest), they no longer appear so ephemeral. The only objection made to deny them the quality of associations is their cyclic occurrence; however, even a terrestrial climax association may be abruptly returned to a new pioneer community if terminated by a catastrophic process (fire, drought, flood, or anthropic intervention): and this is just what occurs to a mature planktonic community when it is disturbed by an external catastrophic influence, like the annual destruction of the stratification.

Whether or not to give the seasonal assemblages the rank of association is a rather academic dilemma. If the temporal changes occur periodically with some regularity, the whole assemblage cycle ('Associationszyclus') may be considered as a sociological entity. The main types of such entities correspond to the types of water bodies based on their trophic degree. Symoens (1951b) already recognized three alliances or groups of associations, respectively, for the plankton of physiologically oligotrophic lakes, morphometrically oligotrophic lakes, and eutrophic lakes. Reynolds (1984) proposes a more elaborate system of assemblages showing their approximate sequential relationships among lakes of different trophic status and morphometry (Figure 1).

Seral successions and the problem of climax

Aquatic biotopes undergo an evolution due to the gradual raising of their bottom towards the water surface by silting or by accumulation of organic debris, thus causing their replacement by terrestrial biotopes ('atterrissement', 'Verlandung'). While they become progressively smaller and shallower, the lakes undergo simultaneously a slow natural eutrophication, already suggested, long ago, by Pearsall (1921). Concomitantly, the aquatic associations succeed each other and, according to the classical view of autogenic succession, each aquatic plant association may be considered as a developmental stage of a *hydrosere*, beginning in water and culminating in a climax association, in a more or less stable equilibrium under the regional climate, often a forest one. An example of such a hydrosere for Western Europe could be represented by the following simplified scheme:

Forest	Quercetum roboris
	↑
Fen wood ('carr')	Alnetum glutinosae
	↑
Fen	Cladietum marisci
	↑
Reedswamp	Scirpo-Phragmitetum australis
	↑
Vegetation with floating leaves	Potameto — Nupharetum lutei
	↑
Submerged vegetation	Ceratophylletum demersi

A possible parallelism between the succession of higher plant associations and algal associations was emphasized by Symoens (1957) for the fens and bogs of the Ardennes and neighbouring districts (Figure 2).

In fact, the view that the vegetation necessarily evolves towards a climatic climax is an oversimplified one. It is now broadly accepted that several climax types may co-exist under the same regional climate.

Some lakes evolve so slowly that — their shallow bays and areas of intense silting excepted — they may be treated as systems as stable as the terrestrial climax formations. In such cases, the succession in the lake culminates in a practically permanent aquatic association.

If some causes preventing the realization of a climax have the same degree of permanence as the climate tending to produce it, this type will not be realized. Since long, it has been shown that the vegetation of raised bogs is in dynamic equilibrium with a complex of climatic, edaphic and topographic factors (Godwin & Conway 1939).

Further, even in natural conditions, e.g. in swampy areas, the succession may be set back to an earlier seral stage by internal processes, like subsidence due to compaction and de-watering of underlying clays (Breen *et al.* 1987).

Superimposed upon the mostly unidirectional autogenic changes are other reversible variations. Examples are the long-term climatic fluctuations which may affect the hydrological and salinity conditions in water basins. Such are also the human activities causing increased nutrient supply, i.e. the man-made eutrophication or cultural eutrophication, stimulating biological production, truncating the euphotic zone and increasing hypolimnetic oxygen uptake and sediment anoxia (see Reynolds 1984). These alterations of the environmental parameters

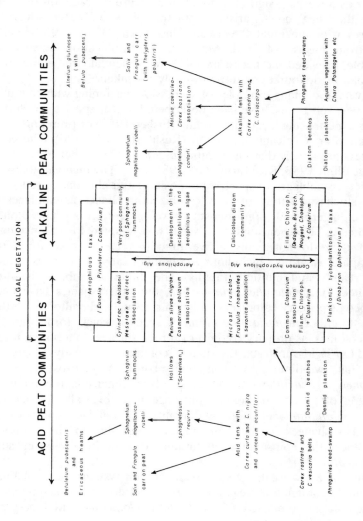

Fig. 2. Succession of fen and bog plant communities of the Ardennes and neighbouring areas. (After Symoens 1957.)

induce long-term changes in the composition of the planktonic associations (see *inter alia* Lund 1972a, 1972b, 1978). The man-induced occurrence of blooms of blue-greens, with all its injurious consequences, is now unfortunately a widespread process. Vallentyne (1974) considers it as a warning sign of a threat of a severe degradation of water quality, the "algal bowl", arising from the misuse of water, which he compares with the great American "dust bowl" of the 1930s which arose from misuse of land! Man-made eutrophication however is reversible, provided the necessary measures, essentially the elimination of the man-made sources of nutrient supply, are taken.

In rivers, too, many examples of succession, beginning with colonization of new substrates, have been described. Pioneer species develop in streams experiencing frequent flooding (Fisher 1983) or presenting new substrata after a volcanic explosion (Rushfort *et al.* 1986). The early colonizers are mostly diatoms (Cattaneo *et al.* 1975, Fisher *et al.* 1982, Korte & Blinn 1983, Rushforth *et al.* 1986, etc.). The high variability of colonizer assemblages in species composition and abundance reflects a certain stochastic nature of early colonization: chance assemblages of early colonizers and opportunistic development in an unexploited environment. However some taxa seem well adapted for colonization, even if they do not really arrive first: this is the case for some monoraphid diatoms, including *Achnanthes* and *Cocconeis* which attach readily and multiply rapidly. The initial assemblage is often dominated by *Achnanthes minutissima*, but other species have been described as primary colonizers: rather surprisingly, Hamilton & Duthie (1984) found *Tabellaria flocculosa*, a species which forms upright zigzag colonies, as primary colonizer on rock surfaces in a boreal forest stream. On rigid substrata, the diatom colonization mostly progresses from a flat, two-dimensional community consisting of prostrate forms (*Cocconeis, Achnanthes*) to an erect, three-dimensional community consisting of species possessing mucilage pads or stalks (*Fragilaria, Synedra, Gomphonema*). On sand grains, the relative abundance of some taxa changes in the same manner, but due to the continual disturbance, the epipsammic community is kept in a pioneer state (Miller *et al.* 1987).

Following a few weeks of succession, stream systems show relatively more stable communities: e.g. blue-green algal dominated assemblages, or assemblages composed of *Cladophora glomerata* and blue-green algae with epiphytic diatoms.

There has been much controversy over the possibility of recognizing algal climax associations. Again this is an academic problem, well summarized by Lhotský (1985).

If we start from the time rhythm given by the ontogenic cycle, we can decide that an algal community without changes in the course of many generations could be characterized as a climax stage. This is certainly the case for the *Spirulinetum platensis* of the African alkaline-saline lakes (soda lakes) where biomass and species composition were found practically constant for months and sometimes even for years (Vareschi 1982; see also Melack 1987).

On the other hand, it is also possible not to search for a climax stage in one association, but in a sequence, in a group of repeating associations, in an association cycle. We emphasized previously the existence of such regular cycles, which then could be treated — on the human time scale — as climaxes.

METHODS OF ALGAL SOCIOLOGY

Sampling

The methods of sampling are very different in the phytosociology of macrophytic and of microphytic vegetation.

The choice of the limits of a phytosociological 'relevé' requires homogeneity and minimum area. For macrophytic associations where the stands can be visualized, this choice raises but limited difficulties. Further, a limited number of visits of the considered plot is generally sufficient.

When dealing with microscopic algae, the association individuals mostly cannot be visualized, which means that the possible horizontal and vertical heterogeneity of the biocoenose will escape the observer's eye. Further, the rapidity of the temporal changes mostly requires repeated sampling at the same place.

The plankton of a lake shows horizontal patchiness, a very complex phenomenon (see Reynolds 1984, pp. 105—112), but sufficient to make a sample far from representative of the whole planktonic association. In case of stable stratification, different algal associations are present in the epilimnion, the metalimnion, and the hypolimnion; further, some algae have well developed ability to regulate their vertical position and to perform vertical migrations (e.g. by buoyancy control in *Microcystis* or by motility in *Ceratium*). Consequently, sampling must be done at different depths in order to obtain a correct view of the composition and the structure of the planktonic community. And finally the rapid temporal changes due to both the short duration of an algal generation and the active or passive movements of the planktonic algae require frequent sampling: planktonic communities cannot be described from infrequent sampling as can most land associations; even "the analysis of dominance should extend (preferably) over a period of years since it will then be possible to distinguish the species which re-appear each year from those which are casuals" (Round 1981).

There is no doubt that the spatial heterogeneity is even greater in communities of attached algae. So, the different faces of a pebble in a torrent bear different epilithic algal assemblages (Cazaubon & Loudiki 1986). In such cases, any structural approach should try to identify the smallest microhabitats and separate their algal assemblages. Our usual methods of sampling algae by pressing wet mosses or scraping sub-merged macrophytes or boulders resemble the sharpening of a pencil with a bulldozer! The advent of scanning electron microscopy (SEM) has permitted a much closer examination of the structure and hetero-geneity of intact communities and revealed distinct microdistribution patterns (Hudon & Bourget 1981, Lamb & Lowe 1981, Hoagland *et al.* 1982, Korte & Blinn 1983). Using SEM, Pringle (1985) showed that in Carp Creek, Cheboygan County, Michigan, chironomid tube-building activities affect diatom microdistribution: a substratum without chironomid tubes is primarily colonized by *Achnanthes minutissima* and *Cocconeis placentula*, exhibiting the lowest species diversity of microhabitats examined; the diatom flora upon sand tubes of *Micro-psectra* is dominated by *Opephora martyi*, as is the flora of sand grains from the stream sediment, whilst on the tubes of *Pseudodiamesa, Nitzschia* and *Navicula* spp. dominate the upperstory and *O. martyi* is located on underlying sand grains.

Identification and assessment of abundance

Algal samples must be considered as equivalent to the sampled areas treated by the phytosociologists. Their analysis is the necessary basis of algal sociology and requires the assessment of their floristic composi-tion by identifying and listing all the taxa present (not only those of a sole taxocene) and the evaluation of their respective abundance.

Initially, abundance was subjectively rated from microscopic exami-nation of the samples and then expressed according to a scale of abundance levels, e.g. very abundant, abundant, rare, very rare, isolated. Such visual evaluation of a species abundance within a sample includes a large degree of unconscious error.

Consequently, ecologists have become increasingly aware of the necessity of using more objective methods based on quantitative data to describe communities. In algal sociology, this implies the counting of cells. In the case of plankton, the method has been made accurate by Utermöhl (1958) and the results may be referred easily to a defined volume of water. In the case of benthic algae, i.e. epiphytic, epilithic, epipelic and epipsammic algae, large variations in the density of algae may occur over quite short distances and replication in sampling is highly advisable. One mostly counts 500 individuals and the results for each species are given as percent of the whole: this may be satisfactory for the description of algal assemblages and assignment to syntaxonomical units. But for the study of the dynamics of algal populations and of temporal changes in the communities, quantitative samples and absolute estimates are necessary (see some methods in Sládečková 1962, Marker & Bolas 1984).

Cell numbers however give but a poor idea of the biomass of the constituent species, because of the very different size of the algal cells: the calculated volume of a cell is about 4 μm^3 for *Aphanothece delicatissima*, 40 μm^3 for *Tetraedron minimum*, 400 μm^3 for *Cyclotella comensis*, 4000 μm^3 for *Closterium aciculare*, 20 000 μm^3 for *Staurastrum paradoxum*, 70 000 μm^3 for *Ceratium hirundinella* (see the tables of cell volumes of bog species in Fetzmann 1956, pp. 732—744; of planktonic species in Nauwerck 1963, pp. 94—97, Findenegg 1969, p. 18, Wetzel 1975, tables 14—6, and Reynolds 1984, pp. 20—23). Consequently, the abundance of a species expressed as its total volume in a sample, should be computed by multiplying the specific cell volume by the number of cells encountered. The problem is somewhat more complicated in the case of colonial or filamentous algae.

Criteria for the recognition and classification of algal associations: character-species or dominant species?

The definition of an association proposed by Flahault & Schröter (1910) gave a more or less balanced importance to the floristic composition, the physiognomy and the ecology.

However the phytosociologists of the school of Braun-Blanquet developed a method where floristic composition is the criterion of primary emphasis. Of paramount importance for recognition of an association is the search for faithful or character-species, i.e. diagnostic species of high fidelity, and of differential species which apparently occur more preferably in the considered association than in other comparable communities. The basic documents are the 'relevés', i.e. the species lists, completed with data on the way of occurrence (abundance, etc.) of each species within the association individual. The analytical characters of the association are those which can be described in one association individual, i.e. those which should be assessed in every relevé: abundance-dominance, sociability, frequency and vitality of the constituent species. The synthetic characters are those which are computed by consideration of all relevés of the association or comparison with those of other associations: presence, constancy, fidelity. In spite of the importance given to fidelity, it has been recognized, even by protagonists of Braun-Blanquet's method, that few species have a total fidelity to one association, and, consequently, that, next to fidelity, dominance and constancy may be usable criteria to define associations. The adaptation of the concepts and the method of Braun-Blanquet to algal sociology has been formally developed by Fetzmann (1956) and Schmitz (1965).

Once recognized, associations are then grouped into higher units, essentially by floristic composition; on the other hand, several categories of differential species groups are used by Duvigneaud (1946) to assess the variations of the associations, i.e. units of lower rank. These principles were largely applied to a hierarchical classification of the algal communities by Symoens (1951b).

A critical analysis of the method of Braun-Blanquet has been made by Whittaker (1962) who pointed out some weaknesses in its theoretical basis and the subjectivity of the approach at several levels, but nevertheless emphasized its merits and achievements and considered the system as "the most successful and most widely applicable means of formal classification of vegetation". Algal associations, as recognized by this method, proved to be useful references for the biological evaluation of waters.

However, algal assemblages may be impoverished or randomly composed and appear as fragmentary, atypical, mixed or transitional when compared to the recognized associations of a formal phytosociological system. Many communities cannot be referred to such associations nor described as such, although they can possibly be assigned to syntaxonomical units of higher rank.

A different approach to the study of plant communities has been developed largely by British authors, initially under the influence of the work of Warming (1895) and its English translation (1909), later in the works of Tansley or inspired by his tradition: recognition of community-types by physiognomy (especially at higher levels) and by dominance (especially at lower levels) and emphasis on vegetation dynamics. The method is characterized by a relative informality, which keeps it rather flexible and better adapted to the description of any algal community. On the other hand, the fact that the dominant taxa are often euryecious and cosmopolitan limits their value in typification: the method is not decisive for this approach, and unsuitable for the elaboration of a world-wide hierarchical classification of communities.

Assessment of species association by multivariate analysis

Until the sixties, recognition and classification of algal associations were achieved by intuitive comparison and manual arrangement of floristic lists of samples, preferably accompanied by some indication of the relative abundance of the species. The associations were named as stated above, after characteristic (if any), differential or dominant species, not without some subjectivity from the authors.

Since the beginning of the sixties, a general trend has appeared in biocenology, based upon the processing of quantitative data by means of multivariate analysis, i.e. that branch of the mathematics that deals with the examination of numerous variables simultaneously. Due to increased access to computers, multivariate analysis made it possible for ecologists to treat efficiently community data involving numerous species, numerous environmental parameters and numerous samples. The application of multivariate analysis techniques is now general in the study of plant and animal communities: see the excellent reviews by Orlóci (1978), Whittaker (1978a, 1978b), Legendre & Legendre (1979), Gauch (1982), Greig-Smith (1983), Orlóci & Kenkel (1985).

Species association: mostly a matrix of association indexes is built up between species, from which clustering or ordination procedures are carried out.

Numerous ways of expressing the degree of association between two species are available (see Mueller-Dombois & Ellenberg 1974, Goodall 1978). Widely used is the phi (ϕ) coefficient, applied e.g. in the study of the attached algal vegetation of the streams of Jämtland, Sweden, by Johansson (1982):

$$\phi_{(X, Y)} = \frac{A \cdot D - B \cdot C}{\sqrt{(A + B)(A + C)(B + D)(C + D)}}$$

where X and Y are the two algae to be compared; A is the number of streams where both algae have been found; B is the number of streams where alga X has been found, but not Y; C is the number of streams where alga Y has been found, but not X; and D is the number of streams where neither X nor Y have been found.

Cluster analysis: cluster analysis (*sensu lato*) is a procedure which assigns objects to groups in such a way that each group reveals some distinct characteristic of the sample. In all clustering procedures, it is essential that the entities within a group show more likeness to each others than to entities of other groups. Assigning objects presenting some similarity to a same group is the basic process of classification. Mere recognition of clusters of points representative of samples produces a non-hierarchical classification: this corresponds mostly to the first step of the treatment of phytosociological relevés and facilitates recognition of vegetation units. Additional arrangement of the clusters into higher units produces a hierarchical classification.

Williams (1971), Goodall (1978), Orlóci (1978) reviewed the numerous clustering techniques. Cluster analysis has been repeatedly applied to algal communities (Symons 1970, McIntire 1973, Colyn &

Koeman 1975, Hillebrand 1977, Johansson 1982, Caljon 1983, Fabri & Leclercq 1984, Dixit & Dickman 1986 etc.).

Widely used for structuring phytosociological tables is the TABORD programme (van der Maarel *et al.* 1978): it is essentially a clustering procedure based on relevé (sample) similarity combined with a repeated relocation procedure for obtaining the larger data matrix values concentrated in blocks along the diagonal of the table. This programme has been used by Johansson (1982).

Ordination: the term ordination was coined by Goodall (1954a, 1954b) to mean the arrangement of units in a uni- or multidimensional order. Ordination techniques are thus used in vegetation research in order to extract from suitable raw data an arrangement of vegetation units in series, in contrast with classification which extracts an arrangement in classes. This arrangement in series reveals usually useful information about the relationships of composition between communities or with environmental variables (see Orlóci 1978, pp. 102 *et seq.*, Whittaker 1978b). Several ordination techniques have been applied to algal ecology.

Factor analysis (FA) was introduced at a rather early stage of the application of multivariate techniques to vegetation analysis (Dagnelie 1960; see also Dagnelie 1975). It has been applied by Symons (1970, 1973) to the study of the ecological relations between species of algae of two Belgian ponds and to the changes in the structures of their algal populations, and by Johansson (1982) to the study of the stream algae of Jämtland.

Principal component analysis (PCA) has been the most commonly used ordination technique in sociology of freshwater algae: it has been applied by Estrada (1972), Reyssac & Roux (1972), Allen & Koonce (1973), McIntire (1973), Descy (1973, 1976a), Johannson (1982), Fabri & Leclercq (1984), Claflin (1987), etc.

These methods proceed by building up an association matrix (usually a correlation or a covariance matrix) between species, each having a score in the set of samples; then, a calculation is made, by means of matrix methods, which enables some 'factors' or 'major axes' to be extracted; these factors correspond to general trends contained in the original data set, and so may have a particular ecological meaning. Species, and then samples, are given scores for these factors, so that their projection on the axes, in a geometrical representation, may present particular patterns and show the linear relationships between variables (species and ecological factors) as well. Such diagrams (ordination scattergrams or plots) provide a general view of the data which is easy to interpret when one main environmental factor is decisive, but very difficult, if not impossible, to obtain from a large data set, with numerous species, ecological parameters and samples.

The reason why PCA became so popular in phytosociology is obvious: the analysis is based on relationships between species or relevés (samples in microphyte phytosociology). In other words, species or samples are arranged according to their ecological similarities.

If we consider only the two first components, two diagrams can be built through PCA to represent the space of these two factors: a species diagram and a sample diagram, which can be interpreted in the same way, but cannot be superimposed to each other.

Alternatively, the reciprocal averaging (RA), also called correspondence analysis, or factor analysis of correspondence (FAC) from the 'analyse factorielle des correspondances' developed a.o. by Benzécri (1973), allows all the points — species and samples — to be represented on the same diagram: for instance, species are placed close to the samples where they reach their optimal score. This feature of correspondence analysis is obviously an outstanding advantage for studies on community structure: species with similar patterns form clusters, which are placed in, or near, clusters of samples with similar species composition.

Furthermore, correspondence analysis provides useful rules for interpretation and allows the processing of many kinds of data, as far as they are properly transformed, without particular requirements usually needed by other methods, e.g. normal distribution of the data. In aquatic ecology, the work of Verneaux (1973) is certainly one of the clearest examples of the value of correspondence analysis in studies of freshwater communities (invertebrates, fish). In algal ecology, RA has been used by Coste (1974), Le Cohu (1974), Descy (1975, 1976b, 1979), Johannson (1982), Gasse et al. (1983). RA applied to the diatom assemblages of running waters in the Ardennes and neighbouring districts, confirmed the basic typology recognized by Symoens (1954, 1957); it showed a series of diatom assemblages from the acid to the alkaline waters, the fundamental 'biotypes' I, III, V, being in close correspondence with the associations previously described (see Figure 3).

RA has still the two main faults of the eigenanalysis methods: the second axis is often an 'arch' or 'horse-shoe' distortion of the first axis, and distances at the ends of the first axis are compressed relative to the middle. Detrended correspondence analysis (DCA) is an improved technique based on RA but correcting these two main difficulties (Hill 1979, Hill & Gauch 1980). The arch effect is removed by adjusting the values on the second axis in successive segments by centring them to zero mean. The variable scaling on the first axis is corrected by adjusting the variance of species scores within the samples to a constant value. The DCA programme DECORANA has been applied to the algal communities of the Flemish lowlands, Belgium, by Caljon (1983).

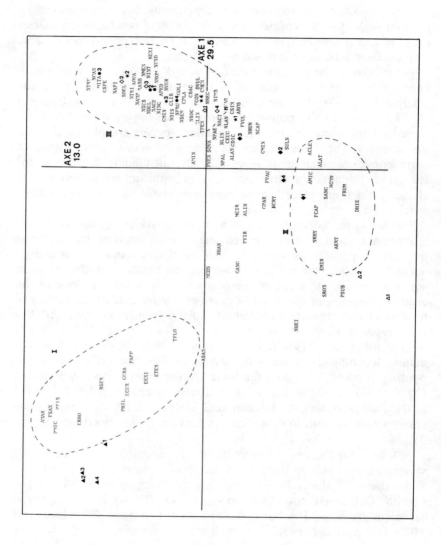

Fig. 3. Factor analysis of correspondence (FAC or RA) on diatom samples collected in the River Meuse basin, Belgium (1983–1984). Six sites (symbols) were sampled at four seasons (1 = spring 1983, 2 = summer 1983, 3 = autumn 1983, 4 = spring 1984);

	mean pH	mean alkalinity (mg/l CaCO$_3$)
Solid triangles ▲	4.4	0.8
White triangles △	6.1	6.7
Solid lozenges ◆	7.2	39.3
White lozenges ◇	7.8	82.8
Circles ●	8.1	222.0
Stars ★	7.9	161.4

Diatom species are identified by four letters. The delimited groups correspond to the associations recognized by Symoens (1954): I. The *Eunotia exigua*-association, dominated by *Eunotia* spp. (e.g. EEXI = *Eunotia exigua*) and other acidophilous taxa; II. The *Diatoma hiemale—Meridion circulare*-association, well characterized by *Diatoma hiemale* var. *mesodon* (DHIE); III. The *Diatoma vulgare—Melosira varians*-association, with many alkaliphilous taxa, like *Diatoma vulgare* (DVUL). The taxa not included in the ellipses are euryecious (taxa near the origin of the axes) or belong to transitional groups.

Multivariate analysis is particularly useful for the study of algal community structure, perhaps more than in any other field of phytosociology, because of the high variability of algal communities in water bodies, resulting from the rapid time changes in their composition and in the environmental factors. Allen & Skagen (1973) gave a demonstration, based on simple geometry that multivariate analysis may be easily understood by means of simple analogies, showing how multidimensional problems are reduced to two or three dimensions. They also show the possibility of emphasizing different aspects of the information, simply by changing the form of the original data (growth rate vs. scores, temporal vs. spatial distribution, environmental parameters, etc.).

However, in spite of their increased objectivity, users of different multivariate analysis techniques have not always recognized the same algal communities from the same data set. Nevertheless, the results yet obtained are promising, considering that most of the work spans less than the past two decades. Multivariate analysis evolves rapidly and there is no doubt that its achievements in algal phytosociology will still be improved by the development and use of new techniques, e.g. the non-metric multidimensional scaling (Shephard & Carroll 1966).

BIOLOGICAL ASSESSMENT OF WATER QUALITY BASED ON ALGAL COMMUNITIES

Ecology was initially defined as the science that considers the relations of living organisms to their environment. But it was known long before the word ecology was created that certain species are characteristic of certain habitat types; in other words, their presence indicates something about the properties of the environment in which they are found. Thus, the concept of the presence of indicator species revealing certain environmental conditions is based on common sense observations.

In principle, every organism has a possible indicator value. Indeed, each species occurs only where several factors have values comprised in a suitable interval. Thus the ecological range ('domaine écologique') of a species is the multifactorial interval which its presence indicates (Gounot 1969, p. 164).

Stenecious species, i.e. those which can live only in a narrow value interval of an environmental factor (e.g. salinity, pH, calcium, flow velocity, heavy metals, etc.) have consequently a high indicator value and may often be used efficiently as indicator species or bio-indicators.

It is not surprising that biologists attempted to estimate the pollution intensity, using the indicator species concept. The classical example of this is the system of saprobic organisms of Kolkwitz and Marsson (1908, 1909), subsequently revised and extended by Liebmann (1951,

1962). The organism lists associated with this system comprise numerous algae as well as macrophytes and animals.

However the user of the indicator species concept may be exposed to two faults: (i) the presence-absence criterion is far too simple, as it does not consider the degree of abundance or vitality of the species within their ecological range; (ii) different species, even sensitive to the same discriminant factor, may have very different indicator-value: stenecious taxa are much better bio-indicators that euryecious. Consequently, the biological estimation method of water quality gave progressively greater consideration to the abundance of the taxa (Pantle & Buck 1955) and to the level of their indicator value (g for 'Gewicht', i.e. indicative weight, in the system of Zelinka & Marvan 1961, improved 1963).

The presence of certain groups of species may often provide a much more effective indication of environmental factors than single species, even abundant or dominant. Fjerdingstad (1964) has clearly emphasized the advantages of a system based on the community level, and proposed a phytosociological system using the sessile algae and bacteria communities for assessing the degree of organic pollution in streams. Examples are the γ-polysaprobic zone with the *Oscillatoria chlorina*-community and the *Sphaerotilus natans*-community; the α-mesosaprobic zone with the *Ulothrix zonata*-community (pollution type with *Oscillatoria* and *Phormidium* spp.), the '*Oscillatoria* benthonicum-community', the *Stigeoclonium tenue*-community, etc. Fjerdingstad (1964) considered the diatoms as rather unreliable pollution-indicators, and he excluded them almost completely from his system, only recognizing a 'pure *Meridion circulare*-community' in the oligosaprobic zone (this meaning that the community can only be assigned to the oligosaprobic zone in the absence of saprophilic and saprobiontic species).

Sládeček (1961) introduced the term system of saprobity in order to extend the previous saprobic system both to biotic and abiotic levels. The value of the saprobity approach is the relatively quick classification of communities which then allows for the identification of zones of impact and recovery from organic pollution, and for deciding on potential water uses. The saprobity system is widely used in Central Europe and has its defensors (see the reviews by Sládeček 1973, 1979). Despite its improvements, it suffers three main drawbacks: (i) consideration on the same scale of all constituents of the biocenosis, whilst bacteria, algae, macrophytes, invertebrates, fishes have very different sizes and generation times and, consequently, should be treated separately; (ii) insufficient knowledge on the ecological characteristics and requirements of many 'indicator' taxa; (iii) confusion for many algal species of their saprobity level with their trophic status.

Therefore, alternative approaches to the saprobity system were developed, based on changes in structure and/or composition of more homogeneous assemblages of organisms, often taxocenes.

The alterations which can occur within a community exposed to water pollution are: changes in the species list, changes in the number of species, changes in the number of individuals per species, and changes in the total number of individuals. All the biological approaches of pollution assessment are based on some or all of these criteria.

Changes in the structure of diatom communities related to water quality have been described by Patrick and her associates (Patrick 1949; Patrick *et al.* 1954). In a natural situation, the curve resulting from a plot of number of species vs. number of individuals per species has the characteristic shape of a log-normal distribution. The response of the community to pollution results in change of shape of the curve, related to the decrease in number of species (elimination of sensitive taxa) and to the increase in number of individuals of a few tolerant species. These structural modifications can also be expressed by diversity indices (e.g. the Shannon-Weaver index) based on the same criteria.

However, it was demonstrated by Archibald (1972) that, for diatom communities, there is no simple relationship between water pollution and diversity. Actually, the range of diversity as well in mildly polluted water as in pure water can be quite wide, as environmental variables other than pollution can influence the community structure. For instance, the structure of a diatom assemblage in acid or in fast-running streams is similar to that observed in very polluted situations. As a consequence, pollution-related diversity changes should only be assessed from a reference structure observed in a similar, but unpolluted situation. The fluctuation index, developed from the stability-diversity concept, can meet this requirement (Kiehm *et al.* 1981).

Alternatively to the structure-based assessment of water quality, changes in floristic composition (i.e. changes in the species list and in the number of individuals per species) have been used in several similar ways for diatoms. The different systems developed use a classification of taxa according to their tolerance to pollution, as well as changes in relative abundance of the taxa of the community. As stated above, pollution causes a decrease of sensitive taxa vs. an increase of tolerant taxa; in extreme conditions, only a few very resistant taxa survive and they usually develop high relative densities.

Lange-Bertalot (1978, 1979), working in the Rhine-Main system (West Germany), assessed that the diatoms can be used as differential species, but not as character-species of the different water quality levels. They have no limits of tolerance towards lower saprobity levels (all species can be found in relatively high numbers of individuals under oligosaprobic conditions), but, on the contrary, have well-defined limits of tolerance towards higher saprobity levels (increase in the saprobity causes the successive elimination of several species groups and replacement of a dominant species group by another one composed of more resistant species). Among stream diatoms, Lange-Bertalot recognized indeed three main groups according to their degree of tolerance.

Two methods based on the multivariate analysis of benthic diatoms communities were developed in the seventies.

Coste (1974, 1978) made a study of the River Seine, France; correspondence analysis allowed him to identify different species groups. As increasing pollution was the main factor accounting for the succession of diatom communities, a classification with regard to pollution sensitivity was possible, as well as distinction between steno- and euryecious taxa. However Coste did not clearly take into account the part of the variation due to natural changes along the river.

The studies carried out at the same period by Descy (1975) in the Belgian basin of the Meuse River yielded similar results, at least as a first step in dealing with two relatively large rivers (Sambre and Meuse), where different degrees of pollution occur: Descy gave a full description of the changes in the basic diatom communities, as well as a classification of the degree of tolerance of about 50 taxa, resulting from processing of quantitative data by PCA.

However the development of monitoring of water pollution based on diatoms, which could be applied to any type of river, requires a preliminary typology of the communities of clean waters in the area considered. We mentioned above the typology assessed for the waters of the Ardennes and neighbouring districts, which served as a reference guide for a study of the changes caused by pollution and an estimation of the water quality in this area (Descy 1979).

The results of these studies, together with the results obtained by other authors, allowed the classification of benthic diatoms according to their degree of sensitivity to pollution, which is summarized elsewhere (Descy *et al.*, in press). Although these works were intended primarily to assess sensitivity indices for the most frequent diatom taxa, it is possible to outline as follows the community changes associated with water pollution:

(a) in slightly to mildly polluted waters, the characteristic species of the different types of clean waters, mostly sensitive towards pollution are progressively replaced by taxa with a wider ecological range, growing in eutrophicated water, such as *Achnanthes lanceolata, Fragilaria vaucheriae, Melosira varians, Navicula cryptocephala, N. exilis, N. gregaria, Nitzschia acicularis, N. linearis, N. paleacea*, etc.

(b) As pollution increases, the species listed above become dominant, and large numbers of more tolerant diatoms may develop, such as

Gomphonema parvulum, Navicula trivialis, Nitzschia amphibia, N. palea, N. filiformis, Surirella ovata, Synedra ulna, etc. Most of these taxa exhibit a large ecological spectrum: they could be referred to as saprophilic in the terminology of the saprobity system.

(c) In heavily polluted waters, only these latter taxa are present, together with the most resistant diatoms; some of them are supposed to be heterotrophic for carbon and nitrogen: *Amphora veneta, Navicula accomoda, N. goeppertiana, N. veneta, Pinnularia brebissonii.*

The 'small Naviculae' (*Navicula atomus, N. cloacina, N. frugalis, N. minima, N. seminulum, N. saprophila,* etc.) are also frequently well-developed under these extreme conditions. These very small diatoms are typical *r*-strategists and are able to develop very high densities in heavily polluted sites, but also in aerophilic conditions, so that their density may largely depend on sampling depth. Their occurrence alone cannot thus indicate a distinct level of saprobity.

As shown previously (Descy 1979), these changes are clearly distinct from the natural succession of the diatom communities associated with the increasing enrichment from acid to alkaline waters, or from oligotrophic to eutrophic waters.

So the development of general systems of water quality monitoring based on algae requires, as for other taxocenes, a reference system which should be obtained by the study of natural communities. The studies in algal sociology, together with the applied research on polluted waters, are a necessary step in order to achieve monitoring methods which could be used over large areas.

ACKNOWLEDGEMENTS

The authors are particularly indebted to Dr. Mary J. Burgis for her kind assistance in the preparation of the English version of this chapter. They are very grateful to Prof. Dr. D. Roggen and D. Van Speybroeck for helpful comments and stimulating discussion. They wish to express their thanks to Blackwell Scientific Publications Ltd. for permission to reproduce Figure 1 from Reynold's paper in *Freshwater Biology.*

REFERENCES

Allen, T. F. H. & Koonce, J. F. (1973) Multivariate approaches to algal stratagems and tactics in systems analysis of phytoplankton. Ecology 54: 1234—1246.

213

Allen, T. F. H. & Skagen, G. (1973) Multivariate geometry as an approach to algal community analysis. Br. phycol. J. 8: 267—287.

Allorge, P. (1921—22) Les associations végétales du Vexin français. Rev. gén. Bot. 33: 481—544, 589—652, 708—810, 34: 71—79, 134—144, 178—191, 251—256, 311—319, 376—383, 425—431, 471—480, 519—528, 564—576, 612—639, 676—701.

Allorge, P. (1925) Sur quelques groupements aquatiques et hygrophiles des Alpes du Briançonnais. In: Brockmann-Jerosch, H (red.), Festschr. C. Schröter. Veröffentl. geobot. Inst. Rübel Zürich 3.H.: 108—125.

Archibald, R. E. M. (1972) Diversity in some South African diatom associations and its relation to water quality. Wat. Res. 6: 1229—1238.

Barkman, J. J. (1958) Phytosociology and ecology of cryptogamic epiphytes. Van Gorcum, Assen.

Barkman, J. J. (1969), Epifytengemeenschappen. In: Westhoff, V. & den Held, A. J., Plantengemeenschappen in Nederland. Thieme. Zutphen, pp. 272—286.

Behre, K. (1956) Zur Algensoziologie des Süsswassers (unter besonderer Berücksichtigung der Litoralalgen). Arch. Hydrobiol. 62(2): 125—164.

Benzécri, J. P. (1973) Analyse des données. Tome 2: Analyse factorielle des correspondances. Dunod, Paris, 619 pp.

Blum, J. L. (1956) The ecology of river algae. Bot. Rev. 22(5): 291—341.

Boudouresque, C. F. (1971) Méthodes d'étude qualitative et quantitative du benthos. Téthys 3: 79—104.

Breen, C. M., Rogers, K. H. & Ashton, P. J. (1988) Vegetation processes in swamps and flooded plains. In: Symoens, J. J. (ed.), Handbook of vegetation science, vol. 15. Vegetation of inland waters. Dr W. Junk. This volume, pp. 223—247.

Budde, H. (1928) Die Algenflora des Sauerländischen Gebirgsbaches. Arch. Hydrobiol. 19: 433—520.

Burgis, M. J. (1971a) An ecological study of the zooplankton in Lake George, Uganda. Ph.D. thesis, Univ. of London.

Burgis, M. J. (1971b) The ecology and production of copepods, particularly Thermocyclops hyalinus, in the tropical Lake George, Uganda. Freshw. Biol. 4: 535—541.

Butcher, R. W. (1940) Studies in the ecology of rivers. IV. Observations on the growth and distribution of the sessile algae in the River Hull, Yorkshire. J. Ecol. 28: 210—233.

Butcher, R. W. (1946) Studies in the ecology of rivers. VI. Algal growth in certain highly calcareous streams. J. Ecol. 33: 268—283.

Butcher, R. W. (1947) Studies in the ecology of rivers. VII. The algae of organically enriched water. J. Ecol. 35: 186—191.

Caljon, A. (1983) Brackish-water phytoplankton of the Flemish lowland. In: Dumont, H. J. (ed.), Developments in Hydrobiology. Junk, the Hague, etc., VIII + 272 pp.

Cattaneo, A., Ghittori, S. & Vendegna, V. (1975) The development of benthonic phytocoenoses on artificial substrates in the Ticino River. Oecologia 19: 315—327.

Cazaubon, A. & Loudiki, M. (1986) Microrépartition des algues épilithiques sur les cailloux d'un torrent corse, le Rizzanèse. Annls. Limnol. 22(1): 3—16.

Chodorowski, A. (1959) Ecological differentiation of turbellarians in Harz-Lake. Polskie Archwm Hydrobiol. 6: 33—73.

Cholnoky, B. J. (1953) Diatomeenassoziationen aus dem Hennops-rivier bei Pretoria. Verhandl. zool.-bot. Ges. Wien 93: 134—149.

Cholnoky, B. J. (1955) Diatomeengesellschaften aus den Donauauen oberhalb von Wien. Verhandl. zool.-bot. Ges. Wien 95: 76—87.

Cholnoky, B. J. (1966) Diatomeenassoziationen aus einigen Quellen in Südwest-Afrika und Bechuanaland. Beih. Nova Hedwigia, 21: 163—273.

214

Cholnoky, B. J. & Schindler, H. (1953) Die Diatomeengesellschaften der Ramsauer Torfmoore. Sitzgsber. österr. Akad. Wiss., math.-nat. Kl. Abt. 1, 162: 597—624.

Claffin, L. W. (1987) Associations between the phytoplanktonic and physicochemical regimes of Lake Michigan. Arch. f. Hydrobiol., Beih. 25 (Ergebn. Limnol. 25): 97—121.

Coesel, P. F. M. (1975) The relevance of desmids in the biological typology and evaluation of fresh waters. Hydrobiol. Bull. (Amsterdam) 9(3): 93—101.

Colyn, F. & Koeman, R. (1975) Das Mikrophytobenthos der Watten, Strände und Riffe um den Hohen Knechtsand in der Wesermündung. Forschungstelle für Insel- und Küstenschutz 26: 53—83.

Coste, M. (1974) Etude sur la mise au point d'une méthode biologique de détermination de la qualité des eaux en milieu fluvial. Trav. Division Qualité des Eaux, Pêche et Pisciculture, C.T.G.R.E.F., Paris, 79 pp. (mimeogr.).

Coste, M. (1978) Sur l'utilisation des Diatomées benthiques pour l'appréciation de la qualité biologique des eaux courantes. Mém. Doct. 3e cycle, Univ. Franche-Comté (Besançon), 145 pp.

Coste, M. & Ricard, M. (1982) Contribution à l'étude des diatomées d'eau douce des Seychelles et de l'île Maurice. Cryptogamie: Algologie 3(4): 279—313.

Dagnelie, P. (1960) Contribution à l'étude des communautés végétales par l'analyse factorielle. Bull. Serv. Carte phytogéogr., sér. B 5: 7—71, 93—195.

Dagnelie, P. (1975) Analyse statistique à plusieurs variables. Presses agron. Gembloux, Gembloux, 362 pp.

den Hartog, C. (1983) Synecological classification of aquatic plant communities. Colloques phytosociol. 10: Les végétations aquatiques et amphibies (Lille, 1981): 171—182.

Denis, M. (1924) Observations algologiques dans les Hautes-Pyrénées. Rev. algol. 1(1): 115—126, (2): 248—266.

Denis, M. (1925) Essai sur la végétation des mares de la forêt de Fontainebleau. Ann. Sci. nat., Bot., 10e sér. 7: 5—163.

Descy, J. P. (1973) La végétation algale benthique de la Meuse belge et ses relations avec la pollution des eaux. Lejeunia, nouv. sér., 66, 62 pp.

Descy, J. P. (1976a) Etude quantitative du peuplement algal benthique en vue de l'établissement d'une méthodologie d'estimation biologique de la qualité des eaux courantes. Application au cours belge de la Meuse et de la Sambre et à la Somme. In: Recherche et technique au service de l'environnement. Cebedoc, Liège, pp. 159—206.

Descy, J. P. (1976b) La végétation algale benthique de la Somme (France) et ses relations avec la qualité des eaux. In: Méthodes biologiques de détection et d'évaluation de la pollution des milieux continentaux. Mém. Soc. r. Bot. Belg. 7: 101—128.

Descy, J. P. (1979) A new approach to water quality estimation using diatoms. Beih. Nova Hedwigia 64: 305—323.

Descy, J. P., Lange-Bertalot, H. & Coste, M. (1987) Algae as water quality assessment characteristics: diatom indices. In: Wetzel, R. G. (ed.), Manual for the assessment of periphyton communities (in press).

Dixit, S. S. & Dickman, M. D. (1986) Correlation of surface sediment diatoms with the present lake water pH in 28 Algoma lakes, Ontario, Canada. Hydrobiologia 131(2): 133—143.

Donat, A. (1926) Zur Kenntnis der Desmidiaceen des norddeutschen Flachlandes. Eine soziologisch-geografische Studie. Pflanzenforschung (Jena) 5, 51 pp.

Drews, G., Prauser, H. & Uhlmann, D. (1961) Massenvorkommen von Synechococcus plancticus nov. spec., einer solitären, planktischen Cyanophycee, in einem Abwasserteich. Beitrag zur Kenntnis der sogenannten "μ-Algen". Arch. Mikrobiol. 39: 101—115.

215

Drury, W. H. & Nisbet, I. C. T. (1973) Succession. J. Arnold Arbor. Harvard Univ. 54: 331—368.

Duvigneaud, P. (1946) La variabilité des associations végétales. Bull. Soc. roy. Bot. Belg., 78: 107—134.

Estrada, M. (1972) Analyse en composantes principales de données de phytoplancton de la zone côtière du Sud de l'Ebre. Invest. Pesquer. 36: 109—118.

Fabri, R. & Leclercq, L. (1984) Etude écologique des rivières du nord du massif Ardennais (Belgique): flore et végétation de diatomées et physico-chimie des eaux. Thèse doctorat Sc., Univ. Liège, 2 vols.

Fetzmann, E. (1956) Beiträge zur Algensoziologie. Sitzungsber. Öst. Akad. Wiss., Mat.-nat. Kl. 165(H.9—10): 709—783.

Fetzmann, E. (1961) Algensoziologische Untersuchungen in Altwässern der Donauauen. Verh. intern. Assoc. theor. angew. Limnol. 14: 466—470.

Fetzmann, E. (1963) Studien zur Algenvegetation der Donau-Auen. Arch. Hydrobiol. 27, Suppl. Donauforschung 1(2): 183—225. See also Kusel-Fetzmann, E.

Findenegg, I. (1969) Expression of populations. In: Vollenweider, R. A. (ed.), A manual on methods for measuring primary production in aquatic environments, including a chapter on Bacteria. Blackwell Scient. Publ., Oxford & Edinburgh, I.B.P. Handb. 12, pp. 16—18.

Fisher, S. G. (1983) Succession in streams. In: Barnes, J. R. & Minshall, G. W. (eds.), Stream ecology: Application and testing of general ecological theory. Plenum Press, New York, pp. 7—27.

Fisher, S. G., Gray, L. J., Grimm, N. B. & Busch, D. B. (1982) Temporal succession in a desert stream ecosystem following flash flooding. Ecol. Monogr. 52: 93—110.

Fjerdingstad, E. (1964) Pollution of streams estimated by benthal phytomicro-organisms. I. A saprobic system based on communities of organisms and ecological factors. Int. Rev. ges. Hydrobiol. 49(1): 63—131.

Flahault, C. & Schröder, C. (1910) Nomenclature phytogéographique. Rapports et propositions. IIIe Congrès international de Botanique (Bruxelles, 14—22 mai 1910). Impr. Zurcher & Furrer, Zürich, 29 + x pp.

Ganf, G. G. (1974) Diurnal mixing and the vertical distribution of phytoplankton in a shallow equatorial lake (Lake George, Uganda). J. Ecol. 62: 611—629.

Gasse, F., Talling, J. F. & Kilham, P. (1983) Diatom assemblages in East Africa: classification, distribution and ecology. Rev. Hydrobiol. trop. 16(1): 3—34.

Gauch, H. (Jr.) (1982) Multivariate analysis in community ecology. Cambridge Univ. Press, Cambridge, etc., x + 298 pp.

Godwin, H. & Conway, V. M. (1939) The ecology of a raised bog near Tregaron, Cardiganshire. J. Ecol. 27: 343—359.

Goodall, D. W. (1954a) Objective methods in the classification of vegetation. III. An essay in the use of factor analysis. Aust. J. Bot. 2: 303—324.

Goodall, D. W. (1954b) Vegetational classification and vegetational continua. Angew. Pflanzensoc. (Wien), Festschr. Aichinger 1: 168—182.

Goodall, D. W. (1978) Sample similarity and species correlation. In: Whittaker, R. H. (ed.), Ordination of plant communities. Junk, The Hague, etc., pp. 99—149.

Gounot, M. (1969) Méthodes d'étude quantitative de la végétation. Masson. Paris, 314 pp.

Greig-Smith, P. (1983) Quantitative plant ecology. 3rd ed., Studies in Ecology, vol. 9. Blackwell Scient. Publ., Oxford, etc., xiv + 359 pp.

Grisebach, A. (1838) Ueber den Einfluss des Climas auf die Begrenzung der natürlichen Floren. Linnaea 12: 159—200.

Guinochet, M. (1938) Etudes sur la végétation de l'étage alpin dans le bassin supérieur de la Tinée. Comm. S.I.G.M.A., no. 59. Bosc. Lyon, 458 pp.

Guinochet, M. (1973) Phytosociologie. In: Coll. d'Ecologie, 1. Masson, Paris, vi + 227 pp.

216

Hamilton, P. B. & Duthie, H. C. (1984) Periphyton colonization of rock surfaces in a boreal forest stream studied by scanning electron microscopy and track auto-radiography. J. Phycol. 20(4): 525—532 (1984).

Happey-Wood, C. M. (1976) The occurrence and relative importance of nanno-Chlorophyta in freshwater algal communities. J. Ecol. 64: 279—92.

Happey-Wood, C. M. (1978) The application of culture methods in studies of the ecology of Small green algae. Mitt. intern. Verein. theor. u. angew. Limnol. 21: 385—397.

Happey-Wood, C. M. (1980) Periodicity of epipelic unicellular volvocales (Chloro-phyceae) in a shallow acid pool. Phycol. 16: 116—128.

Happey-Wood, C. M. & Priddle, J. (1984) The ecology of epipelic algae of five Welsh lakes, with special reference to volvocalean green flagellates (Chlorophyceae). J. Phycol. 20(1): 109—124.

Harris, G. P. (1980) Temporal and spatial cycles in phytoplankton ecology. Mechanisms, methods, models and management. Canad. J. Fish. aquat. Sci. 37: 877—900.

Harris, G. P. (1983) Mixed layer physics and phytoplankton populations: studies in equilibrium and non-equilibrium ecology. Progress in Phycological Research 2: 1—52.

Hartog (den), C. see den Hartog, C.

Harvey, H. W. (1950) On the production of living matter in the sea off Plymouth. Journ. mar. biol. Ass. U.K. 29: 97—137.

Heide (van der), J. see van der Heide, J.

Hill, M. O. (1979) DECORANA. A FORTRAN program for detrended correspond-ence analysis and reciprocal averaging. Cornell Univ., Ithaca, N.Y., 52 pp.

Hill, M. O. & Gauch, H. G. (1980) Detrended correspondence analysis, an improved ordination technique. Vegetatio 42: 47—58.

Hillebrand, H. (1977) Periodicity and distribution of multicellular green algae in two lake areas in the Netherlands. Doktoraatsverh., Vrije Univ. Amsterdam, 178 pp.

Hoagland, K. D., Roemer, S. C. & Rosowski, J. R. (1982) Colonization and community structure of two periphyton assemblages, with emphasis on the diatoms (Bacillario-phyceae). Amer. J. Bot., 69: 188—213.

Hobbie, J. E. (1980) Limnology of tundra ponds, Barrow Alaska. US/IBP Synthesis Series, 13. Academic Press, New York.

Höhn-Ochsner, W. (1978) Beitrag zur Kenntnis der Vegetationseinheiten und Mikro-biozönosen in Moorlandschaften des oberen zürcherischen Glattales. Vierteljahrschr. naturf. Ges. Zürich 123(2): 125—134.

Horn, H. S. (1976) Succession. In: May, R. M. (ed.), Theoretical ecology: Principles and applications. Blackwell, London, pp. 187—204.

Hudon, C. & Bourget, E. (1981) Initial colonization of artificial substrate: community development and structure studied by scanning electron microscope. Canad. J. Fish. aquat. Sci. 38: 1371—1384.

Humboldt (von), A. see von Humboldt, A.

Hutchinson, G. E. (1967) A treatise on limnology. Vol. 2: Introduction to lake biology and the limnoplankton. Wiley & Sons, New York, London, Sydney, xi + 1115 pp.

Jaag, O. (1945) Untersuchungen über die Vegetation und Biologie der Algen des nackten Gesteins in der Alpen, im Jura und im schweizerischen Mittelland. Beitr. Kryptog. Flora Schweiz 9: 8—560.

Johannson, C. (1982) Attached algal vegetation in running waters of Jämtland, Sweden. Acta phytogeogr. suec. 71, 83 pp.

Johannson, C., Kronborg, L. & Thomasson, K. (1977) Attached algal vegetation in running waters. A literature review. Excerpta botanica, sect. B 16: 126—178.

Jørgensen, E. G. (1948) Diatom communities in some Danish lakes and ponds. Det. Kong. Dansk. Vidensk. Selsk., Biol. Skrift. 5 (Nr. 1), 140 pp.

217

Jørgensen, E. G. (1957) Diatom periodicity and silicon assimilation. Dansk. Bot. Arkiv 18(1), 54 pp.
Kalff, J. & Welch, H. E. (1974) Phytoplankton production in Char Lake, a natural polar lake, and Meretta Lake, a polluted polar lake, Cornwallis Island, Northwest Territories. Journ. Fish. Res. Bd. Canada 31: 621—636.
Kalk, M. J. (1979) Zooplankton in Lake Chilwa: adaptations to changes. In: Kalk, M., McLachlan, A. J. & Howard-Williams, C., Lake Chilwa. Studies of change in a tropical ecosystem. Monogr. biol. 35: 123—142.
Kalk, M. J. & Schulten-Senden, C. M. (1977) Zooplankton in a tropical endorheic lake (Lake Chilwa, Malawi) during drying and recovery phases. J. limnol. Soc. South Afr. 3(1): 1—7.
Kiehm, F., Dubois, D. M. & Descy, J. P. (1981) Application de l'indice de fluctuations à l'évaluation de la stabilité de la structure de communautés aquatiques d'eau douce. In: Dubois, D. M. (ed.), Progress in ecological engineering and management by mathematical modelling. CEBEDOC, Liège, pp. 23—27.
Kilham, P. & Kilham, S. S. (1980) The evolutionary ecology of plankton. In: Morris, I. (ed.), The physiological ecology of phytoplankton. Blackwell Scient. Publ., Oxford, pp. 571—597.
Klasvik, B. (1974) Computerized analysis of stream algae. Växtekol. Stud., 5, 100 pp.
Kolkwitz, R. & Marsson, M. (1908) Ökologie der pflanzlichen Saprobien. Ber. deutsch. Bot. Ges., 26a: 505—519.
Kolkwitz, R. & Marsson, M. (1909) Ökologie der tierischen Saprobien. Int. Rev. ges. Hydrobiol. u. Hydrogr. 2: 126—152.
Korte, V. L. & Blinn, D. W. (1983) Diatom colonization on artificial substrata in pool and riffle zones studied by light and scanning electron microscopy. J. Phycol. 19: 332—341.
Kusel-Fetzmann, E. (1979) The algal vegetation of Neusiedlersee. In: Löffler, H. (ed.), Neusiedlersee: The limnology of a shallow lake in Central Europe. Monogr. biol., 37: 171—202.
Lamb, L. A. & Lowe, R. L. (1981) A preliminary investigation of the effect of current speed on periphyton community structure. Micron 12: 211—212.
Lange-Bertalot, H. (1978) Diatomeen-Differentialarten anstelle von Leitformen: ein geeigneteres Kriterium der Gewässerbelastung. Arch. Hydrobiol., Suppl. 51, Algol. Stud. 21: 393—427.
Lange-Bertalot, H. (1979) Pollution tolerance of diatoms as a criterion for water quality estimation. Beih. Nova Hedwigia 64: 285—304.
Laporte, L. J. (1931) Recherches sur la biologie et la systématique des Desmidiées. In: Encycl. Biol., Lechevalier, Paris, t. 9, 150 pp.
Le Cohu, R. (1974) Recherches expérimentales sur l'écologie des algues d'eau douce: Utilisation des enceintes en milieu naturel. Thèse Univ. Rennes, 271 pp.
Legendre, L. & Legendre, P. (1979) Ecologie numérique. Masson, Paris; Presses Univ. Québec. 2 vols.
Leher, K. (1958) Vergleichende ökologische Untersuchungen einiger Desmidiaceengesellschaften in den Hochmooren der Osterseen. Ber. bayer. bot. Ges. (München) 32: 48—83.
Lehn, H. (1968) Litorale Aufwuchsalgen im Pelagial des Bodensee. Beitr. Naturk. Forsch. Südw.-Dtl. 27: 97—100.
Lhotsky, O. (1985) The time factor in the evaluation of algal communities. Verh. intern. Verein. theor. appl. Limnol. 22(5): 2885—2887.
Liebmann, H. (1951) Handbuch der Frischwasser- und Abwasser-Biologie. 1, 1. Aufl. München, 539 pp.
Liebmann, H. (1962) Handbuch der Frischwasser- und Abwasser-Biologie. 1, 2. Aufl. München.

218

Light, J. J., Ellis-Evans, J. C. & Priddle, J. (1981) Phytoplankton ecology in an Antarctic lake. Freshw. Biol. 11: 11—26.

Lorenz, J. R. (1858) Allgemeine Resultate aus der pflanzengeographischen und genetischen Untersuchung der Moore im präalpinen Hügellande Salzburg's. Flora 41: 209—221, 225—237, 241—253, 273—286, 289—302, 344—355, 360—376.

Lund, J. W. G. (1972a) Changes in the biomass of blue-green and other algae in an English lake from 1945—1969. In: Desikachary, T. V. (ed.), Proceedings of the Symposium on Taxonomy and Biology of Blue-green Algae. Univ. of Madras Press, Madras, pp. 305—327.

Lund, J. W. G. (1972b) Eutrophication. Proc. R. Soc. London, B 180: 371—382.

Lund, J. W. G. (1978) Changes in the phytoplankton of an English lake, 1945—1977. Hydrobiol. Journ. 14(1): 6—21.

Maarel (van der), E. see van der Maarel, E.

McIntire, C. D. (1973) Diatom associations in Yaquina estuary, Oregon: A multivariate analysis. J. Phycol. 9: 251—259.

McLachlan, A. J. (1974) Development of some lake ecosystems in tropical Africa, with special reference to the invertebrates. Biol. Rev. 49: 365—397.

Margalef, R. (1947) Limnosociologia. Monogr. Cienc. moderna. Inst. Españ. Edafol., Ecol. y Fisiol. vegetal. Madrid. No. 10, 93 pp.

Margalef, R. (1949) Las asociaciones de Algas en las aguas dulces de pequeño volumen del Noreste de España. Vegetatio 1(4—5): 258—284.

Marker, A. F. H. & Bolas, P. M. (1984) Sampling of non-planktonic algae (benthic algae or periphyton) 1982. In: Methods for the examination of waters and associated materials. Her Majesty's Stationery Office, London, 27 pp.

Melack, J. M. (1988) Aquatic plants in extreme environments. In: Symoens, J. J. (ed.), Handbook of vegetation science, vol. 15. Vegetation of inland waters. Dr W. Junk. This volume, pp. 341—378.

Messikommer, E. (1927) Biologische Studien im Torfmoor von Robenhausen unter besonderer Berücksichtigung der Algenvegetation. Mitt. Bot. Mus. Univ. Zürich 122, 171 pp.

Miller, A. R., Lowe, R. L. & Rotenberry, J. T. (1987) Succession of diatom communities on sand grains. J. Ecol. 75(3): 693—709.

Moss, B. (1979) Algae in Lake Chilwa and the waters of its catchment area. In: Kalk, M., McLachlan, A. J. & Howard-Williams, C., Lake Chilwa. Studies of change in a tropical ecosystem. Monogr. biol., 35: 93—103.

Mueller-Dombois, D. & Ellenberg, H. (1974) Aims and methods of vegetation ecology. Wiley, New York, 547 pp.

Nauwerck, A. (1963) Die Beziehungen zwischen Zooplankton und Phytoplankton im See Erken. Symb. bot. upsal., 17(5), 163 pp.

Nauwerck, A. (1968) Das Phytoplankton des Latnjaure 1954—65. Schweiz. Zeitschr. f. Hydrol. 30: 188—216.

Ong, M. H. & Happey-Wood, C. M. (1979) The seasonal distribution of micro-green algae in two linked but contrasting Welsh lakes. Br. phycol. J. 14: 361—375.

Orlóci, L. (1978) Multivariate analysis in vegetation research. Junk, The Hague, Boston, 2nd ed., ix + 451 pp.

Orlóci, L. & Kenkel, N. C. (1985) Introduction to data analysis with examples from population and community ecology. Intern. Co-op. Publ. House, Fairland, Maryland, xii + 340 pp.

Panknin, W. (1941) Die Vegetation einiger Seen in der Umgebung von Joachimsthal. Bibliotheca botanica, H. 119, 162 pp.

Panknin, W. (1945) Zur Entwicklungsgeschichte der Algensoziologie und zum Problem der 'echten' und 'zugehörigen' Algengesellschaften. Arch. Hydrobiol. 41: 92—111.

Pantle, R. & Buck, H. (1955) Die biologische Überwachung der Gewässer und die Darstellung der Ergebnisse. Gas- und Wasserfach 96.

Patrick, R. (1949) A proposed biological measure of stream conditions, based on a survey of the Conestoga Basin, Lancaster County, Pennsylvania. Proc. Acad. nat. Sci. Phila. 101: 377—341.

Patrick, R., Hohn, M. H. & Wallace, J. H. (1954) A new method for determining the pattern of the diatom flora. Notulae Acad. nat. Sci. Phila., No. 259, 12 pp.

Pearsall, W. H. (1921) The development of vegetation in the English lakes, considered in relation to the general evolution of glacial lakes and rock basins. Proc. R. Soc. London, ser. B 92: 259—284.

Pechlaner, R. (1971) Factors that control the production rate and biomass of phytoplankton in high mountain lakes. Mitt. intern. Verein. theor. angew. Limnol. 19: 125—145.

Poulíčkova, A. (1987) Algae in ground waters below the active stream of a river (Basin of the Morava River, Czechoslovakia). Arch. Hydrobiol., Suppl. 78(1) (Algol. Stud. 46): 65—88.

Priddle, J. & Happey-Wood, C. M. (1983) Significance of small species of Chlorophyta in freshwater phytoplankton communities with special reference to five Welsh lakes. J. Ecol. 71: 793—810.

Pringle, C. M. (1985) Effects of Chironomid (Insecta: Diptera) tube-building activities on stream diatom communities. J. Phycol. 21(2): 185—194.

Rees, A. J. J., Cmiech, H. A. & Leedale, G. F. (1984) Periodicity of phytoplankton over two seasons in a shallow eutrophic lake. Naturalist 109: 81—95.

Reichle, D. E., Dinger, B. E., Edwards, N. T. Harris, W. F. & Sollins, P. (1973) Carbon flow and storage in a woodland ecosystem. In: Woodwell, G. M. & Pecan, E. (eds.), Carbon and the Biosphere. USAEC-CONF-720510, Washington, D.C., pp. 345—365.

Reynolds, C. S. (1980) Phytoplankton assemblages and their periodicity in stratifying lake systems. Holarctic Ecology 3: 141—159.

Reynolds, C. S. (1982) Phytoplankton periodicity: its motivation, mechanisms and manipulations. Rept. Freshw. biol. Assoc. 50: 60—75.

Reynolds, C. S. (1984) Phytoplankton periodicity: the interactions of form, function and environmental variability. Freshw. Biol. 14: 111—142.

Reyssac, J. & Roux, M. (1972) Communautés phytoplanctoniques dans les eaux de Côte d'Ivoire. Mar. Biol. 13: 14—33.

Riley, G. A. (1956) Oceanography of Long Island Sound, 1952—54. IX. Production and utilization of organic matter. Bull. Bingham Oceanogr. Coll. 15: 324—344.

Rodhe, W. (1955) Can phytoplankton production proceed during winter darkness? Verh. intern. Verein. theor. angew. Limnol. 12: 117—122.

Rodhe, W., Vollenweider, R. A. & Nauwerck, A. (1958) The primary production and standing crop of phytoplankton. In: Buzzati-Traverso, A. A. (ed.), Perspectives in marine biology. Univ. of California Press, Berkeley and Los Angeles, pp. 299—332.

Rott, E. (1984) Phytoplankton as biological parameter for the trophic characterization of lakes. Verh. intern. Verein. theor. angew. Limnol. 22: 1078—1085.

Round, F. E. (1957a) Studies on bottom-living algae in some lakes of the English Lake District. Part I. Some chemical features of the sediments related to algal productivities. J. Ecol. 45: 133—148.

Round, F. E. (1957b) Studies on bottom-living algae in some lakes of the English Lake District. Part II. The distribution of Bacillariophyceae on the sediments. J. Ecol. 45: 343—360.

Round, F. E. (1957c) Studies on bottom-living algae in some lakes of the English Lake District. Part III. The distribution on the sediments of algal groups other than the Bacillariophyceae. J. Ecol. 45: 649—664.

Round, F. E. (1960) Studies on bottom-living algae in some lakes of the English Lake District. Part IV. The seasonal cycles of the Bacillariophyceae. J. Ecol. 48: 529—547.

Round, F. E. (1961a) Studies on bottom-living algae in some lakes of the English Lake District. Part V. The seasonal cycles of the Cyanophyceae. J. Ecol. 49: 31—38.

Round, F. E. (1961b) Studies on bottom-living algae in some lakes of the English Lake District. Part VI. The effect on depth on the epipelic algal community. J. Ecol. 49: 245—254.

Round, F. E. (1972) Patterns of seasonal succession of freshwater epipelic algae. Br. phycol. J. 7: 213—220.

Round, F. E. (1981) The ecology of algae. Cambridge Univ. Press. Cambridge, New York, Melbourne, vii + 653 pp.

Rushforth, S. R., Squires, L. E. & Cushing, C. E. (1986) Algal communities of springs and streams in the Mt. St. Helens region, Washington, U.S.A. following the May 1980 eruption. J. Phycol. 22: 129—137.

Rzóska, J. (1966) The biology of reservoirs in the U.S.S.R. In: Lowe-McConnell, R. H. (ed.), Man-made lakes. Academic Press, London, pp. 149—154.

Sakshaug, E. & Olsen, Y. (1986) Nutrient status of phytoplankton blooms in Norwegian waters and algal strategies for nutrient competition. Can. J. Fish. aquat. Sci. 43(2): 389—396.

Satchell, J. E. (1971) Feasibility study of an energy budget for Meathop Wood. In: Duvigneaud, P. (ed.), Productivity of forest ecosystems. UNESCO. Paris, pp. 619—630.

Schmitz, W. (1965) Die Soziologie aquatischer Mikrophyten. In: Tüxen, R. (ed.), Biosoziologie, Ber. intern. Symposium d. intern. Verein. Vegetationsk. (Stolzenau/Weser, 1960), pp. 120—139.

Schroevers, P. J. (1973) Bibliographie der Algen-Gesellschaften (mit Ausschluss derjenigen des Salzwassers). Excerpta botanica, sectio B 12: 241—309.

Shephard, R. N. & Carroll, J. D. (1966) Parametric representation of non-linear data structures. In: Krishnaiah, P. R. (ed.), Multivariate analysis. Acad. Press, New York, pp. 561—592.

Sládeček, V. (1961) Zur biologischen Gliederung der höheren Saprobitätsstufen. Arch. Hydrobiol. 58: 103—121.

Sládeček, V. (1973) System of water quality from the biological point of view. Arch. Hydrobiol., Beih. 7, Ergebn. Limnol. 7, 218 pp.

Sládeček, V. (1979) Continental systems for the assessment of river water quality. In: James, A. (ed.), Biological indicators of water quality. Wiley & Sons, New York.

Sládečková, A. (1962) Limnological investigation methods for the periphyton ('Aufwuchs') community. Bot. Rev. 28(2): 286—350.

Sommer, U. (1981) The role of r- and K-selection in the succession of phytoplankton in Lake Constance. Acta oecologica/oecol. gener. 2: 327—342.

Sörensen, I. (1948) Biological effects of industrial defilements in the River Billebergaån. Acta Limnol. 1, 73 pp.

Stockner, J. G. & Antia, N. J. (1986) Algal picoplankton from marine and freshwater ecosystems: A multidisciplinary perspective. Can. J. Fish. aquat. Sci. 43(12): 2472—2503.

Symoens, J. J. (1951a) A propos d'une association de Desmidiées sphagnophiles. Verh. intern. Verein. theor. angew. Limnol. 11: 392—394.

Symoens, J. J. (1951b) Esquisse d'un système des associations algales d'eau douce. Verh. intern. Verein. theor. angew. Limnol. 11: 395—408.

Symoens, J. J. (1954) Les principales associations algales des eaux courantes de l'Ardenne et des régions voisines. VIIIe Congrès intern. Bot. (Paris, 1954), Rapports et Communic. parvenus avant le Congrès, Sect. 17 (Phycol.): 166—167.

Symoens, J. J. (1957) Les eaux douces de l'Ardenne et des régions voisines: Les milieux et leur végetation algale. Bull. Soc. r. Bot. Belg. 89: 111—314.

Symoens, J. J. (1983) Discussion. In: Mériaux, J. L. Remarques sur la syntaxonomie des Potametea. Colloques phytosociol. 10: Les végétations aquatiques et amphibies (Lille, 1981): 137—138.

Symons, F. (1970) Study of the ecological relations between 30 species of algae by means of a factor analysis. Hydrobiologia 36(3—4): 513—600.

Symons, F. (1973) Study of the changes in the structures of two algal populations: an r-type factor analysis. Hydrobiologia 41(1): 107—112.

Tanaka, N., Nakanishi, M. & Kadota, H. (1974) Nutritional interrelation between bacteria and phytoplankton in a pelagic ecosystem. In: Colwell, R. R. & Morita, R. Y., Effects of the ocean environment on microbial activities, pp. 493—509.

Tansley, A. G. (1916) The development of vegetation. J. Ecol. 4: 198—204.

Tansley, A. G. (1920) The classification of vegetation and the concept of development. J. Ecol. 8: 118—149.

Thunmark, S. (1945) Zur Soziologie des Süsswasser-Planktons. Eine methodolo-gischökologische Studie. Fol. limnol. Scand. 3, 66 pp.

Utermöhl, H. (1958) Zur Vervollkommnung der quantitativen Phytoplankton-Methodik. Mitt. intern. Verein. theor. angew. Limnol. 9, 38 pp.

Vallentyne, J. R. (1974) The algal bowl. Lakes and Man. Dept. Environment, Fisheries and Marine Service, Misc. Publ. No. 22, 185 pp.

van der Heide, J. (1982) Lake Brokopondo. Filling phase limnology of a man-made lake in the humid tropics. Acad. proefschr., Vrije Univ. Amsterdam, 428 pp.

van der Maarel, E., Janssen, J. G. M. & Louppen, J. M. W. (1978) TABORD, a program for structuring phytosociological tables. Vegetatio 38: 143—156.

Vareschi, E. (1982) The ecology of Lake Nakura (Kenya). III. Abiotic factors and primary production. Oecologia 55: 81—101.

Verneaux, J. (1973) Recherches écologiques sur le réseau hydrographique du Doubs. Thèse Doct. Etat, Univ. Besançon, 260 pp.

Verneaux, J. (1982) Expression biologique, qualitative et pratique de l'aptitude des cours d'eau du développement de la faune benthique. Annales scient. Univ. Besançon, Biol. anim., 3: 20 pp.

von Humboldt, A. (1805) Essai sur la géographie des plantes, accompagné d'un tableau physique des régions équinoxiales. Levrault, Schoell & Cie. Paris, 155 pp.

Warming, E. (1895) Plantesamfund. Grundtraek af den økologiske Plantengeografi. Philipsens, Kjøbenhavn, 335 pp.

Warming, E. (1909) Oecology of plants. An introduction to the study of plant-communities. Engl. transl. by Groom, P. & Balfour, I. B. Oxford Univ. Press. Oxford, 422 pp.

Wetzel, R. G. (1975) Limnology. W. B. Saunders Cy, Philadelphia; Eastbourne, Toronto, xii + 743 pp.

White, P. S. (1979) Pattern, process, and natural disturbance in vegetation. Bot. Rev., 45 (No. 3): 229—299.

Whittaker, R. H. (1962) Classification of natural communities. Bot. Rev. 28(1): 1—239.

Whittaker, R. H. (ed.) (1978a) Classification of plant communities. Junk, The Hague, Boston, London.

Whittaker, R. H. (ed.) (1978b) Ordination of plant communities. Junk, The Hague, Boston, London, iv + 388 pp.

Williams, W. T. (1971) Principles of clustering. Annual Rev. Ecol. Syst. 2: 303—326.

Wilmanns, O. (1962) Rindenbewohnende Epiphytengemeinschaften in Südwest-deutschland. Beitr. naturk. Forsch. S.W.-Deutschl. 21: 87—164.

Wright, R. T. (1964) Dynamics of a phytoplankton population community in an ice-covered lake. Limnol. Oceanogr. 9: 163—178.

Zelinka, M. & Marvan, P. (1961) Zur Präzisierung der biologischen Klassifikation der Reinheit fliessender Gewässer. Arch. Hydrobiol. 57: 389—407.

Zelinka, M. & Marvan, P. (1963) Comparisons of methods of saprobial evaluation of water. Vodní hospodářtsvi 13: 291—293 (in Czech).

VEGETATION PROCESSES IN SWAMPS AND FLOODED PLAINS

C. M. BREEN, K. H. ROGERS & P. J. ASHTON

INTRODUCTION

Freshwater wetlands exhibit a continuum of variation in space and time, and categorization results in an artificial classification which although useful, often has little real meaning. Wetzel (1978) noted that definitions vary widely and are commonly useful only in detailed analyses of successional changes of wetland conditions and biota. In this discussion we do not attempt to define precisely the terms floodplain and swamp but consider them to reflect opposite ends of a continuum.

A major feature which locates swamps at one extreme of the continuum is the persistence of standing water among vegetation (Wetzel 1975). Flooded plains lie at the other extreme because standing water is present only during the flood period, and there is a regular alternation of flood and dry phases (Welcomme 1974). Marshes which contain water-saturated sediments but have little or no standing water among the vegetation occupy an intermediate position (Wetzel 1975).

The distinctive hydrogemorphic processes associated with floodplains (Welcomme 1974, Gregory & Walling 1975, Schumm 1977, Hupp & Osterkamp 1985) give rise to a number of fluvial landforms such as oxbow lakes, point bars, meander scrolls, sloughs and back swamp areas. Consequently, flooded plains comprise a mosaic of habitats which may include marshes and swamps.

The floral components of swamps and marshes are derived principally from aquatic sources whereas in flooded plains land-based (terrestrial) and aquatic sources alternate in dominance (Welcomme 1979). This difference has major implications for the structure and functioning of these systems. For example, in swamps the bulk of the primary production occurs in the air and decomposition and consumption occurs later in the water (Howard-Williams & Gaudet 1985). This contrasts with the situation in flooded plains where although the bulk of the production also occurs in the air, consumption occurs both in the air (terrestrial herbivores) and in the water (aquatic herbivores and detritivores).

223

J. J. Symoens (ed.), Vegetation of inland waters. ISBN 90—6193—196—7.
© 1988, Kluwer Academic Publishers, Dordrecht. Printed in the Netherlands.

A conspicuous feature of both swamps and flooded plains is the spatial and temporal separation of production and consumption (Teal 1962, Odum 1971). For aquatic organisms there is always a time lag between production and consumption because the material only enters the water when it collapses (swamps) or when it is flooded (flooded plains). In contrast, large terrestrial grazers are, with few exceptions, almost totally excluded from swamps, whereas they are only temporarily excluded from flooded plains. It is evident therefore that the structure and functioning of these systems is greatly influenced by the nature, strength and duration of these 'barriers' which control the direction and extent of movement of organisms and matter.

Considerable research effort has been directed at nutrient cycling in wetland ecosystems (e.g. Gaudet 1977, Good, Whigham & Simpson 1978, Wetzel 1979, Howard-Williams 1985) and at the elucidation of the rôle of these systems as sources or sinks of energy and nutrients (e.g. Nixon 1980).

In this chapter we examine some of the processes which determine the ways in which swamps and flooded plains function with particular reference to the rôles of the plants in these processes. We do not attempt an exhaustive review of the literature, instead we address selected concepts which, in our opinion, warrant further research.

PROCESSES AT THE COMMUNITY LEVEL

The terms 'community' and 'ecosystem' are sometimes considered to be synonymous although clearly they are not. As den Hartog (1982) has pointed out, a community is 'a structural frame' characterized by its species composition, their arrangement in time and space, the relations between the organisms within the community and their relations with the surrounding biotic and abiotic environment. In contrast, an ecosystem is a functional unit and is characterized by functional parameters such as primary productivity, turnover rates, energy transfer and nutrient cycling.

The aggregation of plants into discrete communities relfects the degree of spatial and temporal discontinuity in the physical and chemical environment. By imposing limits on the distribution of organisms the environment acts to compartmentalize the biota into units within which the organisms interact with both biotic and abiotic factors to modify the community, and amplify its properties. The community therefore reflects both the initial conditions of the environment and the extent to which these have been modified by the physical presence and metabolism of the component species.

Aquatic plant communities usually exhibit distinct zonation (Walker

1970, Cole 1973, Wetzel 1979, Gosselink *et al.* 1981, Spence 1982). Because communities are structurally different and comprized of different species (often with a single dominant species) it can be postulated that they are, to varying degrees, functionally different. The hypothesis would be that spatial and temporal heterogeneity in community distribution leads to functional heterogeneity and that the more distinct the boundaries between communities the stronger would be the mechanisms isolating within community processes. Such an hypothesis suggests also that the properties of a community would tend to conserve resources within the community. Clearly, the extent to which the community is 'isolated' also depends very largely on the existence (and magnitude) of mechanisms which can disrupt isolation i.e. mechanisms which transport material (resources) and information* (Patten & Odum 1981, Solbrig 1981) into and out of the community. This leads to the suggestion that 'isolating mechanisms' and 'transporting mechanisms' oppose one another in the development of boundaries and provide powerful forces in evolution. Different combinations of isolating and transporting mechanisms might therefore be expected to lead to the evolution of communities which have fundamentally different properties and patterns of functioning. The relevance of these processes in swamps and flooded plains deserves special attention.

Isolating mechanisms

It is generally accepted that vegetation changes gradually in space unless there is a discontinuity in the chemical or physical environment (Whittaker 1970, Colinvaux 1973). That aquatic vegetation usually exhibits marked zonation implicates discontinuities in the physical environment as important determinants of community boundaries. We now consider the abiotic factors regulating the development of plant communities in swamps and flooded plains and how they are modified by the biota (biotic regulation).

Abiotic regulation

The water balance (hydrology) is the principal abiotic determinant of the development of all aquatic and semi-aquatic plant communities (Gosselink & Turner 1978). Regulation is effected through the depth,

---

* 'Information' in the ecosystem context includes all those qualitative and natural history phenomena which regulate the timing and rate of conservative energy and matter transfers. They include environmental cues such as temperature change, the structure of surfaces or seed coats and the physical presence of organisms or objects. Negligible amounts of energy are involved in their action of regulating transfers.

timing and duration of flooding and the length of the dry or exposed period (Vesey-FitzGerald 1963, Rees 1978, Furness & Breen, 1980). The interplay between topography and hydrology has a striking influence on the distribution of communities, (Lindsey *et al.* 1961, Cole 1973, Johnson *et al.* 1976, Franz & Bazzaz 1977, Conner *et al.* 1981, Gosselink *et al.* 1981), those intolerant of flooding being closely related to high flood levels and those more tolerant of flooding but less tolerant of prolonged dry periods, showing a closer relationship with more stable water levels associated with river and lake margins. Communities colonizing the zone of periodic inundation have to tolerate the stresses of both submergence and exposure (Robertson *et al.* 1978, Bucholz 1981, Furness & Breen 1982). The zone of periodic inundation, characteristic of flooded plains, is therefore a stressful habitat which is 'unsuitable' for plant growth (Cole 1973). It can be expected to support communities comprising principally stress-tolerators and ruderals (Grime 1979) and to have low species diversity (Colinvaux 1973). Menges & Waller (1983) have demonstrated in the floodplain forests of Wisconsin that the higher elevation areas where the flooding stress is reduced, were dominated by tall competitive perennial forbs. As flooding frequency increased, smaller forbs (stress-tolerant competitors) and tall annuals (competitive ruderals) increased in importance. In areas of greatest flooding frequency, flood tolerant sedges and grasses (stress tolerators) shared the space with fast-maturing, annual forbs (ruderals).

The interaction of soil drought (moisture below wilting point) and flooding have been used to quantify 'habitat unsuitability' in grass-herb swamps, and species diversity was shown to decrease as the habitat became less suitable (Cole 1973). Junk (1970) also implicates the dramatic changes in water level of the Amazon River as a determinant of species numbers colonizing these habitats.

Spence (1982) has recently reviewed the factors controlling the zonation of plants which inhabit perennial freshwater habitats (as distinct from those inhabiting the zone of periodic inundation) and draws a distinction between those species which vary solely in the vertical plane from those having both vertical and horizontal components. Underwater light climate is the principal vertical environmental variable and the factors affecting it have been summarized by Rörslette (1985). Assuming other conditions are favourable, light attenuation by the water column regulates both the depth of colonization by submersed plants and the community species composition. Rörslette (1985) has, however, emphasized that the temporally dynamic nature of aquatic systems can result in 'environmental-induced' mortality.

The precise rôle of light as a determinant of the distribution of emergent hydrophytes is unclear and has been poorly researched.

Although the limit of distribution of submersed plants is usually deeper than for emergents, the assumption that light attenuation by the water column has little significance may not be valid. Seeds and vegetative propagules of aquatic plants (Spence 1964, Sastroutomo 1981, Van Wijk & Trompenaars 1985) are known to have a light requirement for germination which could set the lower limit for colonization by seedlings. Although the limit would be modified to a degree by vegetative growth within the restraints of other environmental factors, it does suggest an important rôle for light attenuation during colonization, and may at least partially account for the restriction of emergent species to fairly shallow water.

Water movement varies vertically and horizontally thereby creating a complex environment which is highly variable in space and time. The rôle of water movement as a causative agent of zonation has been critically reviewed by Spence (1982). Of particular significance for the development of swamps associated with lakes is the formation of suitable habitats through the erosion, transport and deposition of fine sedimentary material in shallow water. This can arise from the deposition of sediment transported by rivers (e.g. delta swamps) or through shoreline and lake bed erosion with deposition occurring in sheltered areas where turbulence is reduced (Weisser 1978, Spence 1982). Thus in any given wind regime the particular physical characteristics of the lake and its inflowing water create depositional zones in which the sediment texture varies spatially because of particle sorting (Spence 1982).

Where sediment deposition occurs in shallow water it can lead to the formation of habitats that are suitable for colonization by swamp and marsh plants. However, suitability is controlled by the balance between sediment deposition and subsidence (de Laune et al. 1971), as well as the texture (Lindsey et al. 1961, Robertson et al. 1978, Hermy & Stieperaere 1981) and chemical composition (Howard-Williams & Walker 1974) of the sediments. Notwithstanding the extensive evidence that textural differences influence species distribution, Hupp & Osterkamp (1985) concluded from a study of a hydrologically dynamic bottomland system, that "vegetation patterns appear to develop more as a result of hydrologic processes associated with each fluvial land form rather than from sediment size characteristics."

This review of the regulation of community distribution by abiotic factors is not exhaustive. Nevertheless, it highlights the significance that must be attached to the regulation of community distribution by abiotic factors and stresses, and emphasizes that because these are frequently both spatially and temporally discontinuous, they lead to the distinct community zonation which is so frequently found in aquatic environments.

Biotic regulation

The colonization of substrata by plants adapted to prevailing environmental conditions amplifies spatial heterogeneity by introducing biotic elements into an otherwise purely physicochemical environment. Once autotrophic organisms are present, succession proceeds in response to physical stresses to plants, competition for resources between plants and interactions with other organisms e.g. herbivores, predators and pathogens (Colinvaux 1973, Drury & Nisbet 1973, Horn 1974; Connel & Slatyer 1977). The result is a cybernetic system (Patten & Odum 1981) which maintains order and, in so doing, conserves resources. It is suggested that plants through their interactions with the physical, chemical and biological environment strengthen existing abiotic (hydrological) isolating mechanisms and create new (biotic) barriers so that the communities become increasingly isolated from one another. In this way they also strengthen internal organization, reduce outputs and tend to become as, if not more, important in determining ecosystem structure and functioning than abiotic processes. We now consider some of the ways in which the biota may effect, maintain and amplify isolation in swamps and flooded plains.

Ultimately, the effectiveness of an isolating barrier is determined by its control over the input and output of resources and information to and from the community. Since hydrology is the principal determinant of input and output, considerable significance can be attached to how the architecture of a plant community (the way in which the community components fill up the available space, den Hartog 1982) influences hydrodynamics.

As plants increasingly fill the available space they offer greater frictional drag and water flow is either dispersed laterally causing the velocity to decline, or the water may be forced into narrower channels in which case velocity rises. The implications for community isolation are quite different: in the former, water enters and flows at a reduced rate through the community; in the latter, both water entry into and movement through the community is reduced. If water bypasses the community by channel flow, the community is isolated by both diminished input and output. If water enters the community and flow rate is diminished, settling of particulates occurs and nutrients may be taken up; resources are therefore acquired by the community. Outflow is reduced through evapotranspiration and water leaves with reduced velocity relative to inflow. The system acts as a sieve (Howard-Williams 1979a, 1979b) trapping or isolating resources and information within the community. Since coarse inorganic material is deposited first in response to a reduction in flow rate, it accumulates forming levées along the margin of the community and where water overflows channel banks. Finer sediments (clay and silts) are deposited further into the community in response to further decreases in flow. As vertical build-up of the levées continues they become stabilized by colonizing vegetation. This restricts further the inflow of water into the community so more water is diverted, inputs and outputs are diminished and are, eventually, restricted to exceptional hydrological events. The community (now transformed into a backwater swamp) becomes isolated from the resources (e.g. nutrients) and information (e.g. water flow and level) formerly supplied by the flow of water.

Without the input of sediments, subsidence due to compaction and de-watering of underlying clays becomes the dominant physical process (Mendelssohn, Turner & McKee 1983). Subsidence is greatest where the rate of supply is lowest and where the sediments are finest; as the centre of the community subsides stagnant waterlogged conditions prevail and plant succession is halted or set back to an earlier seral stage (Buttery & Lambert 1965). These processes can lead to spatial heterogeneity observed within the community.

Nutrient (particularly nitrogen) deficiency is generally considered to be a major factor regulating productivity in swamps (e.g. Valiela & Teal 1974, Gallagher 1975; Richardson *et al.* 1978, Mendelssohn 1979). Dissolved and sediment-adsorbed nutrients (DeLaune, Patrick & Brannon 1976) are transported into the community by flooding waters, and their flux therefore depends on the total volume of water flowing through the community, its velocity and the availability of nutrients to be transported. Brown & Lugo (1982) showed that riverine (hydraulically active) forested wetlands received higher phosphorus loading and were more productive than hydrologically isolated (hydraulically inactive) basin forested wetlands. Supplementation of the nutrient supply to a basin forested wetland by sewage enrichment increased productivity, thereby implicating nutrient supply rather than water balance as the controlling factor. Stagnant (continually flooded) swamps have also been shown to be less productive than seasonally flooded swamps (Conner & Day 1982) and evidence indicates that as the frequency and duration of inundation decreases (i.e. as conditions become less stagnant) productivity increases, but may decrease in sites that are only rarely flooded (Gosselink, Bayley & Turner 1981).

The low productivity of communities growing in stagnant conditions has also been attributed to stress arizing from unfavourable soil conditions that develop during waterlogging (Buttery, Williams & Lambert 1965, Armstrong 1982, Linthurst 1979, Mendelssohn, McKee & Patrick 1981, Mendelssohn, McKee & Postek 1982). Such stressed habitats should characteristically have low species diversity (Colinvaux 1973). There is some evidence to support this view and Gosselink *et al.* (1981) suggested that although peak diversity may occur in seasonally flooded wetlands, structural complexity may be elaborated in systems with long hydroperiods and stagnant water.

Connel & Slatyer (1977) concluded that there was considerable evidence to support the view that during succession "the first occupants preempt the space and will continue to exclude or inhibit later colonists until the former die or are damaged, thus releasing resources." One of the resources affected in this way is light. Because light provides the energy for photosynthesis and also the signals (information) which control processes such as germination and flowering, the manner in which it is attenuated is a significant determinant of the structure and functioning of plant communities.

The rate of carbon assimilation by a plant canopy is the product of the rate of photosynthesis per unit area of photosynthetic tissue and the area of this tissue per unit area of ground or water. We therefore expect that where growth conditions are favourable, the canopy would intercept incident radiation effectively. There is considerable evidence that emergent and floating aquatic plants may intercept so much of the light that very little reaches the substratum (Westlake 1966, Ikusima 1970, Howard-Williams 1972, Jones & Muthuri, 1985).

Efficient interception of incident light provides a mechanism for gaining competitive advantage but it also increases the likelihood of self shading which can reduce overall photosynthetic efficiency. Not surprisingly, stands of submerged and floating leaved plants have been shown to modify their canopy structure to minimise the effects of self shading and attenuation by the water column (Westlake 1966, Adams *et al.* 1974). Interception of available light creates a steep light gradient and provides opportunities for colonization only by specialized, shade-adapted species which have broader, thinner leaves, higher chlorophyll content, lower rates of respiration and lower values for the onset of light saturation (Spence & Chrystal 1970a, 1970b, Lipkin 1979).

Direct effects of the canopy are to intercept both incoming and outgoing radiation and to alter its quality so that green light predominates beneath the canopy in dense weedbeds (Westlake 1964). The filtered light is unfavourable for photosynthesis and also for germination of many plants, since it contains very little red light (600—700 nm) which stimulates germination and much dark red and near infrared (700—800 nm) which inhibits germination. Thus once a plant canopy is developed the indirect effects of modification of the microclimate are manifest.

Many plant communities of swamps and flooded plains exhibit marked seasonality and, in these cases, the barrier formed by the canopy is temporary. In such situations where most, if not all, of the living plant material above the substratum dies back, the rapidity with which canopy space can again be occupied during the growing season will largely determine the magnitude of the isolating mechanism and the success of a particular species in the environment.

There appear to be two major mechanisms by which this is achieved. Some species overwinter as well developed rhizome systems (Thompson 1975, Howard-Williams & Gaudet 1985) which retain control of the volume of substratum space and then permit rapid regrowth through mobilization of stored reserves (Bjork 1957, Bayley & O'Neill 1972, Howard-Williams & Lenton 1975), so that aerial space is quickly reoccupied. An example of this has been provided by Buttery & Lambert (1965) who suggested that in Surlingham Broad, *Glyceria maxima* was more successful than *Phragmites communis* because it grew rapidly in early spring and established a thick vegetative cover before the *Phragmites* shoots appeared above the mud. Verhoeven (1980) observed a similar situation in mesohaline lagoons where early initiation of creeping horizontal rhizomes by *Ruppia cirrhosa* forms a dense mat at or near the bottom which restricts development of *Potamogeton pectinatus*, so that a sparse upper canopy is formed. However, where conditions for growth of *P. pectinatus* are particularly favourable it may form such a dense canopy that development of *R. cirrhosa* is decreased.

Conversely, in deciduous floodplain forests where the dominant species may die back partially each year (Menges & Waller 1983), the canopy layer although temporally quite variable, is persistently occupied. Consequently the changing light climate at the forest floor can separate species spatially and temporally (Menges & Waller 1983).

There is thus considerable evidence supportive of the view that the physical presence of plants through direct effects on light attenuation indirectly controls many aspects of the structure and functioning of the community. In so doing, isolation of the community is favoured. However, the temporal variability of the barrier in temperate systems is such as to render it a weak isolating factor. In the tropics where the canopy may be less variable, seasonally, the degree of isolation afforded by the canopy would be more strongly developed.

The disruption of isolating mechanisms

We have seen how abiotic and biotic factors can, individually or in combination, act as barriers to the transport of resources and information into or out of aquatic plant communities. The strengths or weaknesses of these barriers determine both the extent and properties of the communities in question and control the ways in which the communities function. Therefore, it is important to examine those mechanisms whereby barriers can be disrupted or broken down since their effects also regulate community structure and function.

Clearly, disruptive forces or mechanisms which serve to inactivate or remove barriers function in a manner diametrically opposed to isolating mechanisms and can therefore be conceptualized as 'transporting mechanisms'. In this context, it is evident that most transporting mechanisms are in fact 'extreme' forms of environmental determinants that, under 'normal' circumstances, would promote the formation of barriers. Since hydrological features are the principal determinants of all aquatic communities (Gosselink & Turner 1978), it is logical to expect that they would also be implicated as major transporting mechanisms; this is indeed the case. However, it is important to note that the efficacy of a transporting mechanism depends not only on the magnitude of the event but also on the temporal scale over which the event occurs or is repeated. Irregular or episodic high-energy events (i.e. of large magnitude, for example extreme floods) bring about quite different changes from those induced by regular or seasonal low-energy events. In the former, the intervening period permits gradual recolonization after catastrophic change while the latter promotes the development of more stable communities, for example those that are adapted to an alternating cycle of inundation and exposure. High energy events therefore serve to remove barriers whilst low-energy events merely breach these barriers temporarily.

The rôle of fire as a disruptive mechanism has often been ignored in assessments of aquatic plant communities. In many cases this omission is due to the difficulty of quantifying directly the changes that have occurred. In general, the major impacts of fire on aquatic plant communities are the oxidation (and removal) of that season's above ground (or water) standing stock and the concomitant disruption to consumption, decomposition and nutrient cycling processes (Howard-Williams 1979b). Another, less frequently observed phenomenon, is the occurrence of peat fires within swamp communities. These may effectively remove the below-ground biomass and detritus that has accumulated over long time periods and cause extensive subsidence, often leading to the formation of a new landform (e.g. Lake Liambezi, Eastern Caprivi, Namibia; Seaman et al. 1978, Toerien 1978) or changes in directions of water flow (McCarthy et al. 1986).

Movements and grazing by herbivores provide another means whereby the barriers that isolate aquatic plant communities from each other and from the terrestrial environment may be broken down or their efficacy reduced. For example, in several African situations, the regular movement of Hippopotamus between water and land maintains open channels. Here terrestrial vegetation eaten at night is recycled to the water through faeces during the day, thereby promoting decomposition and nutrient cycling (Viner 1975). Conversely, livestock owned by rural tribespeople living along floodplain margins graze marginal meadows and partially inundated vegetation during the day though only a portion of their faeces is deposited on the flood plain to become available when inundated (Heeg & Breen 1982, Furness & Breen 1985).

Transporting mechanisms that breach or remove barriers therefore disrupt the inherent stability within a plant community, and return it to an earlier successional stage. Clearly, the impact of a particular mechanism will depend on both its magnitude and the frequency of its occurrence. Thus, infrequent catastrophic events may eliminate an entire community and drastically alter the habitat, necessitating re-colonization possibly by different species and resulting in a different plant community. Conversely, recurrent low-intensity events usually cause temporary disturbances and are followed by rapid re-establishment of the original plant community. We now examine some of these processes at the species level and assess their implications for the structure and functioning of plant communities.

VEGETATION PROCESSES IN THE FORMATION AND REMOVAL OF BARRIERS

The development of aquatic plant communities within a framework of spatial and temporal discontinuities in the physical and chemical environment depends upon the ability of each individual species to capture, transform and transmit energy, information and materials (Calow & Townsend 1981, Solbrig 1981). The success of a particular species in a habitat reflects, initially, its genotypic capacity to avoid or overcome competition for the acquisition of resources through its morphological, physiological and reproductive strategies. These resources can then be partitioned between maintenance, growth, storage, repair, defence and alternate modes of reproduction, thus allowing the species to consolidate its position within the habitat. The result is a plant species with distinctive life form, chemical composition and patterns of growth, death and decay. These life history characteristics in turn modify the patterns and rates of information and resource flow and storage in communities and ecosystems. The particular series of mechanisms or processes employed by individual species depends on their ability to respond to or withstand environmental cues, disturbances and stresses (Grime 1979, Solbrig 1981, Mitchell & Rogers 1985). The trend is towards the conservation of acquired resources and information by each plant or species within the community and the maintenance of order within the system via a cybernetic system of feedback responses (Patten & Odum 1981). In so doing, the effects of existing barriers or isolating mechanisms between habitats are maintained and often amplified.

While it is often useful to distinguish between, and discuss separately, the morphological, physiological and reproductive strategies employed by wetland plants, we must emphasize that this categorization is artificial. Nevertheless, whilst taking cognisance of the intricate inter-relationships that exist between these attributes, these three categories provide a convenient basis for discussion. Earlier, we emphasized the principal rôle played by hydrological processes in determining both the extent and character of aquatic and semi-aquatic plant communities. Quite naturally, therefore, hydrological features provide most of the major cues, disturbances and stresses that constitute isolating mechanisms or barriers, thereby eliciting appropriate responses and adaptations from individual species. In this section we examine some of the mechanisms evolved in aquatic plants that form, maintain, extend or overcome barriers and we assess the rôles played by, and the relevance of, these processes to community structure and functioning in swamps and flooded plains.

Morphological features

The almost infinite variety of aquatic habitats created by local climatic conditions and catchment characteristics provide powerful selective forces (Menges & Waller, 1983) that have led to the development of a broad array of morphological attributes and life history strategies in aquatic plants. The survival of a species within a particular habitat depends on the species' attributes which help it to withstand external stresses and disturbances (Grime 1979) while at the same time competing successfully for the available resources. These features are particularly evident in the contrasting morphological features and life history strategies possessed by the component species that constitute the characteristic plant communities of swamps and flooded plains.

Swamps are characterized by the permanence of shallow inundation; water velocities and renewal rates are generally low, though the presence of some form of channel flow does occur in most exorheic swamps (Howard-Williams 1985). The lack of between-flood water stress and the virtual absence of disturbance during floods removes two important barriers that limit plant growth and biomass accumulation. The habitat is therefore less temporally variable than a typical flooded plain and permits colonization by longer-lived plants. Intense inter- and intra-specific competition for space and light, coupled with feedback effects from the physical presence and metabolism of the plants, promotes maximum occupancy of above- and below-ground space. The possession of a clonal or phalanx growth form (Lovett-Doust, 1981) confers great competitive advantages. Thus, swamps characteristically possess a greater plant biomass per unit area (usually composed of large-statured plants) and higher rates of organic matter accumulation than flooded plains (Howard-Williams 1985). These, in turn, provide more feedback effects, amplifying further the environmental constraints on hydrological processes, and leading to further reductions in flow velocity and rates of resource acquisition (Gosselink & Turner 1978, Brown & Lugo 1982; Howard-Williams 1985). Important consequences of these feedback effects for the plants are that resources as well as toxins accumulate within the habitat, the external supply of nutrients is diminished and the sediments play a major rôle in nutrient cycling (Howard-Williams 1985).

Selection for survival in the swamp habitat therefore usually favours long-lived plants of large size that are able to conserve nutrients and resources, and which possess specialized physiological adaptations to cope with anoxic soil conditions and the accumulation of toxic substances. Ultimately, the habitat supports a plant community comprised of a mosaic of mono-specific stands of stress tolerators (*sensu* Grime 1979), such as *Phragmites, Typha* and *Cyperus papyrus*. The development of a dense leaf canopy markedly reduces sub-canopy light levels and lowers the chances of successful propagation by seedlings. Instead, vegetative reproduction becomes the major means whereby individual species maintain or extend their position in the community. Here, the phalanx growth form provides a powerful barrier against encroachment by competing species, or neighbouring clones of the same species. Seasonal cycles of vegetative growth are synchronized to permit rapid occupation of the available space by the developing leaf canopy, followed by the transfer of nutrients and storage compounds to, and accumulation within, underground rhizomes and other vegetative propagules. These reserves can then be mobilized rapidly when they are required either for the re-establishment of aerial plant portions or expansion of the phalanx.

The highly variable hydrological regime of a typical flooded plain exerts far higher degrees of disturbance and stress on the component plant communities than those found in hydrologically inactive swamps. In clear contrast to the situation in swamps, flooded plain plant communities develop distinct patterns of vegetation zonation in response to the interaction of stresses imposed by both the degree and duration of inundation and the degree of exposure (Furness & Breen 1980) and the habitat is temporally unstable in comparison to swamps. The range of morphological and anatomical features and life history strategies possessed by flooded plain plants therefore reflect patterns of resource allocation that are adapted to the variability of the environment.

Arid zone flooded plains that experience intense water stresses imposed by prolonged between-flood desiccation favour colonization by ruderal plants with short life cycles. Survival during stressful periods is achieved by means of dormant underground perennating organs. The production of large quantities of desiccation-resistant seeds provides an important mechanism for surviving longer time-scale catastrophic events such as droughts. Conversely, between-flood desiccation stresses are reduced or absent in more moist tropical and temperate zone flood-plains, allowing the development of slow-growing evergreen and deciduous forests, respectively.

One of the commonest strategies developed by arid-zone floodplain grasses, such as *Vossia* and *Echinochloa* which inhabit the wetter areas (Smith 1976, Furness 1981), combines tolerance of anoxic stress with rapid growth rates during and immediately after floods and little or no growth during dry periods. These plants respond to the onset of flooding by rapid extension of their hollow stolons, thus projecting their photosynthetic tissues above the water surface and permitting the transport of oxygen to their roots. In situations where sustained high water levels are maintained after the initial flooding phase, rapid stolon elongation across the water surface provides a simple yet effective means of surmounting a physical barrier and competing for new habitats. Similarly, this strategy allows plants to occupy newly-exposed, denuded surfaces after flood levels have dropped. The stoloniferous or 'guerilla' growth form (Lovett-Doust 1981) is shared by many aquatic and terrestrial primary colonizers. This growth form provides a high degree of 'parental care' for the offshoot, while still allowing the stolon to be lost with little or no damage to the adult plant if environmental stresses are too severe. A similar strategy possessed by many sub-merged aquatic plants (e.g. *Nymphaea* and *Potamogeton*) is the ability to respond to rising water levels by elongation of their leaf petioles (Sculthorpe 1967). Here, rising water levels provide the cues that trigger the plant's response. Adaptation to signals of forthcoming favourable or unfavourable events is an obvious feature of the regenerative phase of most floodplain plants.

Physiological processes

The permanent inundation of swamp sediments and slow water renewal rates combine to reduce markedly the rates of gas exchange between sediments and the atmosphere. Thus, the metabolism of soil organisms dictates the levels of gases such as nitrogen, oxygen, hydrogen sulphide and methane (Howard-Williams 1985). The most important conse-quences for aquatic plants are that oxygen concentrations are greatly reduced and oxidation-reduction potentials are generally negative within a few centimetres of the sediment surface (Brinson *et al.* 1981).

Under these reduced conditions, high concentrations of iron, manganese and hydrogen sulphide can have toxic effects on plants as can the internal accumulation of the products of anaerobic metabolism (Barber 1961, Mendelssohn *et al.* 1981, Armstrong 1982). These stresses eliminate many plant species and only a few can tolerate the conditions that prevail. However, because of reduced interspecific competition, they become dominant and tend to form monospecific stands. An effective solution to these problems is essential for the survival of plants in flooded habitats, and wetland plants have developed several strategies designed to exclude or tolerate toxins and ensure adequate supplies of oxygen to their root systems.

Under normal soil water conditions, oxygen entering the root from the soil is sufficient to supply the demands of the cortex (Armstrong 1982). However, in the absence of an external supply, oxygen must be transported to the root system via the alternative route of intracellular spaces that connect to the atmosphere via stomata or lenticels. In the case of plants that occupy habitats which are flooded periodically, this route is seldom well developed. Here, strategies are usually based on the development of dormancy or on mechanisms allowing tolerance of toxin accumulation until oxygen supplies are restored during the dry (exposed) season (Crawford 1980).

A variety of altered biochemical transformations have also been identified as means where different aquatic plants prevent the build-up of toxic metabolites, particularly ethanol, produced during anaerobic dark fixation of carbon dioxide (Barber 1961, Sculthorpe 1967). In plant species that are intolerant of flooding, these include the excretion of ethanol into the surrounding anaerobic soil (Crawford 1980) or the rapid induction of alcohol dehydrogenase to detoxify the ethanol (Hook & Scholtens 1980). In contrast, flood-tolerant species tend to prevent the formation of ethanol and accumulate neutral or less toxic compounds such as lactate, malate, succinate, shikimic acid or amino acids, instead (Chirkova 1980, Crawford 1980). These compounds constitute an 'oxygen debt' and can be either accumulated until more favourable conditions prevail, or they can be transported from the roots to the leaves via the transpirational stream (Hook & Scholtens 1980). Oxidative conversions in the leaves then detoxify these compounds (Chirkova 1980, Crawford 1980).

The provision of oxygen to the roots via the stomata has important implications for plant form and ecosystem functioning. Usually, an adequate diffusion path for oxygen takes the form of hollow internodes (e.g. *Phragmites*) or extensive aerenchymatous tissues (e.g. *Typha, Cyperus papyrus*), and porosities may reach 60% (Armstrong 1982). The structure of the aerial portions of herbaceous swamp plants therefore represents a precise compromise between the provision of a diffusion pathway to satisfy the below-ground oxygen demand, and a requirement for structural rigidity which permits the aerial parts to compete for space and light. Consequently, the aerial portions of swamp plants are low in bulk and high in fibre and function as organs that acquire carbon and oxygen from the atmosphere and 'pump' these to the underground organs. These, in turn, act as a reservoir of high-energy carbohydrates and proteins but, being below ground, are generally unavailable to higher trophic levels. In addition, the fibrous nature of the aerial portions of most swamp plants renders them poor food for herbivores (Boyd 1970).

The tall, persistent, fibrous aerial portions of swamp plants (e.g. *Phragmites*) confer a definite advantage in competition for the acquisition of light and space. When allied to a dense clonal or phalanx growth form (Lovett-Doust 1981) the species is able to efficiently exploit all available site resources, suppress competitors by reducing the availability of light, and promote local site domination, often for many years (Hutchings & Bradbury 1986).

Limitations to nutrient supply (for example, by reduced hydrological inputs) can be considered to constitute an important stress that might severely limit primary production in swamps (Brown 1981). However, the very high productivity of most swamp plants (Westlake 1964, Howard-Williams 1985) suggests that the generalization does not, in fact, hold true. The reduction in nutrient inputs due to reduced hydrological inputs, combined with a lack of disturbance, leads to an accumulation of resources over time; thus, the standing stock of nutrients in most swamps is high and nutrient cycling pathways are essentially closed (Howard-Williams 1985). The inherent stability of swamp communities and the efficiency with which nutrients and resources are cycled within the community is reflected in the great age (several thousand years) recorded for swamps in many parts of the world (Thompson & Hamilton 1983, Howard-Williams 1985). Nutrient stresses in swamps should therefore be seen in the context of low rates of supply and not necessarily low availability.

Slow nutrient renewal rates have several implications for plant nutrition and nutrient conservation strategies. In particular, swamp plants such as *Phragmites* and *Typha* can translocate nutrients from senescent to growing shoots, within individual ramets (Dickerman & Wetzel 1985), and possess perennial rhizomes that store nutrients during adverse conditions. Rapid response to environmental cues triggers development of an aerial leaf canopy fuelled by nutrient and carbohydrate reserves translocated from the rhizomes (Dickerman & Wetzel 1985). The development of dense, monospecific stands (phalanxes) through vigorous vegetative growth confers an additional competitive advantage over other species.

In contrast to the situation in swamps, nutrient cycles in flooded plains are generally open; large quantities of resources are brought in and removed during seasonal floods either by physical transport or by initiating decomposition of organs not tolerant of inundation. Thus, relatively few resources are stored within the system. This periodic flushing of dead and living plant biomass from a floodplain during floods, combined with marked dry season water stress, prevents the build up of peat, arrests vegetation successions in immature stages and removes an important feedback loop where plants and peat retard water movements (Heinselman 1975, Gosselink & Turner 1978). Water level fluctuations are often large and rapid (relative to swamps) and impose an additional stress in the form of an alternation between waterlogged (anoxic) and dry conditions. On the flooded plains of semi-arid areas, in contrast to those of more moist tropical or temperate areas, low-growing communities of ruderals and stress-tolerators (e.g. *Cynodon dactylon*; Furness & Breen 1982, Heeg & Breen 1982) develop rapidly and large quantities of nutrients become incorporated in plant biomass. Aerial plant parts are lost during both favourable and unfavourable periods and large quantities of nutrients and energy are allocated to underground storage organs and sexual reproduction strategies. These allocation patterns and the combination of severe disturbance and stress militate against large stature plants, with much supportive and structural tissue, in all but the highest elevation habitats. Short stature and rapid growth rates under fertile conditions provide, in turn, for a rapid turnover of aerial tissues and decrease the need for adaptations designed to prevent the loss of these tissues. Most of the plants growing on semi-arid flooded plains thus provide large nutrient and energy contributions to both the grazing and detrital food chains during wet and dry periods, respectively (Furness & Breen 1982, 1985). By linking these divergent systems the possibilities for retention of resources are limited and the system is dependent upon large inputs in order to maintain productivity. For this reason, upstream impoundments which reduce silt inputs to floodplains can have disastrous consequences.

In contrast, plants occupying the flooded plains of wetter tropical and temperate regions experience less severe water stresses between floods. A flora dominated by large-stature evergreen and deciduous trees develops, for example the Cypress forests of the south-eastern United States of America, and nutrient cycling via grazers is greatly reduced. Instead, the detrital food chain, fuelled by litter and leaf fall, is more prominent and the large quantities of nutrients and resources that are allocated to structural tissues are thus retained within the system (Menges & Waller 1983).

Reproductive strategies

Vegetative and sexual reproduction are essential phases of a plant's life cycle, expressing the outcome of selection pressures that have operated throughout the organism's evolutionary development. Both processes are sensitive to environmental cues and stresses, and combine with morphological and physiological attributes to regulate a plant's response to its environment. Thus, they are largely responsible for the success or failure of the plant to survive in a particular habitat. However, the true adaptation of a plant to a particular environment has been defined by Solbrig (1981) as the capability (of the individual, population or species) to undergo successful sexual reproduction in that environment. Clearly, therefore, growth and multiplication of a species by purely vegetative means does not denote adaptation, rather it indicates tolerance of the environmental constraints. In this context, a given reproductive strategy can be defined by species attributes which reflect the impacts of, and responses elicited by, several different stresses or disturbances, each with possibly conflicting selective pressures (Grime 1979, Menges & Waller 1983). Consequently, a single species can display contrasting reproductive strategies when grown in different environments. Indeed, the extreme variability of most aquatic environments selects for plants possessing a high degree of reproductive plasticity, favouring species with multiple regenerative adaptations (Grime 1979).

Each reproductive stratagem involves a trade-off between the investment of energy and materials in vegetative growth and sexual reproduction (Grime 1979, Solbrig 1981, Menges & Waller 1983). Thus, seed production is favoured at the expense of vegetative growth, and vice versa. Whenever resources are in limited supply, sexual reproduction will reduce the competitiveness and survival ability of the adult (Solbrig 1981). The optimum reproductive stratagem of a plant must therefore balance the options of both sexual and vegetative reproduction; the adoption of a particular process reflects the plant's ability to tolerate, adapt to, or overcome the stresses, disturbances and signals imposed by the environment.

In this context, the stress-tolerant competitive plants that dominate the hydrologically stable, yet highly stressed, habitat of swamps reproduce principally by vegetative reproduction. Sexual reproduction occurs infrequently and both the habitat and the structure of the plant community itself provide strong barriers that reduce success. Reduced water movements into and out of the habitat prevent hydrochory, while unfavourable substratum conditions and unsuitable sub-canopy light levels prevent seed germination within an established community (Howard-Williams 1972). In order to overcome these barriers and thereby extend their range, most swamp plants employ temporarily a ruderal-type sexual reproductive stratagem. Inflorscences are elevated above the canopy level and large numbers of seeds are produced by anemophilous pollination, seed dispersal being accomplished by anemochory (Schulthorpe 1967).

In the highly variable range of habitats that occur on flooded plains, the degree of stress and disturbance experienced by each community dictates the reproductive statagem employed. Typically, repetitive disturbances due to floods arrest the vegetation in immature stages, and promote the development of ruderal plants with a short life cycle and a large sexual reproductive output (Grime 1979). In the case of flooded plains with low topographic relief, these ruderal plants can form large monospecific communities. Many of these short-lived floodplain plants also develop a variety of vegetative propagules such as rhizomes, tubers and turions (Mitchell & Rogers 1985) in addition to large numbers of drought-resistant seeds. The vegetative propagules survive unfavourable seasonal conditions and ensure re-colonization of the habitat. Drought-resistant seeds, on the other hand, are not important for re-colonization of a seasonal basis; rather, they provide the means of surviving catastrophic disturbance such as long time scale droughts (Mitchell & Rogers 1985). Adaptation to signals of forth-coming environmental conditions leads to a shortening of response times and permits flood-plain plants to exploit favourable periods to the full.

SWAMPS AND FLOODED PLAINS: A COMPARISON

The permanently inundated hydrological conditions that characterize swamps favour the establishment of those species possessing life history characteristics which enable biotic isolating mechanisms to dominate the community structure. Feedback effects on both community and ecosystem properties amplify biotic and abiotic barriers and dampen the effects of disruptive mechanisms. Ultimately, the community has a low species diversity and species composition remains relatively stable and homogeneous. In contrast, the disruptive effects of the variable hydrological regime found on most flooded plains restricts the development of biotic isolating mechanisms, thus allowing transporting mechanisms to dominate. The distribution of plant communities within the habitat is largely governed by interactions between topography and the hydrological regime and the species composition of individual communities is dictated by the ability of each species to tolerate the stresses and disturbances imposed by the hydrological regime. Species diversity can be low but species composition changes rapidly in response to irregularities in the hydrological regime and the communities are spatially heterogeneous.

Swamp communities are thus adapted to low rates of resource and information supply and tolerance of toxin-induced stresses brought about by permanent inundation. Resources are conserved within both individual species and the community with the concomitant development of rapid internal nutrient cycling, predominantly via relatively inefficient fungal and bacterial decomposition pathways. Low or weak rates of transport into and out of the community accentuate the conservation of resources within the plant community. Conversely, the plant communities of flooded plains are well adapted to highly variable rates of resource and information supply. In addition, rapid rates of resource transfer both within and out of the community lead to few resources being conserved within the system and the predominance of open nutrient cycles.

Ultimately, swamps and flooded plains can be seen in terms of different combinations of isolating and transporting mechanisms and as occupying two extremes of a hydrological gradient. However, the structure and functioning of these respective ecosystems is largely dependent on the physical presence of plants and their population dynamics and metabolism.

REFERENCES

Adams, D. E. & Anderson, R. C. (1980) Species responses to a moisture gradient in central Illinois forests. Amer. J. Bot. 67: 381—392.

Adams, M. S., Titus, J. & McCracken, M. (1974) Depth distribution of photosynthetic activity in a *Myriophyllum spicatum* community in Lake Wingra. Limnol. and Oceanogr. 19: 457—467.

Armstrong, W. (1982) Waterlogged soils. *In*: Etherinton, J. R. (ed.), Environment and Plant Ecology. John Wiley and Sons, New York.

Barber, D. A. (1961) Gas exchange between *Equisetum limosum* and its environment. J. exper. Bot. 12: 243—251.

Bayly, C. L. & O'Neill, T. A. (1972) Seasonal ionic fluctuations in a *Typha glauca* community. Ecology 52: 714—719.

Björk, S. (1967) Ecologic investigations of *Phragmites communis*. Folia Limnologica Scandinavia 14.

Boyd, C. E. (1970) Production, mineral accumulation and pigment concentrations in *Typha latifolia* and *Scirpus americanus*. Ecology 51: 285—290.

Brinson, M. M., Lugo, A. E. & Brown, S. (1981) Primary productivity, decomposition and consumer activity in freshwater wetlands. Annual Review of Ecology and Systematics 12: 123—161.

Brown, S. (1981) A comparison of the structure, primary productivity, and transpiration of cypress ecosystems in Florida. Ecological Monographs 51: 403—427.

Brown, S. & Lugo A. E. (1982) A comparison of structural and functional characteristics of saltwater and freshwater forested wetlands. *In*: Gopal, B., Turner, R. E., Wetzel, R. G. and Whigham, D. F. (eds.), Wetlands, Ecology and Management. International Scientific Publications, Jaipur.

Bucholz, K. (1981) Effect of minor drainages on woody species distributions in a successional floodplain forest. Canad. J. Forest Res. 11: 671—676.

Buttery, B. R. & Lambert, J. M. (1965) Competition between *Glyceria maxima* and *Phragmites communis* in the region of Surlingham Broad. I. The competition mechanism. J. Ecol. 53: 163—181.

Buttery, B. R., Williams, W. T. & Lambert, J. M. (1965) Competition between *Glyceria maxima* and *Phragmites communis* in the region of Surlingham Broad. II. The fen gradient. J. Ecol. 53: 183—195.

Calow, P. & Townsend, C. R. (1981) Energy, Ecology and Evolution. *In*: Townsend, C. R. & Calow, P. (eds.), Physiological Ecology: An Evolutionary Approach to Resource Use. Blackwell Scientific Publishers, Oxford.

Chirkova, T. V. (1980) Some regulatory mechanisms of plant adaptation to temporal anaerobiosis. *In*: Hook, D. D. & Crawford, R. M. M. (eds.), Plant Life in Anaerobic Environments. Ann Arbor Science Publishers, Ann Arbor.

Cole, N. H. A. (1973) Soil conditions, zonation and species diversity in a seasonally flooded tropical grass-herb swamp in Sierra Leone. J. Ecol. 61: 831—847.

Colinvaux, P. A. (1973) Introduction to Ecology. John Wiley and Sons, New York. 621 pp.

Connel, J. H. & Slatyer, R. O. (1977) Mechanism of succession in natural communities and their role in community stability and organization. Amer. Naturalist 111: 1119—1145.

Conner, W. H. & Day, J. W. (1982) The ecology of forested wetlands in the south eastern United States. *In*: Gopal, B., Turner, R. E., Wetzel, R. G. and Whigham, D. F. (eds.), Wetlands, Ecology and Management. International Scientific Publications, Jaipur.

Conner, W. H., Gosselink, J. G. & Parrondo, R. T. (1981) Comparison of the vegetation of three Louisiana swamp sites with different flooding regimes. Amer. J. Bot. 68: 320—321.

Crawford, R. M. M. (1980) Metabolic adaptations to anoxia. *In*: Hook, D. D. and Crawford, R. M. M. (eds.), Plant Life in Anaerobic Environments. Ann Arbor Science Publishers, Ann Arbor.

DeLaune, R. D., Patrick, W. H. & Brannon, J. M. (1976) Nutrient transformations in Louisiana salt marsh soils, 38 pp., Publication number LSU—J—76—009, Center for Wetland Resources, Louisiana State University, Baton Rouge.

den Hartog, C. (1982) Architecture of macrophyte — dominated aquatic communities. *In*: Symoens, J. J., Hooper, S. S. and Compère, P. (eds.), Studies on Aquatic Vascular Plants. Royal Botanical Society of Belgium, Brussels, pp. 222—234.

Dickerman, J. A. & Wetzel, R. G. (1985) Clonal growth in *Typha latifolia*: population dynamics and demography of the ramets. J. Ecol. 73: 535—552.

Drury, W. H. and Nisbet, I. C. T. (1973) Succession. J. Arnold Arboretum 54: 331—368.

Franz, E. H. and Bazzaz, F. A. (1977) Simulation of vegetation response to modified hydrologic regimes: a probabilistic model based on niche differentiation in a floodplain forest. Ecology 58: 176—183.

Furness, H. D. (1981) The plant ecology of seasonally flooded areas of the Pongola River Floodplain, with particular reference to *Cynodon dactylon* (L.) Pers. Ph.D. Thesis, University of Natal, Pietermaritzburg, South Africa.

Furness, H. D. & Breen, C. M. (1980) The vegetation of seasonally flooded areas of the Pongola River Floodplain. Bothalia 13: 217—231.

Furness, H. D. & Breen, C. M. (1982) Decomposition of *Cynodon dactylon* (L.) Pers. in the seasonally flooded areas of the Pongola River floodplain: pattern and significance of dry matter and nutrient loss. Hydrobiologia 97: 119—126.

Furness, H. D. & Breen, C. M. (1985) Interaction between period of exposure, grazing and crop growth rate of *Cynodon dactylon* (L.) Pers. in seasonally flooded areas of the Pongolo river floodplain. *Hydrobiologia* 126: 65—73.

244

Gallagher, J. L. (1975) Effect of an ammonium nitrate pulse on the growth and elemental composition of natural strands of *Spartina alterniflora* and *Juncus roemerianus* Amer. J. Bot. 62: 644—648.

Gaudet, J. J. (1977) Uptake, accumulation and loss of nutrients by Papyrus in tropical swamps. Ecology 58: 415—422.

Gosselink, J. G. & Turner, R. E. (1978) The role of hydrology in freshwater wetland ecosystems. *In*: Good, R. E., Whigham, D. F. and Simpson, R. L. (eds.), Freshwater Wetlands: Ecological Processes and Management Potential. Academic Press, New York.

Good, R. E., Whigham, D. F. & Simpson, R. L. eds. (1978) Freshwater Wetlands: Ecological Processes and Management Potential. Academic Press, New York.

Gosselink, J. G., Bayley, S. E., Conner, W. H. & Turner, R. E. (1981) Ecological factors in the determination of riparian wetland boundaries. *In*: Clark, J. R. & Benforado, J. (eds.), Wetlands of bottomland hardwood forests. Developments in Agricultural and Managed-Forest Ecology, 11. Elsevier Scientific Publishing Company, Amsterdam.

Gregory, K. J. & Walling, D. E. (1973) Drainage basin form and process. A Geomorphological Approach. Edward Arnold, London.

Grime, J. P. (1979) Plant strategies and vegetation processes. John Wiley and Sons, New York.

Heeg, J. & Breen, C. M. (1982) Man and the Pongolo Floodplain. South African National Scientific Programmes Report No. 56. CSIR, Pretoria.

Heinselman, M. L. (1975) Boreal peatlands in relation to environment. *In*: Hasler, A. D. (ed.), Coupling of Land and Water Systems. Springer-Verlag, New York.

Hermy, M. & Stieperaere, H. (1981) An indirect gradient analysis of the ecological relationships between ancient and recent riverine woodlands to the south of Bruges (Flanders, Belgium). Vegetatio 44: 43—49.

Hook, D. D. & Scholtens, J. R. (1980) Adaptations and flood tolerance of tree species. *In*: Hook, D. D. & Crawford, R. M. M. (eds.), Plant Life in Anaerobic Environments. Ann Arbor Science Publishers, Ann Arbor.

Horn, H. S. (1974) The ecology of secondary succession. Annual Review of Ecology and Systematics 5: 25—37.

Hosner, J. F. & Minckler, L. S. (1963) Bottomland hardwood forests of southern Illinois — regeneration and succession. Ecology 44: 29—41.

Howard-Williams, C. (1972) Limnological studies in an African swamp: seasonal and spatial changes in the swamps of Lake Chilwa, Malawi, Arch. Hydrobiol. 70: 379—391.

Howard-Williams, C. (1979a) The aquatic environment: II Chemical and physical characteristics of the Lake Chilwa swamps. *In*: Kalk, M., McLachlan, A. J. & Howard-Williams, C. (eds.), Lake Chilwa, Studies of Change in a tropical Ecosystem. Dr W. Junk, The Hague.

Howard-Williams, C. (1979b) Interaction between swamp and lake. *In*: Kalk, M., McLachlan, A. J. & Howard-Williams, C. (eds.), Lake Chilwa, Studies of Change in a tropical Ecosystem. Dr W. Junk, The Hague.

Howard-Williams, C. (1985) Cycling and retention of nitrogen and phosphorus in wetlands: a theoretical and applied perspective. Freshw. Biol. 15: 391—431.

Howard-Williams, C. & Gaudet, J. J. (1985) The structure and functioning of African swamps. *In*: Denny, P. (ed.), The Ecology and Management of African Wetland Vegetation. Dr W. Junk, Dordrecht.

Howard-Williams, C. & Lenton, G. M. (1975) The role of the littoral zone in the functioning of a shallow tropical lake ecosystem. Fresw. Biol. 5: 445—459.

Howard-Williams, C. & Walker, B. H. (1974) The vegetation of a tropical African lake: classification and ordination of the vegetation of Lake Chilwa (Malawi). J. Ecol. 62: 831—854.

245

Hupp, C. R. & Osterkamp, U. R. (1985) Bottomland vegetation distribution along Parsage Creek, Virginia, in relation to fluvial land forms. Ecology 66: 670—681.

Hutchings, M. J. & Bradbury, K. (1986) Ecological perspectives on clonal herbs. Bioscience 36: 178—182.

Ikusima, I. (1970) Ecological studies on the productivity of aquatic plant communities: IV Light Conditions and community photosynthetic production. Bot. Magazine (Tokyo) 83: 330—341.

Johnson, W. C., Burgess, B. L. & Keammerer, W. K. (1976) Forest overstory vegetation on the Missouri River floodplain in North Dakota. Ecological Monographs 45: 59—84.

Jones, M. B. & Muthuri, F. M. (1985) The canopy structure and microclimate of Papyrus (Cyperus papyrus) swamps. J. Ecol. 73: 481—491.

Junk, W. (1970) Investigations on the ecology and production biology of the floating meadows (Paspalo—Echinochloetum) on the middle Amazon 1. The floating vegetation and its ecology. Amazoniana 2: 449—495.

Lindsey, A. A., Petty, R. O., Sterling, D. F. & Van Asdell, N. (1961) Vegetation and environment along the Wabash and Tippecanoe Rivers. Ecological Monographs 31: 105—156.

Linthurst, R. A. (1979) The effect of aeration of the growth of Spartina alterniflora Loisel. Amer. J. Bot. 66: 685—691.

Lipkin, Y. (1979) Quantitative aspects of seagrass communities, particularly of those dominated by Halophila stipulacea, in Sinai (northern Red Sea). Aquat. Bot. 7: 119—128. 7: 333—334.

Lovett-Doust, L. (1981) Population dynamics and local specialization in a clonal perennial (Ranunculus repens). 1. The dynamics of ramets in contrasting habitats. J. Ecol. 69: 743—755.

McCarthy, T. S., Ellery, W. N., Rogers, K. H., Cairncross, B. & Ellery, K. (1986) The roles of sedimentation and plant growth in changing flow patterns in the Okavango Delta, Botswana. South Afr. J. Sci. 82: 579—584.

Mendelssohn, I. A. (1979) Influence of nitrogen level, form and application method on the growth response of Spartina alterniflora in North Carolina. Estuaries 2: 106—111.

Mendelssohn, I. A., McKee, K. L. & Patrick, W. H. (1981) Oxygen deficiency in Spartina alterniflora roots: metabolic adaptation to anoxia. Science 214: 439—441.

Mendelssohn, I. A., McKee, K. L. & Postek, M. L. (1982) Sublethal Stresses controlling Spartina alterniflora productivity. In: Gopal, B., Turner, R. E., Wetzel, R. G. & Whigham, D. F. (eds.), Wetlands, Ecology and Management. International Scientific Publications, Jaipur.

Mendelssohn, I. A., Turner, R. E. & McKee, K. L. (1983) Louisiana's eroding coastal zone: Management alternatives. J. Limnol. Soc. Southern Afr. 9: 63—75.

Mendelssohn, I. A. & Seneca, E. D. (1980) The influence of soil drainage on the growth of salt marsh cordgrass Spartina alterniflora in North Carolina. Estuarine and coastal marine Sci. 11: 27—40.

Menges, E. S. & Waller, D. M. (1983) Plant strategies in relation to elevation and light in floodplain herbs. Amer. Naturalist 122: 454—473.

Mitchell, D. S. & Rogers, K. H. (1985) Seasonality/aseasonality of aquatic macrophytes in southern hemisphere inland waters. Hydrobiologia 125: 137—150.

Nixon, S. W. (1980) Between coastal marshes and coastal waters — a review of twenty years of speculation and research on the role of salt marshes in estuarine productivity and water chemistry. In: Hamilton, P. & MacDonald, K. B. (eds.), Estuarine and Wetland Processes, with Emphasis on Modelling. Plenum Press, New York.

Odum, E. P. (1971) Fundamentals of Ecology, 3rd Edn. W. B. Saunders Co., Philadelphia.

246

Patten, B. C. & Odum, E. P. (1981) The cybernetic nature of ecosystems. Amer. Naturalist 118: 886—895.

Rees, W. A. (1978) The ecology of the Kafue Lechwe: soils, water levels and vegetation. J. appl. Ecol. 15: 163—176.

Richardson, C. J., Tilton, D. L., Kadlec, J. A., Chamie, J. P. M. & Wentz, W. A. (1978) Nutrient dynamics of northern wetland ecosystems. In: Good, R. E., Whigham, D. F. & Simpson, R. L. (eds.), Freshwater Wetlands: Ecological Processes and Management Potential. Academic Press, New York.

Robertson, P. A., Weaver, G. T. & Cavanaugh, J. A. (1978) Vegetation and tree species patterns near the northern terminus of the southern floodplain forest. Ecological Monographs 48: 249—267.

Rörslette, B. (1985) Death of submerged macrophytes — actual field observations and some field implications. Aquat. Bot. 22: 7—19.

Sastroutomo, S. S. (1981) Turion formation, dormancy and germination of curly pondweed, Potamogeton crispus L. Aquat. Botany 10: 161—173.

Schumm, S. A. (1977) The Fluvial System. John Wiley and Sons, New York.

Schulthorpe, L. D. (1967) The Biology of Aquatic Vascular Plants. Edward Arnold, London.

Seaman, M. T., Scott, W. E., Walmsley, R. D., van der Waal, B. C. S. & Toerien, D. F (1978) A limnological investigation of Lake Liambezi, Caprivi. J. limnol. Soc. Southern Afr. 4: 129—144.

Sheppey, W. & Osborne, T. (1971) Patterns of use of a floodplain by Zambian mammals. Ecological Monographs 41: 179—205.

Solbrig, O. T. (1981) Energy, information and plant evolution. In: Townsend, C. R. & Calow, P. (eds.), Physiological Ecology: an Evolutionary Approach to Resource Use. Blackwell Scientific Publishers, Oxford.

Smith, P. A. (1976) An outline of the vegetation of the Okavango drainage system. In: Proceedings of the Symposium on the Okavango Delta and its future utilisation. The Botswana Society, Gaberone.

Spence, D. H. N. (1964) The macrophyte vegetation of freshwater lakes, swamps and associated fens. In: Burnett, J. A. (ed.), The Vegetation of Scotland. Oliver Boyd, Edinburgh.

Spence, D. H. N. (1982) The zonation of plants in freshwater lakes. Advances in Ecological Research 12: 37—125.

Spence, D. H. N. & Chrystal, J. (1970a) Photosynthesis and zonation of freshwater macrophytes. I. Depth distribution and shade tolerances. New Phytol. 69: 205—215.

Spence, D. H. N. & Chrystal, J. (1970b) Photosynthesis and zonation of freshwater macrophytes. II. Adaptability of species of deep and shallow water. New Phytol. 69: 217—227.

Teal, J. M. (1962) Energy flow in the salt marsh ecosystem of Georgia. Ecology 43: 614—624.

Thompson, K. (1976) Swamp development in the headwaters of the White Nile. In: Rzoska, J. (ed.), The Nile, Biology of an Ancient River. W. Junk, The Hague, Monographiae Biologicae 29: 177—196.

Thompson, K. & Hamilton, A. C. (1983) Peatlands and swamps of the African continent. In: Gore, A. J. P. (ed.), Mires: Swamps, Bog, Fen and Moor: Ecosystems of the World. Elsevier Publishing Company, Amsterdam.

Valiela, I. & Teal, J. M. (1974) Nutrient limitation in salt marsh vegetation. In: Reimold, R. J. & Queen, W. H. (eds.), Ecology of Halophytes. Academic Press, New York.

Van Wijk, R. J. & Trompenaars, H. J. A. J. (1985) On the germination of turions and the life cycle of Potamogeton trichoides Cham. et Schld. Aquat. Bot. 22: 165—172.

Verhoeven, J. T. (1980) The ecology of *Ruppia*-dominated communities in western Europe. II. Synecological clarification. Structure and dynamics of the macroflora and macrofauna communities. Aquat. Bot. 8: 1—85.

Vesey-FitzGerald, D. F. (1963) Central African grasslands. J. Ecol. 50: 243—274.

Viner, A. B. (1975) The supply of minerals to tropical rivers and lakes (Uganda). *In*: Hasler, A. C. (ed.), Coupling of Land and Water Systems. Springer Verlag, Berlin.

Walker, D. (1970) Direction and rate in some British postglacial hydroseres. *In*: Walker, D. & West, R. (eds.), The Vegetational History of the British Isles. Cambridge University Press, Cambridge.

Weisser, P. J. (1978) A conceptual model of a siltation system in shallow lakes with littoral vegetation. J. limnol. Soc. Southern Africa 4: 145—149.

Welcomme, R. L. (1974) The Fisheries Ecology of African Floodplains. Report FI/FPSZ/74/4. Food and Agriculture Organization of the United Nations, Rome.

Welcomme, R. L. (1979) Fisheries Ecology of Floodplain Rivers. Longman, New York.

Westlake, D. F. (1964) Light extinction, standing crop and photosynthesis within weed beds. Verhandl. intern. Verein. theoret. angew. Limnol. 15: 415—425.

Westlake, D. F. (1966) The light climate for plants in rivers. 2 *In*: Bainbridge, R., Evans, G. C. & Rackham, O. (eds.), Light as an ecological Factor. Blackwell, Oxford.

Wetzel, R. G. (1975) Limnology. W. B. Saunders Cy, Philadelphia; Eastbourne, Toronto, xii + 743 pp.

Wetzel, R. G. (1978) Forward and Introduction. *In*: Good, R. E., Whigham, D. F. & Simpson, R. L. (eds.), Freshwater Wetlands: Ecological Processes and Management Potential. Academic Press, New York.

Wetzel, R. G. (1979) The role of the littoral zone and detritus in lake metabolism. Arch. Hydrobiol. (Ergebn. Limnol.) 13: 145—161.

Whittaker, R. H. (1970) Communities and Ecosystems. Concepts in Biology Series, The MacMillan Company, New York.

THE VEGETATION OF FENS IN RELATION TO THEIR HYDROLOGY AND NUTRIENT DYNAMICS: A CASE STUDY

J. T. A. VERHOEVEN, W. KOERSELMAN & B. BELTMAN

INTRODUCTION

Species-rich quaking fens have become rare in Western Europe. In Great Britain, Germany and The Netherlands they have been drained and turned into agricultural land. In The Netherlands small fens have developed over the last few centuries under conditions of moderate nutrient availability in broads resulting from peat excavations. Due to their scattered location in a landscape of heavily fertilized grasslands, these fen systems are influenced by the agricultural activities in their immediate surroundings [fertilization, spraying (or dumping) of manure] and inlet of polluted river water in dry periods.

This chapter deals with the impact of these activities on these fens and their consequences for fen vegetation. Many agricultural activities result in an increased nutrient availability to fen vegetation. The spraying of manure leads to a high evaporation of ammonia that increases the nitrogen input in precipitation, while leaching and run-off of N, P and K from farmland increase the amounts of these elements brought to the fens by groundwater and surface water. The inlet of polluted river water during dry periods influences the fens in various ways: one, the water carries high loads of nitrate and phosphate; two, it has an ionic composition completely different from the natural fen water; three, penetration of the river water into the fens may enhance the mineralization rate of the large stocks of organic N and P stored in the fen soil, e.g. by changing the soil pH.

The composition of mire vegetation is related to water quality (Sjörs 1950, Moore & Bellamy 1974, Van Wirdum 1980, 1981). Mires mainly influenced by rain are defined as 'ombrotrophic' ('bog') and are regarded as nutrient-poor. 'Minerotrophic' ('fen') mires have inputs of rain as well as groundwater or surface water and can be classified according to the importance and the nutrient load of the water flow ('poor fen' versus 'rich fen', see Du Rietz 1954).

The study presented in this chapter addresses the changes in the fen

249

J. J. Symoens (ed.), Vegetation of inland waters. ISBN 90—6193—196—7.

vegetation as a result of progressive eutrophication. Our initial hypothesis was that eutrophication of the fens leads to an increase in primary productivity, to a decrease in the species density of the vegetation (see Van den Bergh 1979, Grime 1979, Wheeler & Giller 1983) and to shifts in the dominance of species. Further, nutrient availability was thought not only to increase as a result of the increased inputs, but also due to indirect effects, e.g. increased rates of N and P mineralization in the soil.

In a complex of small fen sites surrounded by heavily fertilized grasslands near Utrecht (The Netherlands), correlations between species composition, species density and productivity were determined. Further, the cycling of N, P and K in various types of fens distinguished was studied. The pattern of ground and surface water flows was investigated on a regional and local scale. Finally, an attempt was made to calculate a balance of N, P and K inputs and outputs in two fens under different hydrological conditions.

GEOMORPHOLOGY AND HYDROLOGY OF THE VECHTPLASSEN AREA

Our study area includes several polders (see Figure 1) which are bordered in the North by a sandy glacial moraine (Utrechtse Heuvelrug). These polders have been reclaimed since the 12th century.

In our polders, surface water levels are regulated to make the land suitable for agriculture. Water levels are controlled by the polder board by pumping out water during surplus periods and pumping in water during shortage periods when evapotranspiration exceeds the rainfall.

Generally surface water levels also influence groundwater levels which are mostly within 1—1.5 m of the ground surface. Ditches, 1—1.5 m deep and 1—4 m wide, are the primary channels for water transport. They are connected to canals in which the water level is controlled according to a fixed summer and winter water table which is different for each polder. In the study area, surplus water flows eventually to the river Vecht (Figure 1). In dry periods water from the river Vecht is supplied to the polders.

Rain water infiltrating in the sandy ridge (ca. 40 m above sea level) flows to the lower polders via a thick sand layer (see Figure 2). The groundwater under the polders is under a pressure because of the difference in elevation and because of a peat layer sealing the sandy aquifer. This pressure can be measured as the difference between the groundwater level in a piezometer and the surface water table. Groundwater discharges into turf ponds and ditches because the peat layer is absent there. In areas further from the sandy ridge the

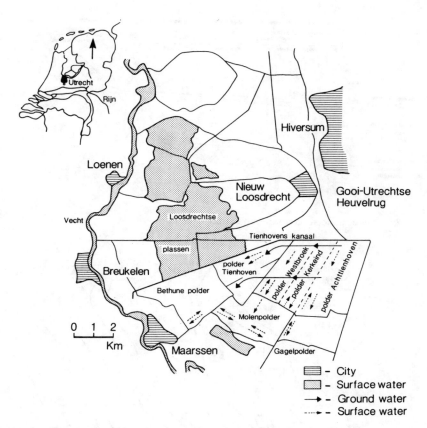

Fig. 1 Study area with different polders and main flow direction of groundwater and surface water.

groundwater level is lower than the surface water level. Here the surface water recharges the groundwater.

The fortnightly measurement of the levels in 100 piezometers and at 40 surface water gauges showed that in most of the polders the groundwater heads are higher than the surface water levels. Further, the piezometers placed 7 m below the soil surface show higher heads than those 2 m below the surface. The results indicate that in most of the area there is groundwater discharge. Upward seepage mainly takes place in the northern and eastern part of the various polders whereas downward seepage occurs in the south-western part.

Maps with groundwater contour lines were drawn to investigate the general flow direction of the groundwater. In Figure 3 such a map is presented for a dry period. The maps for other periods differ only slightly; differences between dry and wet periods are discussed in

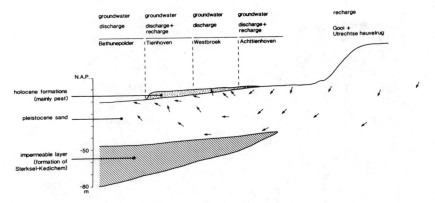

Fig. 2. Geomorphological cross section through the study area.

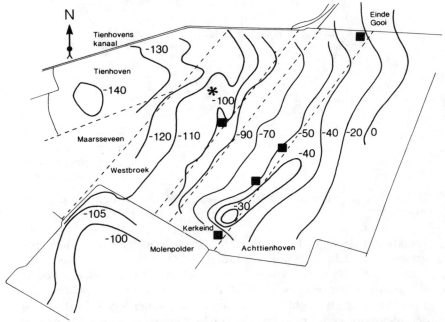

Fig. 3. Groundwater contour map. The flow direction of the groundwater is from high to low water tables perpendicular to the contour lines. * — fen complex area with groundwater discharge, caused by a flow from two sides. ■ — areas with nitrogen (NO_3 and/or NH_4) concentrations above 3 mgN/l. Water tables as cm below sealevel (NAP).

Beltman (1984). The main direction of the groundwater flow is from the east to the west, except for the Tienhoven polder, where the direction of flow is more from the north to the south. These regional flows are directed to the Bethune polder (see Figure 2). This polder is used for drinking water extraction; the water level is kept extremely low so that much groundwater discharges and can easily be extracted.

In the polder Westbroek some remarkable curves in the contour lines are found (see * in Figure 3). The area surrounded by the curves is wet all year round due to groundwater discharge. This is the exact location of one of the major complexes of quaking fens, reed swamps, alder woods and turf ponds. Here the peat cover has been excavated in the 19th century for fuel production resulting in groundwater discharge.

In the Molenpolder the groundwater flows north in the northern part of the polder and south in the southern part. These flow directions seem to be strongly influenced by the presence of canals in the north and west and by the low elevation of the adjacent Bethune polder. The heads of the groundwater (deep and shallow) are continuously lower than the surface water, which makes it a groundwater recharge area.

The water flow patterns can be further illustrated with a cross section of the area (Figure 4). It is clear that soil surface levels and water tables show a discontinuous step-wise decrease in east-west direction, primarily regulated by the water level control in the various polders. The main flow direction is influenced in the north by the flow from the sandy ridge and in the south-east by the low water table of the Bethune polder. If there is a sharp transition in the surface water levels, it is more gradually reflected in the groundwater levels. The slope of the

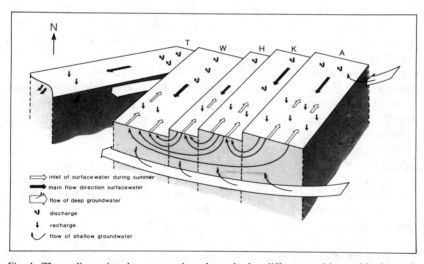

Fig. 4. Three-dimensional cross-section through the different polders with the main groundwater and surface water flow directions. The different standard surface water tables controlled by the polder board result in a discontinual step-wise decrease in east-west direction. T — polder Tienhoven; W — polder Westbroek; H — polder Huis ter Hart; K — polder Kerkeind and A — polder Achttienhoven.

groundwater level near the boundary of the polders is steep, whereas it is very gradual in the middle of the polders. This results in a strong discharge of groundwater at the eastern side of each polder due to the great difference in hydraulic head between groundwater and surface water.

During its flow through the soil the groundwater is influenced by cation and anion exchange processes and by leaching of manure and fertilizers applied on the grasslands and corn fields. A long retention time of water in the soil increases the calcium and the bicarbonate concentration and decreases the hydrogen, potassium, sodium and chloride concentration. The exchange of monovalent for divalent ions is often used to classify the age of different water layers in the soil. Stiff-diagrams (see Figure 5) or the Ionic Ratio *sensu* Van Wirdum (1980, 1981) ($[\frac{1}{2}Ca]/([\frac{1}{2}Ca] + [Cl])$) can be used to characterize water according to the concentrations of major ions. Differences in the ionic composition give information about the type and age of water, i.e. is it atmocline ('rainlike') or lithocline ('groundwater-like') water (Van Wirdum 1980).

To analyse the chemical composition of groundwater and surface water, monthly samples have been taken and analysed with a con-tinuous flow analyser. Four different types of groundwater can be recognised in our study area (see Figure 5). Types IA and IB are characteristic of the discharge areas and mainly found in the polder Westbroek. Both water types are characterized by relatively high concentrations of HCO_3 and low concentrations of SO_4 and Cl. Type IA contains large amounts of Ca, whereas Mg is the predominant cation in IB. Groundwater of type IIA occurs in the sandy ridge where the groundwater is recharged with rain water high in sulphate. Ground-water with a chemical composition of type IIB is found in groundwater recharge areas such as the Molenpolder where polluted river water is supplied during the summer. This water type is characterized by high concentrations of Na, K and Cl.

The land use in the polders (intensive lifestock grazing and crop culture) can be expected to influence the nutrient content of the groundwater. The fertilization of the grasslands does not lead to increased phosphate levels in the groundwater, because phosphorus adsorbs readily to soil particles (Nichols 1983). Orthophosphate con-centrations in groundwater are below 0.5 mg/l. In most of the area the ammonium-N and nitrate-N concentrations in groundwater are very low as well (< 1.0 mg/l). Locally, concentrations higher than 3 mg/l are found (see Figure 3). Where such high concentrations occur in recharge areas, they are directly due to the fertilization and spreading of manure on the pastures or corn fields at the same spot. An example are the fields located near 'Einde Gooi' in the north-west where high concen-

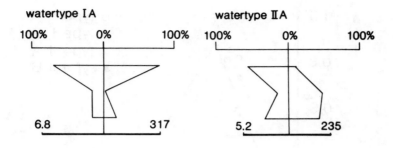

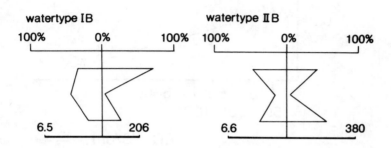

Fig. 5A. Stiff diagrams of the main water types. In each diagram the percentual contribution of [Ca], [Mg] and [Na] + [K] to the total sum of cations are expressed at the left side and the percentual contribution of [HCO₃], [SO₄] and [Cl] to the total sum of anions at the right side.

$$
\begin{array}{l}
\text{Ca} ------|--- \text{HCO}_3 \\
\text{Mg} -----|--- \text{SO}_4 \\
\text{Na} + \text{K} --|-- \text{Cl} \\
\text{pH} -----|-- \text{conductivity (mS/cm)}
\end{array}
$$

trations (between 12 and 25 mg/l nitrate-N) were measured in several piezometers at different depth below soil surface. As the groundwater flows from here eventually to the quaking fens in the polder Westbroek (see Figure 3), the very high N fertilization has a eutrophicating effect beyond the immediately surrounding environment. During its flow through the sandy aquifer, the groundwater becomes anaerobic and nitrate is converted into ammonium. The N content will decrease due to denitrification (up to 90% loss under suitable conditions (ICW 1985)) and due to mixing of different groundwater streams. However, the concentrations of ammonium-N at 'the end of the stream' (e.g. in the groundwater discharging in the polder Westbroek) are in the range of 3 to 5 mg/l which is distinctly above background values. Hence,

256

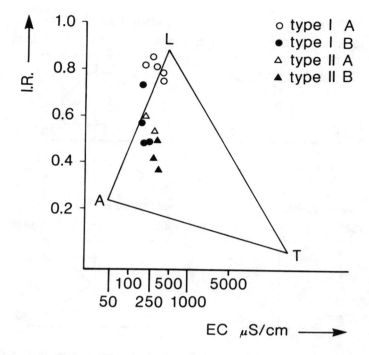

Fig. 5B. Ionic ratio ($[\frac{1}{2}Ca]/([\frac{1}{2}Ca] + [Cl])$) plotted versus electrical conductivity. L, reference point lithocline (ground) water; A, reference point atmocline (rain) water; T, reference point thalassocline (sea) water. Data points for the 4 water types distinguished are indicated.

eutrophication via groundwater is a serious threat to mesotrophic fens in discharge areas.

Another very important source of eutrophication of the fens should be mentioned here. The polder boards supply water to the polders in periods of high evaporation and low rainfall. This water originates from the river Vecht that carries (particularly during such periods) strongly polluted Rhine water high in sodium, chloride, phosphate, nitrate and organic toxins. There is a strong difference between the polders with primarily groundwater discharge and those with recharge with respect to the amount of river water that has to be supplied. In most of the discharge areas, upward groundwater seepage compensates for the losses due to the evapotranspiration and only small amounts of river water have to be supplied. In recharge areas, however, large supplies of river water lead to high inputs of nutrients, sodium and chloride. The high chloride concentrations in the shallow groundwater in polders with recharge (see Figure 5A and B) clearly illustrates the effect of such supplies.

GROUNDWATER AND SURFACE WATER FLOWS
ON A LOCAL SCALE

A more complex situation is found if we descend from the regional scale (polders) to the local scale (fens). The quaking fens have a distinct vertical structure: the vegetation forms a 40 cm thick floating mat consisting of dead and living roots and peat; underneath the mat there is a water layer. This water is mixed with muck and peat to a variable extent; a roughly 10 cm thick solid sapropel layer covers the sandy bottom that is mostly about 1 m below the surface (see Figure 6). Depending on the degree to which the water layer is filled up with peat, the mat floats more or less freely and responds to fluctuations in the water table. The vegetation in the fens is mown each year in July to prevent the succession to alder woodland.

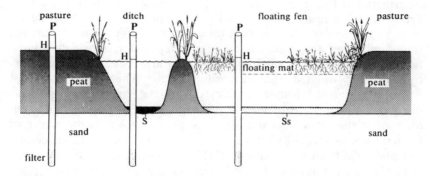

Fig. 6. Cross section through set of landscape elements including a quaking fen, a ditch and a pasture. P — piezometer; H — piezometer head; S — sapropel layer; Ss — sapropel layer of solid peat.

The hydraulic heads measured in the different landscape elements where groundwater discharge occurs (see Figure 6) show slight differences. Under the pasture the groundwater head is higher than under the ditch. In the ditch the peat cover is absent, resulting in groundwater discharge into the ditch. The upward seepage results in a decrease in the pressure and thus in a lower groundwater head. In the quaking fen most of the peat layer has been removed, but there still remains a thin solid sapropel layer with a low permeability. This increases resistance to flow of upwelling groundwater. This results in a groundwater head between those under the pasture and the ditch. Thus, even within the same polder, local differences in hydrology may occur which are related to the current land use as well as land use history.

In our study of eutrophication, special attention was given to surface water infiltration and the movement of polluted water within fens. At two sites (Westbroek I and Molenpolder) water samples were collected at several depths along transects parallel with the water flow direction, and chemically analyzed. The dry and warm summer of 1986 provided excellent conditions for such a study. To maintain sufficiently high water tables in the polder, the polder board pumped in large amounts of polluted river water in both the Molenpolder and polder Westbroek (an exceptional amount of river water was supplied to the latter polder). The river water has, as stated before, a very poor quality. It can be chemically distinguished from 'fen water' by its extremely high sodium and chloride contents, and a very high electric conductivity. Hence, conductivity versus distance diagrams clearly demonstrate the process of surface water infiltration (see Figure 7).

As can be seen from Figure 7 A—D, the Molenpolder fen can be penetrated by surface water from two sides (ditch and turf pond). From June 1 onwards, river water was supplied to raise water levels to a height suitable for agricultural practices. Infiltration of polluted water enhanced the conductivity of the fen water, especially at the borders. However, a small spot in the central part of the fen appears hydrologically isolated from this polluted water. This spot stays almost exclusively rain-fed (ombrotrophic) during the entire summer. This phenomenon can be explained by assuming that rain water that accumulated in the fen during the winter is 'pushed' towards the centre of the fen from two sides by infiltrating surface water, under influence of a low hydraulic head at the centre of the fen and higher hydraulic heads at the borders. It apparently takes more than a dry summer to replace all rain water piled up in the fen during the winter by polluted surface water. The ombrotrophic character of the central part of the fen is reflected in the species composition of the vegetation, especially that of the bryophyte layer. While in the largest part of the fen the *Sphagnum* carpet is dominated by *Sphagnum squarrosum* and *S. fimbriatum*, species indicative of eutrophic conditions, in the central part *S. magellanicum*, a species indicative of more ombrotrophic conditions, is abundant.

In this fen, surface water infiltration mainly takes place below the soil surface. This is because the floating mat rises with rising water levels and is seldom inundated. Sheet flow over the mat therefore seldom occurs; the predominant type of water flow in this fen appears subsurface flow.

Surface water infiltration in the Westbroek I fen is illustrated in Figure 7 E—H. The hydrological setting of this fen differs distinctly from that of the Molenpolder fen. Infiltration occurs from one side only (ditch), and the floating mat is anchored on baulks of solid uncut peat

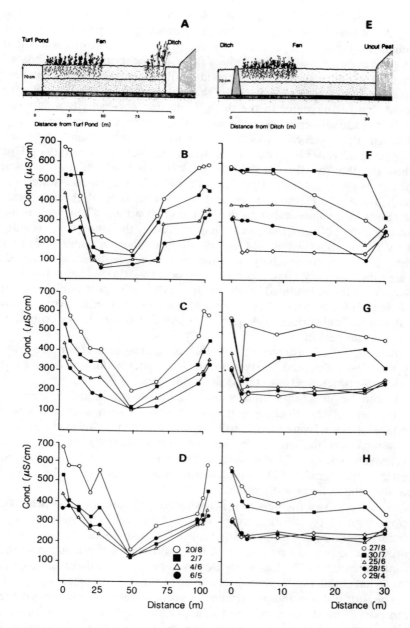

Fig. 7. Electric conductivity of water samples along transects in the Molenpolder fen (A—D) and the Westbroek I fen (E—H). A, E: Schematic representation of the transects. B, F: Conductivity at the soil surface. C, G: Conductivity 20 cm below soil surface. D, H: Conductivity 60 cm below soil surface.

at all sides. Thus, when the water level in the ditch rises due to the supply of river water (from May 22 onwards) the peat mat is very limited in its response, and is consequently flooded by ditch water that spreads over the mat by rapid sheet flow. The peat mat stays inundated during the entire summer. As a result of the occurrence of rapid sheet flow, the conductivity of water at the soil surface of the fen tends to equal that of the water in the ditch. Note that conductivity at the soil surface is already enhanced one week after the first river water supply, although the conductivity of the water in the ditch remains the same (the polluted water has not yet penetrated that far into the polder). This illustrates that overflow of the peat bank between the fen and the ditch due to rising of the water level causes rapid inflow of surface water: the ditch 'bursts over' into the fen and the conductivity of the fen water increases to a value similar to that of the ditch water. When, after some time lag, the polluted river water has arrived at the study site, conductivity of the ditch water and subsequently of the fen water is further enhanced. The conductivity of water at greater depths also increases over time, but only after a certain time lag. This shows that subsurface flow in the fen is inhibited by the presence of a solid peat bank between the fen and the ditch. The observed decrease in conductivity of fen water at the soil surface at August 27 is correlated with heavy rainfall in the previous period.

Although the border of the fen that is furthest away from the ditch appears less influenced by river water, differences within the fen are not nearly so pronounced as in the Molenpolder fen.

It can be concluded that a true floating fen will only rarely be inundated, as the floating mat rides up and down with the water surface. Hence, water flow will be predominantly subsurface, and certain parts of the fen may be isolated from surface waters and almost exclusively rain-fed. Under such circumstances it might be expected that a certain gradient in trophic status develops along a transect that parallels the water flow direction.

However, when the peat mat is anchored on baulks of solid peat, the upper limit of the floating range will be frequently exceeded. Consequently, rapid sheet flow of water over the mat dominates over subsurface flow, providing a source of both nutrients and pollutants to the plant communities. As this water flow is very rapid, gradients in trophic status that parallel water flow direction are not to be expected.

This study shows the importance of surface water infiltration for the transport of nutrients as related to the floating capacity of the root mat. It should be kept in mind, however, that the location of the Westbroek I fen in a groundwater discharge area practically shuts it off from river water, except for very dry periods, whereas the Molenpolder fen is influenced by river water each summer.

SPECIES COMPOSITION OF THE
PLANT COMMUNITIES

A phytosociological analysis of the phanerogam and bryophyte vegetation in the fens resulted in the recognition of three main vegetation types that are correlated with distinct sets of hydrological conditions (see also Vermeer 1985, Vermeer & Verhoeven 1987).

A summary of the most important differential species, the mean biomass production and the hydrological situation of the fen types is given in Table I.

Table I. Differential species, biomass production and hydrological characteristics of fen types.

fen type	differential spp. phanerogams	differential spp. bryophytes	biomass g/m²	hydrology
1a	*Carex diandra* *Carex rostrata* *Pedicularis palustris* *Caltha palustris* *Ranunculus lingua* *Menyanthes trifoliata*	*Calliergon cordifolium* *Plagiomnium affine* *Bryum pseudotriquetrum* *Calliergonella cuspidata*	250—500	seepage of lithocline groundwater
1b	as type 1a, dominance of *Carex acutiformis* and *Carex paniculata*	as type 1a	500—700	seepage of lithocline groundwater, eutrophicated
2	*Carex lasiocarpa* *Carex tumidicarpa* *Carex panicea* *Erica tetralix* *Drosera rotundifolia* *Myrica gale*	*Kurzia pauciflora* *Cephalozia connivens* *Sphagnum flexuosum* *Sphagnum papillosum* *Campylium stellatum* *Fissidens adianthoides*	200—400	isolation from ground and surface water
3	*Carex acutiformis* *Carex paniculata* *Carex disticha*	*Sphagnum squarrosum* *Sphagnum fimbriatum* *Aulacomnium palustre*	500—700	inflow of polluted river water

Phytocoenoses of type 1, a vegetation type with high diversity, occur in areas with upward seepage of 'old' groundwater (water quality type IA and IB in Figure 5). Where fens in such areas are being eutrophicated or not regularly cut, the biomass production is higher and tall *Carex* species more and more dominate the vegetation (type 1b).

Plant communities of type 2 are also very diverse and contain many rare species. They can be found in areas that are more or less isolated from ground and surface water and are, therefore, mainly fed by rain water (water quality type IIA in Figure 5). These communities have a low biomass production and show a thick bryophyte layer dominated by a number of *Sphagnum* species.

On the contrary, the communities of type 3 are few in species and show a high biomass productivity. There is a thick layer of bryophytes primarily dominated by *Sphagnum squarrosum*. Fens with this vegetation type occur in areas with a downward groundwater flow where the water deficit occurring in the summer is suppleted by the inlet of polluted water from the river Vecht (water quality type IIB in Figure 5).

The communities described here may be attributed to previously described plant associations. Vegetation type 1 can be assigned to the *Scorpidio-Caricetum diandrae* (Koch 1926) Westhoff 1968, vegetation type 2 to the *Sphagno-Caricetum lasiocarpae* (Gadeceau 1909) Steffen 1931 and vegetation type 3 is a transition between the *Magnocaricion* and the *Phragmition* (see Westhoff & den Held, 1969).

The species composition and productivity of the fens described here shows large differences with that described by Wheeler (1980a, 1980b, 1980c) and Wheeler & Giller (1983) for fens in the British Norfolk area. There, most fens are cut only once every 4 years and they mostly have lower groundwater tables than in our area; hence they show a much higher biomass production and stronger dominance by tall species such as *Cladium mariscus*, *Phragmites australis*, and *Peucedanum palustre*.

CORRELATION BETWEEN SPECIES DENSITY, PRODUCTIVITY AND THE AMOUNTS OF N, P AND K IN THE VEGETATION AND IN THE SOIL

A general relationship between species density and biomass production for herbaceous vegetation has been proposed by Al-Mufti *et al.* (1977) and Grime (1979). Species density in low-productive stands is indicated to be low; when productivity increases there is initially a rise in the species density until a maximal annual above-ground biomass value of about 510 g/m²; a further increase in productivity is accompanied by decreasing species density. This picture was confirmed by Vermeer & Berendse (1983) who compared fens, reed marshes and grasslands: in their study the biomass range with increasing species density (200—450 g/m²) was associated with the fens, and the range with decreasing species density (450—1000 g/m²) with the reed marshes and grasslands. Wheeler & Giller (1983) found a decreasing species density with increasing biomass production in the Norfolk fens (biomass range

500—3000 g/m²); they stressed the strong variability of species density in the first part of their range (500—1000 g/m²).

In our study we investigated the correlation between species density and biomass production in 37 fen sites in the study area. All sites sampled were subject to annual cutting in the summer. At each site, biomass and species density were determined in 10 replicate 20 × 20 cm samples just before cutting (July); the biomass range (converted to g/m²) was 300—1000. All sampling methods are described in more detail in Vermeer (1985).

There was no significant correlation between the two variables (see Figure 8). Low as well as high species densities are found over the complete biomass range. The variation in species density is probably due to variation in environmental conditions at the site; apparently, there is no direct relation to the nutrient availability. An important factor is that all sites are cut each year, so that litter accumulation is relatively small. This apparently permits a high species density, also in

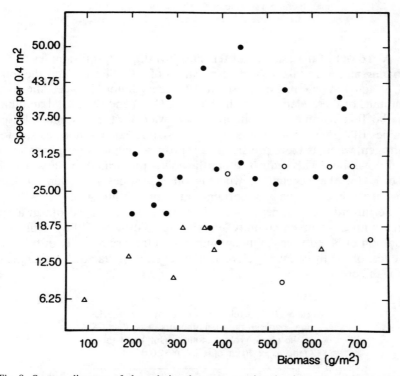

Fig. 8. Scatter diagram of the relation between species density and maximum shoot biomass (July) in 37 fen sites in the Vechtplassen area. ●, vegetation type 1a; ○, vegetation type 1b; △, vegetation type 2 (Vermeer & Verhoeven, 1987).

relatively high-productive situations. There are indications that species density is low at sites where *Sphagnum* spp. form an important component of the vegetation (fen types 2 and 3).

The correlations between the above-ground biomass and the total amount of N, P and K contained therein were investigated for the same sites as indicated above. In Table II the correlation coefficients are given. Significant positive correlations exist for N, P as well as K. This means that a higher productivity is associated with a higher uptake of all three elements.

Table II. Correlation coefficients for correlations between maximum above-ground biomass (July) and the amounts of N, P and K contained therein. *, $p <$ 0.05; **, $p < 0.01$.

max. biomass *versus* total N in biomass	0.348*
max. biomass *versus* total P in biomass	0.628**
max. biomass *versus* total K in biomass	0.478**

At 16 out of the total set of fen sites investigated, soil samples of the floating mat were taken for determination of bulk density, and the total N, P and K content. For these sites the correlation between the total amounts of these elements in the biomass (July) and their total amounts in the floating mat (in peat, litter and water) were determined (see Table III). It is remarkable that no significant relation exists for nitrogen, whereas there are significant positive relations for phosphorus and potassium. This indicates that the soil is the primary reservoir of P and K for the vegetation, whereas other sources (e.g. rain, surface water) are at least equally important for the availability of N.

Fertilization experiments with N, P and K were carried out in a fen area with a plant community of type 1a (Vermeer 1986). Only the addition of N led to a significant increase in the productivity of this fen. Hence, of the three macro-elements, nitrogen is the factor limiting plant growth here.

Table III. Correlation coefficients for correlations between the total amounts of N, P and K in the above-ground biomass (July) and those in the upper 40 cm of soil. *, $p < 0.05$

N in biomass *versus* N in soil	not sign.
P in biomass *versus* P in soil	0.579*
K in biomass *versus* K in soil	0.683*

CYCLING OF N AND P: THE INFLUENCE OF LITTER QUALITY AND ENVIRONMENTAL FACTORS ON N AND P MINERALIZATION

Fens are peat-accumulating ecosystems. Decomposition of the plant litter produced is incomplete and humic compounds together with certain amounts of N and P are stored in the fen soil. This does not mean that remineralization of organic N and P is an insignificant source of nutrients to the fen vegetation; net mineralization per quantity of organic matter may be slower than in terrestrial systems, but the much larger amounts of organic matter in the fen soil may well lead to higher total mineralization per area.

Mineralization of organic N and P is associated with the bacterial decomposition of plant litter and peat. Mineralization rates are dependent on environmental conditions (temperature, pH, redox potential, see Godshalk & Wetzel 1978c) and on the chemical composition of the organic material. Just after death, plant material loses N and P due to leaching of dissolved organic compounds (Godshalk & Wetzel 1978a). Subsequently, the decomposition of the plant litter is accompanied by bacterial immobilization of N and P until the C/N and C/P ratios of the litter have dropped to 'critical' values; from then on the further decomposition is accompanied by mineralization (see Parnas 1975; Godshalk & Wetzel 1978b).

Plant litter in the fens is essentially of three categories: phanerogam shoot material, phanerogam root material, and bryophyte (mainly *Sphagnum*) material. The importance of the phanerogam shoot material is relatively low due to the annual cutting of the fens in July. Bryophyte material is of quantitative importance only in the fens of vegetation types 2 and 3, where a thick *Sphagnum* carpet is found. Plant roots are known to be less degradable than shoots due to a higher fibre and lignin content and lower N and P contents (Sikora & Keeney 1983). *Sphagnum* material is also relatively resistant to decomposition (Clymo 1978).

Mineralization rates were determined *in situ* at 6 fen sites, together covering all vegetation types indicated in Table I. Three of the sites studied are located in the 'Vechtplassen' area, the other three sites are located along a gradient in a fen in 'De Weerribben', another area of fens located near Emmen, The Netherlands. An indication of the vegetation type, productivity and hydrology of the various fen sites is given in Table IV.

The three sites along the gradient ('Stobbe' I to III) are located in a rectangular fen bordering a ditch (Figure 9). Ditch water flows continually into the fen, influencing strongly the water quality at site III. Sites I and II are, however, not reached by this water flow and are mainly fed by rain water.

Table IV. Some characteristics of the fen sites where N and P mineralization was studied.

Site	Location	veg. type	biomass max (g/m²)	hydrology
Molenpolder	Vechtplassen	3	543	downward seepage, river water
Westbroek I	Vechtplassen	1a	295	upward seepage, groundwater
Westbroek II	Vechtplassen	1a	369	upward seepage, groundwater
Stobbe I	Weeribben	2	590	downward seepage, rain water
Stobbe II	Weerribben	2	493	downward seepage, rain water
Stobbe III	Weerribben	1b	1378	downward seepage, lithocl. surface water

Soil material collected at a depth of 20 cm was incubated for 4 to 6 weeks in plastic bottles in the field at the exact site of collection. Available N and P was determined by shaking the soil samples for 1 h with 1M KCl (ammonium extraction) and demineralized water (phosphate and nitrate extraction). By comparing available N and P in fresh soil samples with those in incubated samples, mineralization or immobilization of these elements can be established. Further, an indication of the mineralization rate can be obtained. For further description of the methods see Verhoeven & Arts (1987). The incubation procedure was repeated during the period April-November.

The results of the experiments for the three sites in the 'Vechtplassen' area are given in Figure 10. It is clear that N as well as P mineralization are much faster in the fen in the Molenpolder than in the two fens in Westbroek. In Table V the amounts of N and P mineralized are totalled for the two periods of growth before and after the cutting of the vegetation in July and compared with the amounts contained in the above-ground parts of the vegetation. The amounts of N and P mineralized are similar to or smaller than those contained in the vegetation in the two fens in Westbroek. In the fen in the Molenpolder, however, more N and P is mineralized than is taken up in the above-ground biomass.

Figure 11 gives the results for the gradient in the 'Weerribben' area. It is clear that N and P mineralization are faster at site I and II than at site III. When the total amounts mineralized are compared with the amounts present in plant biomass at the end of the growing season

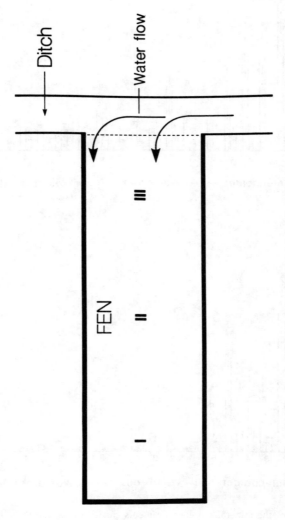

Fig. 9. Location of the sampling sites in the fen at 'De Weerribben'. The fen is bordered by impervious peat walls on three sides and by a ditch on one side.

268

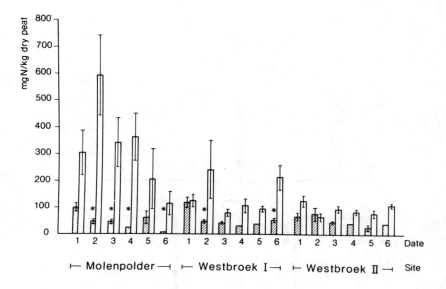

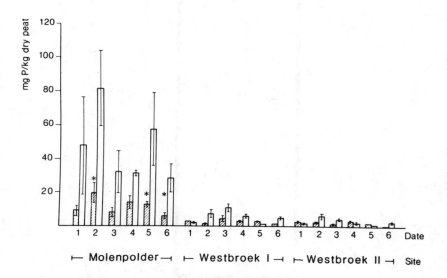

Fig. 10. Results of the mineralization experiments carried out at the three fen sites in the Vechtplassen area. Extractable inorganic N and P, respectively, in fresh soil samples are compared with that in samples that had been incubated for 6 weeks at the site of collection. Bars indicate mean values of 5 samples, vertical lines indicate standard errors. The significance of differences was tested after a two-way analysis of variance. *, $p < 0.05$; ** $p < 0.01$. Dates: 1 — March 27; 2 — May 10; 3 — June, 20; 4 — August 1; 5 — September 19; 6 — November 11.

Table V. The amounts of N and P mineralized compared to the amounts taken up by the above-ground plant parts. The mineralization rates measured at a depth of 20 cm were generalized for the upper 20 cm of the mire soil.

locality	N mineralized (g/m²)		N in plant biomass (g/m²)	
	prior to	after cutting	prior to	after cutting
	April—July	July—Nov.	April—July	July—Nov.
Westbr I	3.4	4.5	4.5	4.5
Westbr II	1.8	2.9	5.5	3.2
Molenp	15.8	8.6	7.2	6.5
	April—August		April—August	
Stobbe I	8.4	—	4.6	—
Stobbe II	7.5	—	3.8	—
Stobbe III	1.4	—	14.3	—

locality	P mineralized (g/m²)		P in plant biomass (g/m²)	
	prior to	after cutting	prior to	after cutting
	April—July	July—Nov.	April—July	July—Nov.
Westbr I	0.17	0.06	0.31	0.25
Westbr II	0.12	0.01	0.24	0.14
Molenp	2.19	1.22	0.66	0.46
	April—August		April—August	
Stobbe I	0.65	—	0.23	—
Stobbe II	0.16	—	0.16	—
Stobbe III	−0.04	—	0.89	—

(Table V) it is remarkable that the amounts of N and P mineralized exceed the amounts present in the above-ground plant parts at the low-productive sites I and II, whereas the amount of N mineralized is low and P is even immobilized at the high-productive site III.

It can be concluded that there is no relation between N and P mineralization rate and productivity or N and P uptake by the vegetation. The results show that there is a striking difference in mineralization rate between fens with a *Sphagnum* carpet and those with a vegetation of primarily phanerogams. N as well as P mineralization are distinctly more rapid in fens with a *Sphagnum* layer. This is apparently due to the different behaviour of *Sphagnum* litter as compared to phanerogam root and shoot litter. *Sphagnum* litter is known to contain a high proportion of refractory compounds primarily associated with the cell wall structures (Clymo 1965) but to release rapidly the N and P

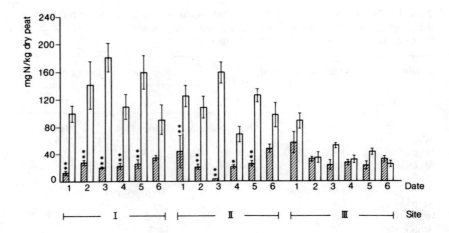

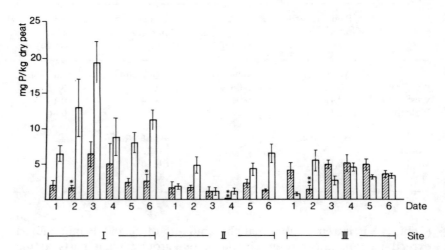

Fig. 11. Results of the mineralization experiments carried out along the gradient in the fen at 'De Weerribben'. Samples were incubated for 4 weeks here. Dates: 1 — May, 21; 2 — June 15; 3 — July 12; 4 — August 8; 5 — September 4; 6 — October 2. For further legends see Fig. 10.

present in the cytoplasm (Sikora & Keeney 1983). The presence of a *Sphagnum* carpet probably influences the decomposition process in an indirect way as well: the pH at the fen surface is around 5.5 in the presence of *Sphagnum* and 6.5 where there is no carpet. The solubility of phosphorus is strongly influenced by the pH (Patrick & Mikkelsen 1971); the high P mineralization in fens with *Sphagnum* carpets may be partly due to the low pH. The results given here give further evidence for the conclusions of Waughman (1980) that N as well as P availability is higher in ombrotrophic bogs than in minerotrophic fens.

A surprising result of our study is the absence of a correlation between N and P mineralization and plant productivity and nutrient uptake. The only conclusion can be that phanerogam growth is limited by other factors than the N and P availability when a *Sphagnum* carpet is present. The limitation might be imposed by an inhibition of nutrient uptake due to the lower pH or to the presence of toxic phenolic compounds produced by *Sphagnum* (Sikora & Keeney 1983). The high productivity at the Stobbe III site is probably due to the continuous input of nutrients via the inflowing ditch water.

NUTRIENT BUDGETS FOR TWO FENS UNDER CONTRASTING HYDROLOGICAL CONDITIONS

Water regime and nutrient availability

In the previous part of this chapter it has been shown that there is a relation between the species composition of fen plant communities and hydrological conditions (see Table I). It is well known that nutrient dynamics in wetlands are a function of both water movement and water chemistry (e.g. Carter *et al.* 1979, Howard-Williams 1985, Carter 1986, LaBaugh 1986). The water budget of wetlands affects their structure and function because it controls the input and output of nutrients, and their retention time in the system. Nutrient input-output budgets provide a framework for analyses of wetland nutrient dynamics. The difference between annual input and output for a given nutrient tells whether that nutrient is being accumulated within the system, is being lost from the system, or is passing through the system. From this, long-term developments of the system, such as eutrophication, may be predicted.

Apart from its direct effects on the nutrient balance of wetlands, the hydrologic setting affects nutrient availability in an indirect way. Phosphate availability, for example, is strongly controlled by adsorbance to Ca, Fe and Al complexes. The extent of adsorption depends on the water and soil chemistry (e.g. pH, contents of Ca, Fe and Al) (Nichols 1983, Richardson 1985, Stumm & Morgan 1981, Reddy & Rao 1983). As the ionic composition of water from both possible sources, rain and groundwater, is essentially different (see Verhoeven 1986), it is clear that the water regime not only affects nutrient dynamics of wetlands in a direct way, but that it also may have important indirect effects on the soil nutrient availability.

Despite the critical role of hydrology in the determination of chemical balances for wetlands, so far only a limited number of studies have been based on a comprehensive water balance study (Carter 1986, LaBaugh 1986, Howard-Williams 1985). With regard to mires a few nutrient budgets have been published for bogs (Crisp 1966, Rosswall & Granhall 1980, Hemond 1983) but there are no data available for a nutrient budget for fens (Verhoeven 1986). This is probably due to the more complex hydrological regime of minerotrophic mires in general, as compared with ombrotrophic mires.

In the following section of this chapter we present preliminary results of a study on the hydrological and nutrient budgets of two quaking fens in the 'Vechtplassen' area. Two fens of different trophic status were selected for the study. One site (Westbroek I) belongs to vegetation type 1a (see Table I), indicative of mesotrophic soil conditions. The vegetation at the other site (Molenpolder) was classified as type 3 indicative or more eutrophic soil conditions. The fens are located in polder Westbroek and the Molenpolder, respectively (see Figure 1). In both fens, the internal cycling of nutrients within the system was studied as well. The fens will be referred to as 'mesotrophic' (Westbroek I) and 'eutrophic' (Molenpolder) below.

It should be kept in mind that the data presented here are preliminary, so that the ultimate budgets may vary from those presented below. However, we do not think that future refinements of the balance will seriously alter the trends shown.

Water and nutrient budgets

An estimated water budget for 1985 is given in Table VI. In both fens the major input of water is precipitation, and the major output is due to evapotranspiration in the summer and surface water outflow in the winter. In the mesotrophic fen there is groundwater discharge, whereas in the eutrophic fen groundwater recharge occurs. It is clear that groundwater flow direction has a marked effect on surface water input. Upward groundwater seepage isolates the mesotrophic fen from surface water infiltration during the summer. Surface water inputs in the fens will be higher on average, as the 1985 summer was relatively cold and wet. Calculations of the 1985 N, P and K budget (Table VII) illustrate that the eutrophic fen receives considerably more nutrients via surface water inflow. This is due to the poor water quality of the surface water in the Molenpolder where this fen is located. In this polder, supply of polluted river water takes place during the summer to compensate for losses due to evapotranspiration and downward seepage, in order to maintain high surface water levels suitable for agricultural practices. The activities by the polder board lead to a deterioration of the surface

Table VI. Estimated water budget (mm/y) for 1985 for 2 fens.

Site	Westbroek I (mesotrophic)	Molenpolder (eutrophic)
Input		
surface water[1]	48	172
groundwater[2]	124	0
precipitation[3]	798	798
Output		
surface water[1]	488	380
groundwater[2]	0	108
evapotranspiration[4]	482	482

[1] Calculated by difference over periods of one month, taking differences in storage over time into consideration.
[2] Estimated from measurements of hydraulic heads at different depths, and measurements of the hydraulic conductivity of soil layers in between.
[3] Measured with open rain gauges at the field site.
[4] Estimated from water loss from lysimeters incubated at the field site.

water and have a pronounced effect on the annual P and K budget of the eutrophic fen. Although the surface water N input is also enhanced in the eutrophic fen, the extremely high N input in precipitation (dry and wet deposition) reduces the overall significance of surface water N inputs. It is to be expected that surface water nutrient inputs will be several times higher in the dry 1986 summer, due to larger surface water inputs.

Some remarks should be made concerning the N input via precipitation. We emphasize that the extremely high N load of precipitation is primarily a result of air contamination with nitrogen-oxides and ammonia. It is estimated that natural background N inputs via bulk precipitation are as low as 0.7—3.0 kg N/ha/y (see van Aalst 1984), whereas a bulk precipitation input of 17.7 kg N/ha/y was measured at our study sites. It is well known that plants have a high filtering capacity, and may readily adsorb atmospheric gasses and aerosols (e.g. Hosker & Lindberg 1982). Thus, only a fraction of the total dry deposition on the vegetation will be collected in open ('bulk') rain gauges. Total deposition rates in Table VII are estimated from bulk precipitation measurements, using a conversion factor (see note 2 in Table VII for further details of this calculation). From this table it can be seen that roughly half of the total precipitation nutrient input will be collected in open rain gauges. Although adjustments to our present estimates of total deposition may have to be made, preliminary results

Table VII. Estimated nitrogen, phosphorus and potassium budget (kg/ha/y) for 1985 for 2 fens.

Site	Westbroek I (mesotrophic)			Molenpolder (eutrophic)		
	N	P	K	N	P	K
Input						
surface water[1]	0.17	0.02	0.29	1.28	0.72	12.41
groundwater[1]	2.52	0.03	0.74	0.00	0.00	0.00
precipitation[1 2]	39.90	1.56	29.34	39.90	1.56	29.34
total	42.59	1.61	30.37	41.18	2.28	41.75
Output						
surface water[1]	0.63	0.12	5.37	2.39	0.70	17.77
groundwater[1]	0.00	0.00	0.00	0.65	0.09	3.89
cutting[3]	32.85	2.25	27.00	21.94	1.69	20.63
total	33.48	2.37	32.37	24.98	2.48	42.29
Net gain or loss	+9.11	−0.76	−2.00	+16.20	−0.20	−0.54
(% of input)	(21)	(47)	(7)	(39)	(9)	(1)

[1] Mass flow obtained by the product of the nutrient concentration and discharge of water.
[2] Includes estimated dry deposition, based on the following assumptions: — total dry deposition on the vegetation is 2 × wet-only deposition (van Aalst 1984). — wet-only deposition is 0.75 × bulk deposition (Ridder 1984). Bulk precipitation of N, P and K is 17.7, 0.69 and 13.0 kg/ha/y, respectively.
[3] Estimated from chemical analyses on biomass samples taken just before and after the cutting.

of field measurements seem to confirm the gross correctness of the estimates.

Groundwater flow causes an input of ammonium in the mesotrophic fen (see also Verhoeven *et al.* 1983). Groundwater contour maps and analyses of water samples show that the ammonium originates from manured and fertilized grasslands in adjacent polders (see Figure 3). It is transported towards the fen by groundwater flow over a considerable distance (Figure 1). Until now, cation exchange processes in the lowest solid peat layer prevent the ammonium from reaching the root zone. In the long run, however, saturation of this layer is likely to occur (Verhoeven 1983).

Export of nutrients occurs mainly due to harvesting of the vegetation in July. Differences in export between the two fens are not caused by differences in nutrient stock at the time of mowing, but are a consequence of differences in the vegetation structure. At the eutrophic site a large part of the nutrient stock (approx. 30%) is located in the

Sphagnum carpet that is not removed during the cutting and subsequent haying.

It appears that both fens accumulate N (Table VII), the eutrophic fen about twice as much as the mesotrophic fen. It should be noted, however, that N_2 fixation and denitrification were not considered in the calculations. Investigations showed that both fens exhibit significant potential denitrification and N_2 fixation rates. The importance of both processes under natural conditions will be investigated.

The fens differ significantly in P budget. P outputs broadly balance inputs at the eutrophic site, whereas there is a net P removal from the mesotrophic site. This is correlated with a notably lower soil water phosphate concentration in the latter fen. The very low phosphate concentration in the mesotrophic fen (approx. 0.06 mg ortho-PO_4/l) will partly be caused by the input of lithocline groundwater and surface water. The water penetrating the fen is rich in Ca and Fe, and these elements are known to form complexes with phosphate, thereby reducing the phosphate availability in the soil (Nichols 1983). The extreme importance of upwelling lithocline groundwater for the persistance and survival of mesotrophic fen plant communities is discussed in detail by van Wirdum (1979, 1981, 1982) and Grootjans (1985).

The sum of inputs and outputs for K is approximately in balance, but there exists a distinct difference in the total amounts of input and output between the fens. As fluxes of K are considerably larger at the eutrophic site, K availability is expected to be better too (Howard-Williams 1985).

It can be concluded that the regional groundwater regime is of major importance to the nutrient dynamics of fens that lay scattered in a matrix of pastures that are heavily fertilized (approx. 300 kg N/ha/y). Manured pastures and quaking fens are interrelated by means of groundwater flow. This may cause undesirable nitrogen inputs via upward groundwater seepage. The protection of fens against eutrophication should therefore include measures to prevent eutrophication of the groundwater in 'upstream' agricultural areas, that may well be remote. Except for a possible N addition, groundwater input must certainly be considered as essential for the survival of mesotrophic fens in the long run. The constant supply of Ca and Fe reduces phosphate availability in the soil. Furthermore, in polders that are discharge areas for groundwater there is less need for additional river water supply during the summer months to maintain desired water levels. Hence, water in the ditches in discharge areas consists of a mixture of rainwater and groundwater and is, in general, of good quality. In polders that are recharge areas for groundwater, the addition of polluted river water leads to eutrophication of surface waters and subsequently enhances surface water nutrient inputs on fens.

DISCUSSION

The water level fluctuations in the fens studied are small because the water tables of the polders are controlled by the water authorities. Further, most fens have a floating vegetation that rises and falls with the rising and falling water table. As a result, the position of the water table is continually near the top of the floating mat, leading to very constant soil moisture conditions. These fens are, therefore, very suitable systems to study the influence of water quality on the species composition and structure of the vegetation; in most other wetlands (including most fen systems) the variations in the water table have strong effects on the vegetation that obscure the effects of differences in water quality (see Gosselink & Turner 1978).

Another important forcing function in the fens studied is the fact that they are being cut each year in the summer. This leads to a significant increase in the output of nutrients, as is clearly shown in the nutrient budgets presented here. Further, cutting prevents the accumulation of litter and thereby keeps the vegetation sufficiently open to permit high species densities (Bakker & De Vries 1985, Van den Bergh 1979). The difference in cutting regime is one of the main reasons why the vegetation in the fens in East Anglia and Norfolk, Great Britain, shows much higher standing crops and smaller species densities; there the fens are being cut once every two to four years, mainly to stimulate the growth of sedge (*Cladium mariscus*) and reed (*Phragmites australis*) (Wheeler 1980a, 1980b, 1980c, Wheeler & Giller 1983).

In our study area, the ionic composition of the fen water has a strong influence on the species composition of the vegetation. The ionic composition of the fen water is the result of mixing of water from three main sources: rain, groundwater and surface water. Various ways of characterizing the water quality in relation to the species composition of the vegetation have been proposed (Van Wirdum 1980, 1981, Kemmers & Jansen 1985a, Sjörs 1950). These studies mainly focus on the major cations and anions in the water, not on plant nutrients. In our study we want to specificially address the question if the effect of water quality on species composition can be explained by an effect on nutrient availability to the vegetation; further, we want to analyze the effects of increasing nutrient loads of the inflowing (rain, ground and surface) water.

An important factor to consider in this respect is the development of the fen systems in time. From an analysis of aerial photographs it was established that most fens in our area have terrestrialized in the 1930—40s. The period that they have been annually cut differs from about 15 to about 25 years. Some fens (Het Hol, Kortenhoef) have terrestrialized already at the beginning of the century and have been annually cut for more than 40 years.

Fens of the first age category that are located in polders with discharge of old groundwater contain plant communities of type 1a. Plant growth in these fens is limited by the availability of N. Prolonged enrichment of such sites with N will lead to an increase in the productivity and to a higher degree of dominance of highly productive species such as *Carex acutiformis* and *C. paniculata*, whereas the total species composition is only slightly influenced (type 1b). Due to the very high N inputs from precipitation, there is a net gain of N on the nutrient budget, despite of the annual mowing of the vegetation. This extra N accumulates mainly as organic N in the fen soil. As the total amount of organic N present in the soil is one of the factors determining the mineralization rate (see Kemmers and Jansen, 1985b) N will become more available on the longer term as a result of this accumulation process. When the amounts of N mineralized during the vegetation season are compared with the amount of N supplied by the hydrological inputs (Table VIII) it is clearly shown that mineralization is a very important N source. Part of the total N available is removed by mowing, part may be denitrified and the rest accumulates in the soil.

The nutrient budget further shows that P is lost from the system in significant quantities. Table VIII shows that mineralization is by far the most important source of phosphate in all the fens studied. In the fens with groundwater discharge part of the phosphate mineralized or supplied by input sources is adsorbed or precipitated with Ca and Fe compounds in the inflowing groundwater. As a result of the net annual removal of P in hay, the P content of the soil will gradually decrease. A shift in the factor limiting plant growth from N to P is to be expected on the longer term. The peat soil of the 'old' fens at 'Het Hol' has a significantly lower total P content than those of the younger fens in the area, which strongly supports this conclusion.

Table VIII. The contribution of input sources and recycling to the total availability of inorganic N and P. (values as kg/ha, year).

source	Westbroek I	Molenpolder
Nitrogen		
atmospheric deposition	39.9	39.9
surface water	0.2	1.3
groundwater	2.5	0.0
mineralization in soil	79.	244.
Phosphorus		
atmospheric deposition	1.56	1.56
surface water	0.02	0.72
groundwater	0.03	0.0
mineralization in soil	2.3	34.2

So far, the influence of N and P availability on the productivity and species composition of the fens is obvious; the productivity is initially limited by N availability (type 1a), increases gradually as a result of the high N inputs from precipitation (type 1b) and finally decreases again due to P limitation brought about by the annual removal of hay (type 2).

A complicating factor, however, is that the vegetation layer of the fen becomes more and more isolated from the groundwater as peat accumulation proceeds. At a certain stage in this development the water quality in the upper layer of the fen is sufficiently atmocline to permit the establishment of *Sphagnum* species. As soon as the *Sphagnum* layer forms an important component of the vegetation, the mineralization rates of N and P are substantially increased. Surprisingly enough this does not result in a higher productivity by the phanerogams. Apparently, the *Sphagnum* layer brings about other changes as well (pH and possibly toxins) that prevent an increase in phanerogam productivity (see Sikora and Keeney, 1983). The exact moment of the invasion by *Sphagnum* spp. in the temporal development described is difficult to indicate; it does follow the change from type 1 into type 2 judging from the fact that *Sphagna* are absent in type 1 communities and form mats of varying thickness in type 2 communities.

The fact that ombrotrophic mires have higher levels of available N and P in the soil than minerotrophic mires was stressed by Waughman (1980) who compared South German fens and bogs. He points to nutrient uptake problems rather than their availability as such to explain the lower phanerogam productivity in bogs. Our results in the 'ombrotrophic' sites in 'De Weerribben' accord with this view.

Fens located in polders with groundwater recharge are much more under the influence of surface water, in our area mainly polluted river water. It is remarkable, however, that also under such circumstances the main input source for N remains atmospheric deposition. Inputs of P and K are higher than in the fens in the groundwater discharge areas. A faster accumulation of N takes place and P inputs and outputs are more or less in balance. The vegetation in these fens is dominated by tall sedge and grass species and by eutraphent *Sphagnum* species (type 3). The N and P mineralization rates in the soil are extremely high. Again, the negative influence of the *Sphagnum* layer on the phanerogams is distinct: here shoot density rather than shoot size decrease with increasing importance of *Sphagnum squarrosum*. N as well as P availability are markedly increased as a result of the surface water inflow. In the absence of a *Sphagnum* layer this would likely result in high biomass production.

The completely different species composition (bryophyte as well as phanerogam layer) in the fens with surface water inflow cannot be explained by a difference in N and P dynamics. It is very probable that there are direct influences from the higher concentrations of Na and Cl in the surface water that originates from the Rhine.

In this study, nutrient availability has been shown to be a key factor explaining the differences in species composition between the various fens. There are, however, also other, mainly biotic factors involved that relate to the water quality aspects other than nutrients *per se*. The invasion of *Sphagnum* species in the vegetation is of great importance. Because of the completely different behaviour of dead *Sphagnum* material compared to phanerogam litter in decomposition and mineralization, a *Sphagnum* carpet causes important changes in the nutrient dynamics of the fens. The break-down of organic matter is slowed down, whereas the mineralization of N and P is accelerated. Further, the growth of phanerogams is for some reason suppressed. The moment of *Sphagnum* invasion is different for fens in groundwater discharge and groundwater recharge areas: in the former category of fens *Sphagnum* can only invade when the fen is filled with peat below the floating mat sufficiently to block the groundwater stream, i.e. in a late stage of the development. Fens of the latter category develop a *Sphagnum* carpet already in an earlier stage.

Another factor of great importance is the source of the surface water that is supplied to the fens with groundwater recharge. If this surface water originates from rivers high in NaCl (Rhine) the species composition is drastically influenced and much of the typical fen communities is lost.

FUTURE DEVELOPMENTS

Because of the persistently high levels of N in atmospheric deposition in The Netherlands, it is to be expected that the fens with a type 1a community will quickly develop towards type 1b (increasing biomass production and changes in the dominance of species; estimated time period 10 years). The further development towards type 2 (decreasing biomass production due to persistent summer mowing) will take place slower (20—40 years). For the conservation of type 1a communities it will be necessary to start a regime of summer mowing on recently terrestrialized turf ponds in groundwater discharge areas. Upon the start of the mowing regime, the vegetation will change towards type 1a in 10 years.

Fen communities of types 1a, 1b and 2 can only be conserved if the water regime in the polders guarantees a continuous supply of lithocline

280

water. In our study area, this means that the water table in the polders with groundwater discharge should be kept at a relatively low level, whereas water tables of polders upstream and in the glacial hills should be kept high. Water management should be directed to the minimization of surface water inflows.

The inlet of water originating from the Rhine should be totally avoided.

ACKNOWLEDGEMENTS

We are indebted to Prof. Dr. P. J. M. van der Aart and Prof. Dr. A. G. van der Valk for critically reading and improving this chapter.

REFERENCES

Bakker, J. P. & De Vries, Y. (1985) The results of different cutting regimes in grassland taken out of the agricultural system. Munstersch. Geogr. Arbeiten 20: 51—57.

Beltman, B. (1984) The role of water balance studies in environmental planning. Proc. First Int. Seminar of the IALE, (Roskilde, Denmark, October 1984), 4: 131—136.

Carter, V., Bedinger, M. S., Novitzki, R. P. & Wilen, W. O. (1979) Water resources in wetlands. In: Wetland function and values: the state of our understanding. Greeson, P. E., Clark, J. R. & Clark, J. E. (Eds.). American Water Res. Ass., Minneapolis 344—376.

Carter, V. (1986) An overview of the hydrologic concerns related to wetlands in the United States. Can. J. Bot. 64: 364—374.

Clymo, R. S. (1965) Experiments on the break-down of Sphagnum in bogs. J. Ecol. 53: 747—757.

Clymo, R. S. (1978) A model of peat bog growth. In: Heal, O. W. & Perkins, D. F. (Eds.), Production ecology of British moors and montane grasslands. Springer, Berlin 187—223.

Crisp, D. T. (1966) Input and output of minerals for an area of Pennine moorland: the importance of precipitation, drainage, peat erosion and animals. J. appl. Ecol. 3: 327—348.

Du Rietz, E. (1954) Die Mineral Bodenwasserzeigergrenze als Grundlage einer natürlichen Zweigliederung der Nord- und Mitteleuropäische Moore. Vegetatio 5: 571—585.

Godshalk, G. L. & Wetzel, R. G. (1978a) Decomposition of aquatic angiosperms. I. Dissolved components. Aquat. Bot. 5: 281—300.

Godshalk, G. L. & Wetzel, R. G. (1978b) Decomposition of aquatic angiosperms. II. Particulate components. Aquat. Bot. 5: 301—327.

Godshalk, G. L. & Wetzel, R. G. (1978c) Decomposition of aquatic angiosperms. III. Zostera marina and a conceptual model of decomposition. Aquat. Bot. 5: 329—354.

Gosselink, J. G. & Turner, R. E. (1978) The role of hydrology in freshwater wetland ecosystems. In: Good, R. E., Whigham, D. F. & Simpson, R. L. (Eds.), Freshwater wetlands. Ecological processes and management potential. Acad. Press, New York, pp. 63—78.

Grime, J. P. (1979) Plant strategies and vegetation processes. Wiley, New York.

281

Grootjans, A. P. (1985) Changes of groundwater regime in wet meadows. Thesis, State University of Groningen, 146 pp.

Hemond, H. F. (1983) The nitrogen budget of Thoreau's bog. Ecology 64: 99—109.

Hosker, R. P. & Lindberg, S. E. (1982) Review: atmospheric deposition and plant assimilation of gasses and particles. Atmosph. Environm. 16: 889—910.

Howard-Williams, C. (1985) Cycling and retention of nitrogen and phosphorus in wetlands: a theoretical and applied perspective. Freshw. Biol. 15: 391—431.

ICW (1985) Nitraatproblematiek bij grondwaterwinning. Inst. v. Cultuurtechniek en Waterhuishouding, Wageningen. Rapport 12: 1—49.

Kemmers, R. H. & Jansen, P. C. (1985a) Waterhuishouding in relatie tot de beschikbaarheid van vocht en voedingsstoffen voor natuurlijke vegetaties. Cultuurtechn. Tijdschr. 24(4): 195—211.

Kemmers, R. H. & Jansen, P. C. (1985b) Stikstofmineralisatie in onbemeste halfnatuurlijke graslanden Inst. Cultuurtechniek en Waterhuishouding, Wageningen. Rapport 14: 1—20.

LaBaugh, J. W. (1986) Wetland ecosystem studies from a hydrologic perspective. Water Res. Bull. 22: 1—10.

Moore, P. D. & Bellamy, D. J. (1974) Peatlands. Elek Science, London, 221 pp.

Nichols, D. S. (1983) Capacity of natural wetlands to remove nutrients from wastewater. J. Water Poll. Control Fed. 55: 495—505.

Parnas, H. (1975) Model for decomposition of organic material by micro-organisms. Soil Biol. Biochem. 7: 161—169.

Patrick, W. H. & Mikkelsen, D. S. (1971) Plant nutrient behaviour in flooded soils. Adv. Agron. 1971: 187—212.

Reddy, K. R. & Rao, P. S. C. (1983) Nitrogen and phosphorus fluxes from a flooded organic soil. Soil Sci. 136: 300—307.

Richardson, C. J. (1985) Mechanisms controlling phosphorus retention capacity in freshwater wetlands. Science 228: 1424—1427.

Ridder, T. B. (1984) Vijf jaar metingen van de samenstelling van de neerslag. In: Zure regen, oorzaken, effekten en beleid. Adema, E. H. & Van Ham, J. (Eds.), Pudoc, Wageningen, pp. 55—58.

Rosswall, T. & Granhall, U. (1980) Nitrogen cycling in a subarctic ombrotrophic mire. Ecol. Bull. (Stockholm) 30: 209—234.

Sikora, L. J. & Keeney, D. R. (1983) Further aspects of soil chemistry under anaerobic conditions. In: Gore, A. J. P. (ed.), Mires: swamp, bog, fen and moor. A. General Studies. Ecosystems of the World, vol. 4A, Elsevier, Amsterdam, pp. 247—256.

Sjörs, H. (1950) On the relation between vegetation and electrolytes in North Swedish mire waters. Oikos 2: 239—258.

Stumm, W. & Morgan, J. J. (1981) Aquatic Chemistry. John Wiley and Sons Inc., New York, 780 pp.

Van Aalst, R. M. (1984) Depositie van verzurende stoffen in Nederland. In: Zure regen, oorzaken, effekten en beleid. Adema, E. H. & Van Ham, J. (Eds.), Pudoc, Wageningen, pp. 66—70.

Van den Bergh, J. P. (1979) Changes in the composition of mixed populations of grassland species. In: Werger, M. J. A. (ed.), The study of vegetation. Junk, The Hague, pp. 57—80.

Van Wirdum, G. (1979) Dynamic aspects of trophic gradients in a mire complex. CHO-TNO Proc. and Inform. 25: 66—82.

Van Wirdum, G. (1980) Eenvoudige beschrijving van de waterkwaliteitsverandering gedurende de hydrologische kringloop ten behoeve van de natuurbescherming. CHO-TNO Rapp. en Nota's 5: 118—143.

Van Wirdum, G. (1981) Linking up the natec subsystem in models for the water management. CHO-TNO Proc. and Inform. 27: 108—128.

282

Van Wirdum, G. (1982) The ecohydrological approach to nature protection. RIN Leersum, Annual Report: 60—74.

Verhoeven, J. T. A. (1983) Nutrient dynamics in mesotrophic fens under the influence of eutrophicated ground water. Proc. Inst. Symp. Aquat. Macrophytes (Nijmegen, 18—23 September 1983), pp. 241—250.

Verhoeven, J. T. A. (1986) Nutrient dynamics in minerotrophic peat mires. Aquat. Bot. 25: 117—138.

Verhoeven, J. T. A. & Arts, H. H. M. (1987) Nutrient dynamics in small mesotrophic fens surrounded by cultivated land. II. Nutrient uptake by plant biomass in relation to the mineralization of soil organic matter. Oecologia (Berlin) 72: 557—561.

Verhoeven, J. T. A., Van Beek, S., Dekker, M. & Storm, W. (1983) Nutrient dynamics in small mesotrophic fens surrounded by cultivated land. I. Productivity and nutrient uptake by the vegetation in relation to the flow of eutrophicated ground water. Oecologia (Berlin) 60: 25—33.

Vermeer, J. G. (1985) Effects of nutrient availability and ground water level on shoot biomass and species composition of mesotrophic plant communities. Thesis, University of Utrecht, 142 pp.

Vermeer, J. G. (1986) The effect of nutrients on shoot biomass and species composition of wetland and hayfield communities. Oecol. Plant. 7(21): 31—41.

Vermeer, J. G. & Berendse, F. (1983) The relationship between nutrient availability, shoot biomass and species richness in grassland and wetland communities. Vegetatio 53: 121—126.

Vermeer, J. G. & Verhoeven, J. T. A. (1987) Species composition and biomass production of mesotrophic fens in relation to the nutrient status of the organic soil. Oecol. Plant. 8(22): 321—330.

Waughman, G. J. (1980) Chemical aspects of the ecology of some South German peatlands. J. Ecol. 68: 1025—1046.

Westhoff, V. & den Held, A. J. (1969) Plantengemeenschappen in Nederland. Thieme, Zutphen, 324 pp.

Wheeler, B. D. (1980a) Plant communities of rich-fen systems in England and Wales. I. Introduction, tall sedge and reed communities. J. Ecol. 68: 365—395.

Wheeler, B. D. (1980b) Plant communities of rich-fen systems in England and Wales. II. Communities of calcareous mires. J. Ecol. 68: 405—420.

Wheeler, B. D. (1980c) Plant communities of rich-fen systems in England and Wales. III. Fen meadow, fen grassland and fen woodland communities, and contact communities. J. Ecol. 68:761—788.

Wheeler, B. D. & Giller, K. E. (1983) Species richness of herbaceous vegetation in Broadland, Norfolk, in relation to the quantity of above-ground plant material. J. Ecol. 70: 179—200.

WATER FLOW AND THE VEGETATION OF RUNNING WATERS

F. H. DAWSON

INTRODUCTION

The vegetation of running waters is defined and determined by the effect and interactions of a single physical factor, water flow, which governs plant-form, dominates the growth-controlling factors and defines the habitat. Thus very high or turbulent flows will directly determine the presence or absence of instream vegetation and even light becomes subordinate to its influence. At high but less extreme flows, aquatic vegetation may be confined to the margins and to islands where it is in direct competition with terrestrial vegetation. Very low flows allow the development of a vegetation characteristic of static waters, i.e. pond and lakes. Between these extremes of water flow, the physical and chemical factors can control growth, interact with fluviatile plants to regulate seasonal biomass and to create flowing water communities; plants are still subject to the broad environmental limits of high and low temperature etc. (Whitton 1972). Such communities can vary during the year influenced either by seasonal changes in their physical environment, particularly by water flow, or by competition between species. The plant community at a site therefore reflects the balance achieved between the physicochemical environment and the plants tolerance, adaption to or their modification of these conditions by their presence. The species present within the community reflect those available and able to colonise. The dominance of a species may occur through the rapidity of its development during favourable growth periods which may result in a change of dominant species even seasonally. The spatial variation of fluviatile aquatic vegetation is naturally high and mosaics of 'phytocoenoses' often result, primarily, from the turbulent nature of water flows in combination with the immediate or long-term history, of its interaction with the bed or underlying rocks. Interaction by man particularly mechanical 'weed' control and channel reconstruction often increases the uniformity of the habitat. Such changes may increase the tolerance of recovery of the system to

283

J. J. Symoens (ed.), Vegetation of inland waters. ISBN 90—6193—196—7.
© *1988, Kluwer Academic Publishers, Dordrecht. Printed in the Netherlands.*

frequent but not to catastrophic events. The vegetation which results is more likely to be limited to the more invasive and fast growing plants and is likely to be less varied in its species composition. Plants themselves exhibit a wide range of growth habits which together with much plasticity in somatic adaption, often produces a wide range of phenotypic expression including submerged and arial leaf forms. Plants, despite belonging to many taxonomic groups, respond to environmental forces or selection to produce a common simplicity or uniformity of form between members of different taxa in particular habitats when strongly influenced by environmental variables particularly water flow.

TERMINOLOGY FOR VEGETATION AND HABITAT

It is convenient to have an agreed terminology for the vegetation of waters of similar type but it should be simple, comprehensive and clearly understandable. Such definitions however lie within a framework which is fundamentally confused by the wide taxonomic mixture of species, the great diversity of habit and the plasticity of the organisms involved. Thus for example the simple group 'water plants' is not useful, for it is frequently assumed to be just higher plants, but can reasonably include small algae, fungi and lichens. Taxonomic grouping can also be inappropriate for aquatic vegetation because plants from several phyla frequently coexist and interact as a group.

A pragmatic approach to definitions is adopted here, without instigating a new system (see Wiegleb 1988). Thus, firstly, water plants are simply defined as those which grow in, on or close to water provided that they are not more dominant in another habitat. From this the first division is purely a functional one of size, reflecting the technical aspects of study and accepts that there is often some overlap in practice. Thus the larger water plants are termed macrophytes and include Angiosperms, Gymnosperms, Pteridophytes, Byophytes and large algae, whereas the smaller water plants, particularly unicellular algae, are microphytes. Freshwater macrophytes (cf. aquatic macrophytes, to include marine and brackish waters) are divided into emergent (occasionally joined with marginal), surface-floating, floating-leaved and submerged plants. The latter may be further sub-divided e.g. to caulescent, rosette and thalloid (Sculthorpe 1967, Westlake 1976). The floating-leaved and submerged macrophyte groups are frequently joined in other classifications as 'euhydrophytes' or true water-plants for they normally are either completely submerged, or anchored to the substratum or sometimes having both surface and submerged leaves. These approaches together with life-form classification in general has been clearly discussed by Denny (1985). The origins of this approach are distant, probably commencing with J. J. Dillenius in the early 1700s and can be followed to Schenck's work in the 1880s before being split between the phytosociological approach of Raunkiaer and Braun-Blanquet and the more ecological approach associated with Britain (Tansley 1911, Arber 1920).

There have been, and doubtless will continue to be, attempts to refine existing definitions but, however, the subtleties of such terms are rarely warranted by the current knowledge of aquatic botany (and eco-physiological explanations) at the general level. The more the divisions, the less the differences between them and the greater the possibilities for overlap. Notwithstanding that they often may appear adequate at specific sites for particular macrophytes; thus the extended definitions of habit or life form (and habitat) are thus viewed by some as exercises in semantics and etymology. The architecture of an aquatic community i.e. 'the way in which the community components fill the available space' (volume), is a useful criteria for classification as it attempts to illustrate in a functional way the spatial interrelationships of members of the community rather than just cover by the addition of structural elements (den Hartog 1982) but this approach still requires data on biomass and productivity before general application. Hence, freshwater macrophytes of running waters are defined here as 'fluviatile macrophytes' but acknowledging that such a definition can never be precise because, for example, many grow in lakes, whereas other 'river' plants may, in addition, grow near the inlet or outlet of lakes because they are influenced strongly by the water flows present. Lake shores are subject to turbulence and wave action more than unidirectional flow and are therefore not included. The definition here is thus based upon water flow and not directly on habitat.

The vegetation of running water can be defined to include plants regulated primarily by the direct and indirect conditions of water flow. Defining such a grouping however is difficult for overlaps with other aquatic situations are often found such as along or in irregularities of the margins of running waters; the latter includes such situations as reed-swamps and backwater or oxbow communities but probably should not include wet-meadow communities despite their intermittent inundation. Alternatively if a broad approach is taken than intermittent streams, 'lakes' with either high through-flows or short retention times together with flood plains, would be included in this grouping; such examples however question the basis of the definition of 'aquatic', for is a plant aquatic merely if it is adapted to withstand water or water movements? Contrast, for example, the Podostemaceae which typically live aerially but on boulders in rivers and are subject to occasional to frequent inundation and may have roots trailing in the water, with a terrestrial plant of river margins in sub-arctic regions which has spiral

286

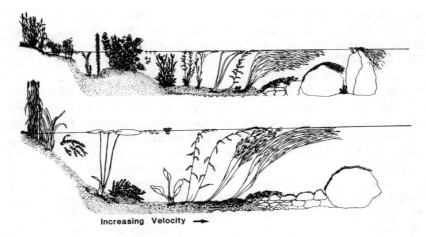

Increasing Velocity ➔

Fig. 1. Typical growth forms of fluviatile macrophytes in slow (left) to very fast flows (right) in shallow and deeper flowing waters.

spring-like roots to resist loss by the growth of ice crystals from the soil. Alternatively there are plants which can dominate some river situations but are not adapted to flowing conditions and survive only by rapid colonisation from marginal semi-terrestrial areas and brief periods of rapid growth.

THE FLUVIATILE HABITAT AND
ITS CLASSIFICATION

The physical interaction of plants with the energy possessed by the water flowing past governs and selects for suitable plants species through their vegetative hydraulic resistance (Dawson & Charlton, in press). A successful plant must be able to resist this fluviatile energy. Thus the plants size, form and stand structure, together with the strength of its stems and the security of its method of anchoring to the substratum, are major factors in maintaining its presence. In waters with seasonal changes in flow e.g. snow melt or tropical dry season, an herbaceous perennial habit is frequently advantageous. This habit is characterised by dieback of the plant or the shedding of the majority of its shoots to reduce its resistance to flow, thus avoiding the wash-out of its basal stems and particularly its roots or rhizomes. The power of the flowing water is governed by gradient which in combination with the surface rocks creates the environment into which plants invade and frequently modify. Such interactions include the moderation of flows through growth which increases the water levels thus reducing fluviatile power and the stabilisation of silt banks but through this moderation and encroachment, the plant may destroy the habitat for the plant itself.

Fluviatile habitats have growth forms but not always taxa in common. The variety of habitat is large and ranges from sources with rapid runoff such as mountain torrents, snow melt, or constantly flowing springs, through streams and rivers with slow, rapid or alternating flow to large deep sluggish sometimes turbid rivers of enormous size; water velocities in smaller streams are frequently more turbulent and thus appear faster than those placid but higher flow of large rivers with fully-developed velocity profiles.

The studies of fluviatile habitats have been undertaken in various ways and to varying degrees of detail. These studies can be grouped from geographic regions down to detailed sections of small reaches. Examples of studies by continent, include the USSR (Zhadin & Gerd 1961), Australia (Bayly & Williams 1973), Africa (Beadle 1974, Rzóska 1976), by region river basins of South America (Bonetto 1975), mountain torrents of Tien Shan (Taubaev 1970, Brodsky 1980), by country, Britain (e.g. Haslam & Wolseley 1981, Holmes 1983), Germany; by river, Volga (Ekzertzev 1979), Danube (Szemes 1967), Lot (Dècamps 1978), Wye (Edwards & Brooker 1982), Sungai Gombak (Bishop 1973) and more recently in a series edited by Davies and Walker (1986); and by section, for example, the sites of the IBP Programme (Le Cren & Lowe-McConnell 1980). There are common factors between habitats but the great variety of other interacting physical and chemical factors are frequently incidental and serve only to confuse fundamental studies of distribution, community and growth controlling factors. However generalising about habitats can lead to some insight, thus for example fast flows characterised by substrates of stones or boulders are frequently colonised by the smaller plants particularly Bryophytes; the latter are characterised by being small in size, with numerous tenacious rhizoids in mosses, flattened shoots in liverworts and thin layers in algae and aquatic lichens. Bryophytes, in particular, are very adaptable to stress and some may tolerate factors such as low light levels, drying out and freezing, abrasion by particles or the accumulation of toxic metals (Glyme & Carr 1974, Lewis 1973, Wehr & Whitton 1986). Small lowland streams originating in springs or seepages are, by contrast, frequently dominated by emergent macrophytes (Michaelis 1976). At the other extreme of the habitat range, large rivers which are characteristically deeper with relatively less turbulence allowing the development of higher water velocities, have algal dominated communities more frequently. This is in part due to the length of time (retention time) available for their multiplication and development but also due to the lower light levels at the substratum through physical absorption and turbidity of the water preventing macrophyte growth. The seasonal deposition of silts or soft sediments also inhibit colonisation and affect the permanence of macrophytes. Fluviatile macrophytes having rhizomes are thus limited to the margins, backwaters or shallows, depending upon the seasonal flows and morphology of the river and its substrate. The characteristics of large rivers range from those in Europe which are frequently channelised or constrained and may have erosion controls, to those enormous rivers of South America, China or India, in which the macrophyte vegetation may be minimal either through deeply coloured waters or seasonally high sediment loads over unstable substrates or even constancy of bed-position. Some of these large rivers may have mobile mats of reedswamp-like communities which are not necessarily connected to the bed or bank, but form by rapid growth in uncharacteristically slow downstream flows and may achieve substantial size even if only seasonally.

A particular fluviatile habitat is not limited to large characteristic areas but may occur spasmodically in other areas. Thus, the very high energy or mountainous-torrent habitat may occur as rapids in otherwise large sluggish rivers, for example the Podostemaceae, which characteristically grow on boulder torrents in tropical regions. Other habitats are difficult to even equate as fluviatile for example the seasonal or very high energy ones of the sides of steep gorges or the extreme energy ones of waterfall (Gimmingham & Birse 1957, Dawson 1973). The fitness of a particular growth-form for a particular type of habitat and the associated features of the structure of the community, may reappear under recurring habitat conditions almost irrespective of the particular species present. The habitat is restricted by physical factors, particularly flow, such that at high energies only larger stones and boulders are stable and remain but at low energies in slow water, then sediment may fill the crevices but, unless dramatic climatic changes have occurred, boulders are unlikely to be present. Thus in summary the volume of water flowing regulates fundamentally the parameters of width, depth, velocity and turbulence, which in combination with the constraints of surface geology determines the permissible vegetation-range.

The grouping or classifying of plants can lead to a better understanding of the competitive and other interrelationships of the species and identification of common growth controlling factors through an understanding of the interaction of associated plant groups with their environment. It should be remembered that classifications are often functional, by the method of study, by the habitat group, by association of species and, as such, may be made without a basic understanding of the interrelationships i.e. they may only be superficial. The classification of freshwater macrophyte habitats has produced a variety of systems but only the ecological zonation developed for fluviatile freshwater macrophytes by Butcher (1933) and modified by the addition of seasonal changes in discharge as the frequency and severity of spates by Westlake (1975) will be given here. This classification was chosen for it applies more extensively than Britain, for which it was developed, although some species will undoubtedly change (Table I). The next modification envisaged is the addition of water depth. This type of classification originated when it was realised that divisions on the basis of physicochemical environment or edaphic factors were an inevitable extension of the plant community or association approach, if progress in understanding the ecological habitat were to be achieved. In early studies of British freshwaters, Tansley (1911, 1953) tentatively considered that the aquatic formation for a region could be subdivided on the basis of the aeration of the water and on the amount and nature of dissolved mineral salts. Even in this simple classification three further divisions were required to accommodate the effects of water flow in determining the dominance of submerged and floating-leaved plants. Arber (1920) elaborated upon this approach to include the effects of substratum in determining the distribution of aquatics and the range of altitude of their occurrence. Examples include *Ranunculus trichophyllus* to 2500 m in the European Alps, *Myriophyllum*, *Lemna* and *Callitriche* spp. to above 2400 m near Chimborazo, S. America and particularly *Potamogeton pectinatus* from sea-level to 5000 m in both Venezuela and Tibet. Thus the Butcher-Westlake classification of the fluviatile habitat originated in the community approach which still continues, but adds the ecological perspective. Phenological aspects have been less emphasised but are becoming more widely recognised as useful incorporations and as greater understanding is currently bringing together the sociological and the functional relationships of growth controlling mechanisms of river plants (Castellano 1977, Thomen & Westlake 1981).

COMMUNITIES, SUCCESSION AND CLIMAX

Fluviatile communities reflect water conditions. In the large long rivers planktonic algal communities develop but such rivers are also usually deep and subject to particular conditions not found in smaller rivers, especially severe floods which preclude the establishment and growth of macrophytes (Kofoid 1908). In shallow small running waters, communities are more dependent for high biomass upon nutrient conditions in particular. Increasing or high nutrient-status favours epiphytic algal growth or attached algal communities in preference to a submerged macrophyte community, although large rhizomatous plants with surface

Table I. Typical dominating species of some major types of British river (from Westlake 1975, with modification from Butcher 1933).

Approx. range of Ca++ & conductivity μmho cm⁻¹	Source of river; spates	Substratum and approximate range of current velocity				
		Torrential <0.6 m s⁻¹ Boulders and shingle	Fast 0.7—0.25 m s⁻¹ Gravel and sand	Moderate 0.3—0.15 m s⁻¹ Sand and silt	Slow 0.2—0.05 m s⁻¹ Silt and mud	Sluggish-marginal <0.1 m s⁻¹ Mud or peat
<10 ~60	Mountains Frequent & high	Fontinalis squamosa & other mosses	Ranunculus fluitans	Potamogeton × sparganifolius	Juncus bulbosus Eleocharis acicularis	no examples
	Hills Moderate	Hypnum palustre	R. fluitans Myriophyllum sp.	Potamogeton nitens Callitriche spp.	Juncus bulbosus Callitriche hamulata	rare
	Lowland bogs Slight	no examples	Myriophyllum alterniflorum Ranunculus fluitans	Potamogeton polygonifolius P. alpinus	Potamogeton gramineus	Juncus bulbosus
5—50 ~40—300	Mountains Frequent & high to moderate	Fontinalis antipyretica Lemanea fluviatilis	Myriophyllum sp.	P. alpinus Sparganium emersum	Callitriche spp. Elodea canadensis	no examples
40—100 ~250—500	Mountains Frequent & high	Fontinalis antipyretica	Myriophyllum spicatum Ranunculus fluitans	S. emersum Potamogeton crispus	E. canadensis Potamogeton crispus	no examples
	Hills Moderate	no examples	Myriophyllum spicatum Ranunculus spp.	Potamogeton perfoliatus Ranunculus peltatus Sparganium emersum	Callitriche spp. Elodea canadensis Potamogeton pectinatus	Sparganium erectum
<80 ~400	Mountains Frequent & high	F. antipyretica	Ranunculus penicillatus Apium nodiflorum	Potamogeton perfoliatus Sparganium emersum	Callitriche spp. Elodea canadensis	no examples
	Hills Moderate	no examples	Ranunculus penicillatus Apium nodiflorum	Ranunculus peltatus Potamogeton pectinatus P. perfoliatus	Callitriche spp. Elodea canadensis Nuphar luteum	S. erectum
	Hills Slight	no examples	Ranunculus calcareus Apium nodiflorum	P. lucens Rorippa agg. Sparganium emersum Nuphar luteum	Elodea canadensis Nuphar luteum Potamogeton lucens P. pectinatus	S. erectum Glyceria maxima

leaves or emergent stems may overcome this (e.g. Golterman & Kouwe 1980).

The validity of the individual concepts of succession and climax for fluviatile aquatic plants depends upon the selection of a fixed point defined either geographically or on the river. The former considers that a community is similar to that of the hydrosere of a pond and which is assumed to progress to the extinction of the water body, but this is at variance with the concept of the river, which continues to flow, but elsewhere. Water flow will continue regardless and the habitat will remain available except in the case of subterranean flow or massive tectonic change. Thus redefining the question, is the climax of a river succession still a river? Tansley (1939) overcomes this dilemma by invoking the colonisation of shallow quiet water near the bank by fringing reed swamp plants and whose further colonisation is prevented by the water current and deeper water, resulting in a cessation of the progress of succession ('arrested succession'). It is becoming recognised that plant growth does affect water flows and influencing the direction of flow but at the same time being controlled by these flows resulting in a metastable state or dynamic stability (e.g. Dawson *et al.* 1978). However the results of an extreme event e.g. severe flood, causing a change in channel position is of little real relevance here, provided water continues to move in the same general direction. The main philosophical difference which remains between rivers and lakes is that in rivers, water will continue to flow somewhere in the valley and provide habitats for fluviatile aquatic plants in contrast to lake succession which will reach a climax and then cease to exist, until geological activity creates a new one.

There are generally few stages in the progressive change of species in fluviatile aquatic vegetation, regardless of which succession concept is adopted. Water flow is probably the prime reason for this as its seasonal or intermittent changes stresses the vegetation and may severely reduce the population. An example of seasonal progression and changes in the seasonal dominance of species together with year to year interactions resulting in an oscillating cyclic succession has been observed for *Ranunculus calcareus* and *Nasturtium officinale* (Dawson *et al.* 1978). There is circumstantial evidence that certain coloniser species are required before others can invade; the former e.g. *Ranunculus calcareus* tends to modify the environment before the latter especially emergent species can invade or ever colonise the beds e.g. *Nasturtium officinale* (Dawson *et al.* 1978). By comparison with the general terrestrial situation although there are colonisers or invasive species and the climax series is frequently short. Other types of succession are seen but for fluviatile aquatic plants few have been documented e.g. the seasonal downstream progression of plant development and invasion; one study has demonstrated a downstream ecocline in the development-time of flowers in *Ranunculus* (Dawson 1980b).

Terrestrial riparian vegetation rather than marginal fluviatile vegetation, may directly influence the aquatic environment and its succession. Shade, for example, must be a prime example of a direct effect; decreased light controls or selects for the slower growing fluviatile macrophytes e.g. *Callitriche* spp. in Europe. An alternative influence can be the encroachment of terrestrial vegetation, especially tree roots, which can reduce velocities in the margins but may increase flow towards the centre of the stream.

PLANT GROWTH AND ENVIRONMENTAL GROWTH CONTROLLING FACTORS

Fluviatile macrophytes in common with other plants, have certain general requirements for growth. Although several other authors had considered the general characters of the aquatic environment for water plants (Glück 1905), it was left to Butcher (1933) to review them and to extend this line of approach to introduce many of the special or particular characteristics of the fluviatile habitat (Table II), the fundamental characteristic of which is the normal existence of a continuous flow of water in one direction. The interactions with this lead to the selection of a community of plants which can tolerate the conditions prevailing. Butcher, in addition, outlined several topics of importance to the study which was the basis of study for half a century (Table III).

The maximum biomass which fluviatile macrophytes achieve at a particular site, is the result of a balance between the conditions available for growth [basically light, inorganic nutrients and carbon (carbon dioxide and carbonates), water temperature and water velocity] and the plants' physiological responses at its current state of growth (Westlake 1967, Mitchell 1974). The turbidity and light penetration of the water has been noted as particularly important (e.g. Ham *et al.* 1981) particularly during the early growth of submerged plants before they reach the surface. Plant growth also requires the supply of several other elements in suitable chemical forms in addition to inorganic carbon of an acceptable type (Raven *et al.* 1982), particularly the macro-nutrients nitrate, phosphate and potassium together with other minor and trace elements which are often available in the water and/or sediments (Denny 1983); the absence of toxic or growth-inhibiting compounds is also a requirement. If the general growth conditions are met, then plants, either large or small, will be able to grow in the majority of aquatic sites. For example in the smaller rivers and streams fluviatile macrophytes are more likely to be abundant in shallower

Table II. A general comparison of (a) the aquatic and terrestrial environment and (b) the special conditions for flowing water plants (based on Butcher 1933).

(a) Aquatic *vs* Terrestrial environment
   (1) Superabundance of water
   (2) Reduced light intensities underwater by absorption in proportion to
      (i) depth of water
      (ii) quantity of suspended material
   (3) Less rapid and smaller fluctuations in temperature
   (4) Lower concentrations of dissolved oxygen and carbon dioxide (unless carbonates present); plant metabolism can result in oxygen deficiencies and change the pH of the water
   (5) Increased nutrient availability from solution in addition to substratum

(b) Flowing *vs* static water
   (1) Unidirectional water flow
      (i) supplying/removing dissolved and suspended mineral and organic material
      (ii) further reduction in the variation in the physicochemical conditions e.g. temperature
   (2) Further reductions in quantity and quality of underwater light especially by increased suspended material through rainfall etc.
   (3) Further relative reductions in the daily variation of dissolved oxygen and carbon dioxide, dependent upon turbulence and current speed

Special conditions

   (4) Current affects
      (i) substratum for rooting of plants by redeposition or by input from land especially during floods
      (ii) sources of mineral salts from stream bed, from settlement around and within plant stands, and by precipitation and absorption of colloids
   (5) Changes in current affect
      (i) bed stability, nature of bed and turnover and introduction of differing materials especially during floods (plants may increase silt deposition by their presence, if not dislodged)
      (ii) water levels and plant growth and stability
   (6) Mechanical effect of current
      continuing downstream stress and strain on the plants allowing only 'rooted' plants to remain, encourages
      (i) strong stems and leaves
      (ii) efficient rooting systems
      (iii) growth habit of low resistance to flow
   (7) Man
      — plant and channel management e.g. weed cutting, dredging
      — changes in flow, water transfer, abstraction and discharge
      — changes in flow pattern, channelisation, land drainage
      — fisheries
      — navigation
      — water power

Table III. Major topics in the study of river vegetation, based upon Butcher (1933)

A. Impermanence of position of vegetation and movement to unoccupied (or more suitable) areas
B. Seasonal growth varies with:
   1. Growth patterns. Species show either inherent or environmentally controlled patterns unless so stressed by the environment that their seasonal growth cycle is limited only to their presence with little seasonal difference in biomass
   2. Nature of habitat
   3. Changes in substrate
   4. Temperature regime
C. Colonisation
D. Effect of vegetation on water current and depth
   Significant retardation of the current occurs with dense growths and is associated with deposition of most of the transported materials
   Increases in flow variation and in eddy production
E. Macrophyte vegetation as a habitat and food for organisms
   1. Food source, directly and indirectly
   2. Protection from predation — shelter and habitats
   3. Increases surface area for epiphytic algae
F. Relationship of vegetation to river fertility (productivity)
   1. Increases in nutrient supply from
      (i) temporary, flow-dependent retention or deposition of silt and fine organic particles
      (ii) release by death and decay of plant material
[2. Transfer of mineral salts from substratum to water for algal growth   ]
   3. Bank protection and bed stabilisation during floods
   4. Possible oxygenation agent, balances dependent upon downstream loss and not through respiration *in situ*; and daily oxygen balance with atmosphere

---

locations (< 3 m; Dawson 1976), but may be limited in their ability to invade, become established or remain because of an unsuitable stream bed or water flow during either the whole or part of the year at particular sites. Invasion is often by turion or by vegetative propagule, i.e. pieces of plant stem or 'root', but less frequently by seed (e.g. Coffey & McNabb 1974, Castellano 1977, Sastromoto *et al.* 1979). The slow flow of water may reduce the potential for spread by propagule and favour seed dispersing plants, or plants which can reinvade from the banks or marginal areas. Once established however plants may be unable to survive competition with other macrophytes or algae. Many fluviatile macrophytes overwinter as buried stems or small rhizomes as in Batrachian Ranunculaceae, or as tubers in *Potamogeton* spp. (Dawson 1976, van Wijk 1983, Madsen 1986).

Planktonic algae are favoured in downstream sections of large rivers as mentioned above. This is partly by increasing scour, and partly by limiting large plants through the increase of depth as river channel dimensions increase, but also because sufficient time elapses for the development of an algal population to develop during the downstream passage of water. The length of the period available for development is also of importance to large plants for, unless large food reserves are available, even fast growing plants have modest and finite growth rates (Westlake 1966). As plants increase in bulk, self-shading develops and the supply of inorganic carbon and nutrients may become progressively more difficult. When considering individual plant biomasses, space for growth and growth strategy may become of paramount importance (Castellano 1977, Howard-Williams 1982). The biomass achieved and annual production in various aquatic sites together with errors in their determination, have been reviewed again by Westlake (1982).

Seasonal cycles. Changes in biomass in the late winter or spring are often initiated by an herbaceous-type development from overwintering storage organs, or by redevelopment of over-wintering stems, to produce the seasonal biomass which is then acted upon by the environment. The latter has for decades been discussed by authors e.g. Butcher (1933), who have produced sets of factors which are undoubtedly likely to control plants particularly in rivers, but few have produced quantitative evidence on single or even related factors (Table II). This work followed a detailed study of sections on two lowland English rivers and illustrated the importance of several factors to the growth of river plants and the interaction between species and the physico-chemical environment. Amongst the features he introduced, he emphasised particularly the persistence of some species of plants throughout the year whilst others decay in the autumn and, also, the impermanence of aquatic vegetation. The latter was demonstrated by the constant movement of vegetation to unoccupied areas, leaving other areas barren.

Studies on fluviatile macrophytes include those of the seasonal changes in the biomass of the emergent macrophyte *Nasturtium officinale* R. Br. (Castellano 1977) and of the submerged macrophyte *Ranunculus penicillatus* var. *calcareus* (R. W. Butcher) C. D. K. Cook (Dawson 1976a) and their year-to-year interactions in relation to discharge cycles (Dawson *et al.* 1978). The latter, with their seasonal variation, is probably one of the most important of the between-site variables governing biomass difference; such a relationship was also demonstrated by Brooker *et al.* (1978) who showed an inverse relationship between the flow of water in the April—June period and the intensity of plant growth. A more extreme case of discharge variation for a shaded section of a lowland chalkstream has been described by Ham *et al.* (1982a) as affecting species of the genera *Ranunculus, Berula* and *Callitriche.* Some limited studies on the growth rate and seasonal cycle have also been undertaken in artificial systems in defined conditions for example by Eichenberger (1983) to estimate the uptake of nutrients of *Ranunculus fluitans* Lam. and the effects of grazing by invertebrates.

Year to year variation of either the mean rate of growth or the maximum biomass achieved can occur, dependent upon the time for which there is sufficient light available to the plant because of fluctuations in light-penetration through the water (Westlake 1966b). The effects of rainfall and turbidity of the water are often far more prolonged in larger and deeper streams; seasonal boat traffic also has a similar effect (Murphy & Eaton 1982). If the light levels are below the light compensation of growth balance point for extended periods, then growth is not possible and the biomass achieved by the time of flowering is reduced; other species which develop later, or have large over-wintering storage organs and short growing seasons, may become dominant if this situation continues for several years e.g. *Nuphar* spp. in larger rivers.

Even under relatively similar discharge cycles, changes occur, as shown during the 4-year monitoring of the biomass of *Ranunculus calcareus*, in an unshaded section of a stream, in order to determine the growth cycle (Dawson 1976). This study showed that the maximum summer biomass in the four seasons declined to half that initially found, but this resulted primarily from the plants not being subjected to annual mid-summer plant removal by cutting. It was suggested that the decline occurred as much from the suppression of autumnal regrowth by moribund plant material overlying the shoots giving a low overwintering biomass, as from the release from the synchronising influence of regular cutting on plant regrowth. This suggestion was confirmed by the study of Ham *et al.* (1981), who also found a similar response to cutting. In related work over several seasons on this same site different species and cycles of biomasses were found by Ham *et al.* (1982) between shaded and unshaded sites and illustrate the effects of competition between an aggressively growing species *Ranunculus* and other less aggressive ones *Callitriche* and *Berula*; these differences also indicate the degree of shade tolerance or adaption amongst these species. A long-term study of the biomass variations found by determining the total weight of weed removed from a managed river, divided the causes equally between environment in the early part of the year, the timing and extent of management in the previous year, and the inherent growth cycle of the plant (Westlake & Dawson 1982). In a subsequent study on the river in which pre-emptive removal in the previous autumn was undertaken, the importance of management was confirmed because reductions of up to a third were found in comparison with the expected biomass (Westlake & Dawson 1986).

Environmental Growth Controlling Factors

Despite aquatic plants, both large and small, having been discussed and reviewed regularly by a series of authors either wholly or in part flowing waters (Pond 1905, Kofoid 1908, Tansley 1911, 1953, Arber 1920, Glück 1924, 1936, Gessner 1955, 1959, Hoehne 1955, Sculthorpe 1967, Hynes 1970, Westlake 1975, Haslam 1978), there is as yet little fundamental understanding of the distribution of species. However several growth controlling factors have been isolated and their effects discussed qualitatively, these effects still require quantification. Eco-physiological data on the likely controlling factors is increasing and is assisting in the interpretation of growth and distribution. Much basic survey work on flowing waters has been undertaken at a variety of levels (Rawlence & Whitton 1976, Ham *et al.* 1981, 1982, Holmes 1983) but much only indicates probable explanations of distribution whilst a few give data on correlated physico-chemical factors from relatively infrequent determinations. Thus most of the studies so far are of an observational or functional, rather than causal nature.

An integrated approach is desirable when studying the vegetation of flowing waters because of the numerous potential interactions of the biotic, physical and chemical environment. Such an approach should seek to combine factors affecting the vegetation and include such things as the population composition and structure, the stand structure and physiological responses of the species present together with the levels and the short- and long-term temporal variations of controlling physical factors at the plant's surfaces; the latter would include water velocity, nutrient concentration, water or water and air temperatures and irradiance level and duration. Water flow influences the plant habitat, through its force grading the substrate where possible, and though seasonal changes in the supply, deposition and removal of materials. The interaction of the above factors, with the plants' responses or tolerances results in plant associations and in their short-term seasonal or long-term changes.

Water Flow

Water flow affects fluviatile macrophytes both directly by the force and turbulence of its movement, and indirectly through their metabolism by affecting the supply of nutrients and the removal of by-products. Water flow frequently varies with season through changes in the rate of discharge, but the seasonal growth of plants can modify the speed or velocity of its flow. Major changes in flow may eliminate or strongly suppress the growth of a plant population. Plants adapt to such conditions by shedding their above-substratum parts or by rapid reinvasion from less vulnerable areas through the production of numerous propogules or seeds. The dynamic nature of the fluviatile environment makes the predictability of the presence of a particular species difficult although representatives of a group of species whose flow-tolerance limits lie within the field range are likely to be present.

The *direct effects* of water force on fluviatile macrophytes has mainly been studied from the viewpoint of the resistance to water flow resulting in the interaction of vegetation and flow which is created in river engineering although there have been a few studies related to macrophyte ecology e.g. flow damage to plants, (Haslam 1978); these topics together with others related to vegetative hydraulic roughness are included in a recent bibliography (Dawson & Charlton, in press). The presence and seasonal growth of vegetation in lowland flowing waters may result in significant reductions in water velocity with corresponding increases in water level and even over-bank flows. The reduced velocities encourage the deposition of sediment and in combination with the plants, may cause changes in the magnitude or direction of currents and may erode banks or create new channels. Conversely presence of vegetation particularly at or on the margins does offer protection to banks and this technique is frequently used in many regions of the world but may not be reported in scientific literature; the original studies of vegetative resistance concerned the selection of grasses to line farm-channels subject to intermittent flows (this work should currently have application in erosion control in Third World countries). Submerged macrophytes, although not productive in terms of dry weight in comparison to terrestrial plants, have a much higher water and frequently air content, and can also rapidly change their position in the water column and their form, to create a 'blockage' to water flow channels (Westlake 1981, Hydraulics Research Ltd. 1985). While these applied studies have been restricted primarily to effects on the channel, they have included aspects relating to the vegetation such as the spacing of vegetation in channels and in flood plains in relation to its hydraulic resistance and sedimentation studies in vegetated channels and the development of better equations (Dawson & Charlton, in press).

The interaction of water flow with various species of fluviatile macrophyte is complex and varies not only with water velocity depth of flow and turbulence but with the seasonal state of growth, size and condition of the macrophyte. In essence, the resistance offered by the plant in maintaining its presence, increases more rapidly than the increase in water velocity. Macrophytes have developed various basic strategies to overcome water force, greater structural strength, greater flexibility, growth in thin layers and by the restriction of their growth to non-critical seasons of water flow. Thus the more flexible plants e.g. *Batrachium Ranunculi*, adopt a more stream-line and compact form with increasing velocity until they either become damaged or prone, as a compensation mechanism to minimise the forces acting upon them. In

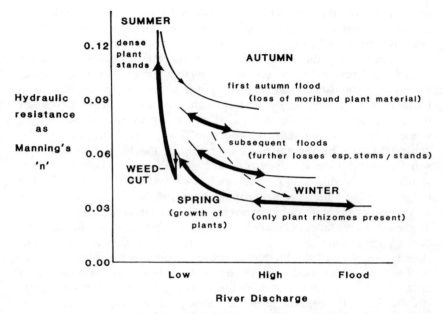

Fig. 2. The seasonal relationships of hydraulic roughness to discharge for a lowland river (1—1.5 m in depth) with dense summer stands of attached fluviatile macrophytes: thin lines indicate ranges of discharge values, thick lines indicate the predominant values (simplified from Dawson & Robinson 1984).

contrast, other species groups have a low initial resistance and can tolerate small increases in velocity by, for example, having leaves at the water surface (Nymphaeaceae), but hydraulic resistance rapidly increases as leaves tip and become submerged. Quantitatively vegetative hydraulic resistance is determined by the surface in contact with the water and the number and form of plant stands which includes their height, width and flexibility (and buoyancy) together with their component stems and leaves (Dawson & Charlton, in press). Stem frequency and the permeability of stands of various species with closely packed stems (*Ranunculus, Nasturtium*) have been illustrated by Marshall and Westlake (in press) in which they found water velocity within stands were only a tenth of that in the main flow in a lowland river site. Another study has measured the water force acting upon individual plant stands of a *Ranunculus* spp. at their natural sites in a section of river and confirmed that realistic forces were being measured by multiplying the average force by the number of plants in the population, that this alone could account for about half of the flow resistance when directly measured as a difference in water height (Dawson & Robinson 1984). A simple calculation suggests that in temperate regions, the maximum biomass achievable by submerged plants growing near their maximum tolerated water velocity, can raise the water level in a low-land river by 0.5—1.0 m.

In summarising the directs of flow on vegetation in flowing systems in order of decreasing gradient or of fluviatile power, then macrophytes act as opportunists in the higher energy areas rarely reaching high biomasses; they survive by finding niches protected from full water force but are vulnerable to loss, if not compact or without strong mechanisms of attachment. In moderate flows, high biomasses can be achieved partly by vegetative modification of the habitat at least sea-sonally by the plant. Plants also resist higher flows by adaption of their stand forms. In slow flows a wide variety of macrophytes can exist and ranges overlap with 'lake' macrophytes, but whilst increased velocities may cause near catastrophic removal, other growth conditions are often of greater importance. Channel modification frequently causes vegeta-tion changes not only in the modified reach but also further down-stream (Brookes 1985). Such changes result directly from the objective of the engineering work i.e. increasing the rate of loss of water from upstream, but schemes normally fail to recognise the effect on down-stream channel morphology or on subsequent changes in vegetation. Water storage dams dependent upon this method of operation affect vegetation downstream but may increase vegetation by preventing floods e.g. Lowe (1979); dams also provide, through the intermittent release of water, a technique for the control of excessive growths of fluviatile macrophytes (Dawson et al. 1976).

The *indirect effects* of water flow operate on the environment of the plant and can affect its metabolism through altering the light available for growth, the temperature of the water or the supply of inorganic car-bon or nutrients. The plant's metabolism can also be affected reflexively by its own processes, for example, the increase of dissolved oxygen within the plant stand which enhances respiration at the expense of net photosynthesis.

Light

Data is sparse on the effects on biomass of light within the water column of flowing waters although data are available on light levels reaching the water surface in streams and rivers in varying degrees of marginal shade (Dawson & Kern-Hansen 1979). Initially, these authors demonstrated that the biomass of three problem species of submerged fluviatile macrophytes was proportional to the light at the stream surface and subsequently that this was true for both artificially shaded sections of stream and naturally shaded areas, when compared to the biomass in adjacent open and shaded sections of stream; this was despite the possible effects of light saturation and self shading. This information was used to describe the shade given by some types of marginal terrestrial vegetation to streams of differing width and orientation, for use as an alternative management technique for streams with excessive quantities of aquatic macrophytes. In the subsequent year of this study, the biomass of macrophytes decreased in both the experimental and control sections (see above). Speculation suggests that in regularly managed streams, the more vigorously growing macrophytes are selected for, and in combination with late summer regrowth, allows a higher overwintering biomass and a good start to the next seasons growth (cf. preemptive control, Westlake & Dawson 1986). The use of shade as a long-term alternative control may select for less vigorous plants which are attempting merely to maintain their presence, rather than grow vigorously. Observation of shaded stream areas suggests that it is not uncommon for infrequent large plants to develop with many uncolonised or bare areas in preference to many smaller plants; this phenomenon remains to be explained.

Chemistry

Water flow makes possible the near continual supply and availability of nutrients to fluviatile macrophytes directly through the replenishment of water and indirectly through the supply and refreshment of sediments; conversely however pollutants are equally easily available and distributed by flow to macrophytes. This changing environment means that nutrients in particular are frequently not a limiting factor to growth, but their presence may lead more easily to epiphyte growths in water with high nutrient levels than in static water situations. Of specific macronutrients the level of nitrate is frequently sufficient, or in excess, in many vegetated running waters and in general it is rare for nutrients to be thought limiting in lowland rivers e.g. Casey & Westlake (1974). In studies of lowland chalkstreams in southern England only on one occasion has a phosphate phosphorus level (of 5 μgl^{-1}) occurred which could have possibly been considered limiting (Casey & Westlake 1974). Nutrient limitation is often proposed as a control for plant growth but it would require to be very low in both water and sediment, to be effective although in channel precipitation (e.g. ferric hydroxide) or adsorption could reduce levels prior to critical stages in plant growth (Kern-Hansen & Dawson 1978).

Metabolism

Water flow affects submerged macrophyte metabolism through the exchange of dissolved carbon dioxide and oxygen in particular, as well as determining the exchange of the latter with the air through the water surface. Macrophytes respond to partial pressures i.e. the concentrations of these dissolved gases whereas the degree of saturation indicates the direction of its movement at the air-water interface. It is commonly stated that macrophytes oxygenate water i.e. they are 'oxygenators'; it is also true that they consume a significant portion of oxygen overnight. Diel cycles of dissolved oxygen frequently show an asymmetric distribution of concentration which with results of preliminary studies of the accentuated oxygen concentrations within stands, is likely to lead to gas bubble release; this can result in low oxygen concentrations and other non-desirable effects e.g. fish kills with excessive growth of macrophytes. Analysis of diel oxygen curves has produced a variety of results, for instance Dawson *et al.* (1982) showed that significantly smaller ranges could be found in unmanaged than in managed streams whilst Thyssen (1982) showed that growth of a periphyton community could compensate for the partial removal of macrophytes in a relatively nutrient poor natural stream of relatively low macrophyte biomass (<150 g dry wt m^{-2}). Jorga & Weise (1977) showed photosynthetic oxygen production was a function of biomass and light intensity to *c.* 250 g dry weight m^{-2} after which the oxygen level became critical to the system in the early morning hours. Decomposition by macrophytes at the end of the growing season can, if *in situ* also increase oxygen consumption; emergent plants in general do create an obvious oxygen demand when they decompose after fall into water (Szczepanski 1976). Gaseous exchange is faster in more turbulent water and thus less diel variation may be noted in fast streams; ice cover inhibits exchange and leads to low oxygen concentration through demand by the benthic communities, although algal growth can occur on the underside of ice sheets. Low flows are often associated with summer conditions, and greater exchanges of heat with the water elevate water temperatures and also macrophyte metabolism in addition to reducing the oxygen holding capacity of the water (see e.g. Jorga & Weise 1977). A combination of these factors with decreasing light levels after mid summer, probably leads to decline in the seasonal cycle of growth of many macrophytes.

THE ROLE OF MACROPHYTES IN THE FLUVIATILE ECOSYSTEMS

In common with other ecosystems, fluviatile macrophytes provide a variety of habitats and become food for many levels of the food chain; the role of aquatic macrophytes has been reviewed by Gaudet (1974) and Marshall & Westlake (1978). Fluviatile macrophytes provide substrate or shelter for all ages and sizes of fauna together with many micro-organisms, algae, etc. (Whitcome 1968, Harrod 1964, Böttger 1978, Mills 1981, Scott 1985). Their presence influences their surroundings and frequently results in an even greater variety of habitats through lower mean velocities, sediment deposition and release of nutrients or organic compounds. On death, fluviatile macrophytes provide an important source of easily degradable organic material (Westlake *et al.* 1972), especially in shallower flowing waters; this can be equivalent on an areal basis to the terrestrial inputs such as trees leaves (Dawson 1976b). The breakdown of macrophyte material in lowland chalkstreams is typically twice as fast as terrestrial leaves; hence the former is infrequently seen moving downstream. Water flow distributes materials for breakdown at rates faster than in static situations. This redistribution and mixing of materials is most apparent in situations with mixed terrestrial and macrophyte input because the latter retard the downstream passage of the former materials by keeping it near its site of production and in a dispersed and suitably oxygenated state for rapid fragmentation (Dawson 1980). In forests, with few fluviatile macrophytes, leaf material may accumulate in deep anaerobic layers in areas of low flow or is lost downstream from the ecosystem (Slack & Feltz 1970, Dawson 1976a, Böttger 1978). At the other extreme of unshaded flowing waters, excessive quantities of macrophyte, often monospecific stands, in lowland areas are removed to reduce water levels or flood-risk. Apart from loss of organic material, the water velocity produced oscillates, seasonally, between extremes and requires a very tolerant or rapidly invasive flora (and fauna) to accommodate these changes. Other aspects of the role of fluviatile macrophytes has been demonstrated by studies on macrophyte dominated systems subjected to various perturbations. Examples include the problems especially of the depression of dissolved oxygen at night in stream dominated by *Potamogeton pectinatus* which received sewage effluent; the increase in the biomass of *Ranunculus fluitans* with increasing eutrophication and the judicious use of a control technique, shade, to avoid catastrophic changes in macrophyte populations especially towards an algal dominated one (Ska & van der Borght 1986).

Future research on fluviatile macrophytes should not only incorporate a sound understanding of the species, their seasonal growth cycles and associated changes in morphology and stand structure but the ways in which plants respond to their respective natural environments and to man's intervention with them, through management (Westlake 1968). Such basic studies assist in the introduction of management in its broadest form or other human perturbation of the environment. Such is the extent of the latter that in many countries it is

304

now difficult to differentiate between natural plant populations and the accepted normal situation which has resulted from decades or centuries of management.

REFERENCES

Arber, A. (1920) Water Plants. University Press, Cambridge, 436 pp.

Bayly, I. A. E. & Williams, W. D. (1973) Inland Waters and their Ecology. Longman, Camberwell, Victoria, Australia, 316 pp.

Beadle, L. C. (1974) The inland waters of Africa. Longman, London, 365 pp.

Bishop, J. E. (1973) Limnology of a small Malayan River Sungai Gombak. Monographiae Biologicae 22 Dr. W. Junk, The Hague, 385 pp.

Bonetto, A. A. (1975) Landscapes of River Basins (South America). In: Hasler, A. D. (ed.), Ecological Studies 10. Coupling of land and water systems. Springer, New York. 173—197.

Böttger, K. (1978) Ökologischer Gewässerschutz eines norddeutschen Tieflandsbaches. Schr. Naturwiss. Ver. Schleswig-Holstein 48: 1—12.

Brodsky, K. A. (1980) Mountain torrent of the Tien Shan. Monographiae Biologicae 39, Dr. W. Junk, The Hague, 311 pp.

Brooker, M. P., Morris, D. L. & Wilson, C. J. (1978) Plant flow relationships in the R. Wye catchment. Proc. EWRS 5th Symp. on Aquatic Weeds 1978, pp. 63—70.

Brookes, A. (1985) River channelization: traditional engineering methods, physical effects and alternative practices. Progress in Physical Geography 9: 44—73.

Butcher, R. W. (1933) Studies on the Ecology of Rivers. I. on the distribution of macrophyte vegetation in the rivers of Britain. J. Ecol. 21: 58—91.

Capblancq, J. & Dauta, A. (1978) Phytoplankton et production primaire de la rivière Lot. Annls Limnol. 14: 85—112.

Casey, H. & Westlake, D. F. (1974) Growth and nutrient relationships of macrophytes in Sydling Water, a small unpolluted chalk stream. Proc. Eur. Weed Res. Counc. 4th int. Symp. Aquatic Weeds 1974, pp. 69—76.

Castellano, E. (1977) Productividad de Rorippa Nasturtium aquaticum (L.) Hayek. Thesis Agreg., Department of Biology, Universidad de los Andes, Venezuela.

Coffey, B. T. & McNabb, C. D. (1974) Eurasian water-milfoil in Michigan. Michigan Bot. 13: 159—165.

Davies, B. R. & Walker, K. F. (1986) The Ecology of River Systems, Editors Monographiae Biologicae 60, Junk, Dordrecht, 793 pp.

Dawson, F. H. (1973) Notes on the production of stream bryophytes in the High Pyrenees (France). Annls Limnol. 9 (3): 231—240.

Dawson, F. H. (1976a) The organic contribution of stream edge forest litter fall to the chalk stream ecosystem. Oikos 27, 13—18.

Dawson, F. H. (1976b) The annual production of the aquatic macrophyte Ranunculus penicillatus var. calcareus (R. W. Butcher) C. D. K. Cook. Aquat. Bot. 2: 51—73.

Dawson, F. H. (1978) The seasonal effects of aquatic plant growth on the flow of water in a stream. Proc. 5th Eur. Weed Res. Counc. Symp. Aquat. Weeds, pp. 71—78.

Dawson, F. H. (1980a) The origin, composition and downstream transport of plant material in a small chalk stream. Freshwat. Biol. 10, 419—435.

Dawson, F. H. (1980b) The flowering of Ranunculus penicillatus var. calcareus in the River Piddle, Dorset. Aquat. Bot. 9, 145—157.

Dawson, F. H., Castellano, E. & Ladle, M. (1978) The concept of species succession in relation to river vegetation and management. Verh. int. Verein. theor. angew. Limnol. 20: 1451—1456.

305

Dawson, F. H. & Charlton, F. G. (in press) A bibliography on the hydraulic resistance or roughness of vegetated watercourses. Occasional Publication of Freshwater Biological Association No. 25.

Dawson, F. H. & Kern-Hansen, U. (1979) The effect of natural and artificial shade on the macrophytes of lowland streams and the use of shade as a management technique. Int. Rev. ges. Hydrobiol. Hydrogr. 64 (4): 437—455.

Dawson, F. H., Kern-Hansen, U. & Westlake, D. F. (1982) Water plants and the temperature and oxygen regimes of lowland streams. In: Symoens, J. J., Hooper, S. S. & Compère, P. (eds.), Studies on Aquatic Vascular Plants. Royal Botanical Society of Belgium, Brussels, pp. 214—221.

Dawson, F. H. & Robinson, W. N. (1984) Submerged macrophytes and the hydraulic roughness of a lowland chalkstream. Verh. int. Verein. theor. angew. Limnol. 22 (3): 1944—1948.

Dècamps, H. (Ed.) (1978) Hydrobiologie de la Rivière Lot. Annls Limnol. 14 (1—2): 1—180.

den Hartog, C. (1982) Architecture of macrophyte-dominated aquatic communities. In: Symoens, J. J., Hooper, S. S. & Compère, P. Studies on Aquatic Vascular Plants. Royal Botanical Society of Belgium, Brussels, pp. 222—234.

Denny, P. (1980) Solute movement in submerged angiosperms. Biol. Rev. 55: 65—92.

Denny, P. (1985) Submerged and floating-leaved aquatic macrophytes (euhydrophytes). In: Denny, P. (ed.), The ecology and management of African wetland vegetation. Geobotany 6: 19—42.

Edwards, R. W. and Brooker, M. P. (1982) The ecology of the Wye. Monographiae Biologicae 50, Dr. W. Junk, The Hague, 164 pp.

Eichenberger, E. (1983) The effect of the seasons on the growth of Ranunculus fluitans Lam. Proc. Int. Symp. Aquat. Macrophytes (Nijmegen, 18—23 September 1983), pp. 63—67.

Ekzertzev, V. A. (1979) The higher aquatic vegetation of the Volga. In: Mordukhai-Boltovskoi, Ph.D. (ed.), The River Volga and its life. Monographiae Biologicae 33, 271—194. Dr. W. Junk, The Hague.

Gaudet, J. J. (1974) The normal role of vegetation in water. In: Mitchell, D. S. (ed.), vegetation its use and control. UNESCO, Paris, pp. 24—37.

Gessner, F. (1955) Hydrobotanik. Die physiologischen Grundlagen der Pflanzenverbreitung im Wasser. I. Energiehaushalt. Deutscher Verl. der Wissenschaften, Berlin, 517 pp.

Gessner, F. (1959) Hydrobotanik. Die physiologischen Grundlagen der Pflanzenverbreitung im Wasser. II. Stoffhaushalt. Deutscher Verl. der Wissenschaften, Berlin, 701 pp.

Gimmingham, C. H. & Birse, E. M. (1957) Ecological studies on growth forms in Bryophytes I. Correlations between growth form and habitat. J. Ecol. 45: 533—545.

Glück, H. (1905) Biologische und morphologische Untersuchungen über Wasser- und Sumpfgewächse. I. Die Lebensgeschichte der europäischen Alismaceen. Fischer, Jena.

Glück, H. (1906) Biologische und morphologische Untersuchungen über Wasser- und Sumpfgewächse. II. Untersuchungen über die Mitteleuropäischen Utricularia-Arten, über die Turionenbildung bei Wasserpflanzen, sowie über Ceratophyllum. Fischer, Jena.

Glück, H. (1911) Biologische und morphologische Untersuchungen über Wasser- und Sumpfgewächse. III. Die Uferflora. Fischer, Jena.

Glück, H. (1924) Biologische und morphologische Untersuchungen über Wasser- und Sumpfgewächse. IV. Submerse und Schwimmblattflora. Fischer, Jena, 746 pp.

Glyme, J. M. & Carr, R. E. (1974) Temperature survival of Fontinalis novae-angliae Sull. The Bryologist 77: 17—22.

306

Golterman, H. L. & Kouwe, F. A. (1980) Chemical budgets and nutrient pathways. *In*: Le Cren, E. D. & Lowe-McConnell, R. H. (eds.), The Functioning of freshwater ecosystems. IBP No. 22. Cambridge Univ. Press, pp. 85—140.

Gopal, B. & Sharma, K. P. (1981) Water Hyacinth (*Eichhornia crassipes*) the most troublesome weed of the world. Hindasia Publications, Delhi, India.

Harrod, J. J. (1964) The distribution of invertebrates on submerged aquatic plants in a chalkstream. J. anim. Ecol. 33: 335—348.

Haslam, S. M. (1978) River Plants. Cambridge University Press, Cambridge, 396 pp.

Haslam, S. M. & Wolseley, P. A. (1981) River Vegetation, its Identification, Assessment and Mangement, Cambridge University Press, Cambridge, 154 pp.

Holmes, N. T. H. (1983) Typing British rivers according to their flora. Focus on Nature Conservation No. 4 Nature Conservancy Council, Huntingdon, Cambridgeshire, 194 pp.

Ham, S. F., Wright, J. F. & Berrie, A. D. (1981) Growth and recession of aquatic macrophytes on an unshaded section of the River Lambourn, England, from 1971 to 1976. Freshwat. Biol. 11: 381—390.

Ham, S. F., Cooling, D. A., Hiley, P. D., McLeish, P. R., Scorgie H. R. A. & Berrie, A. D. (1982a) Growth and recession of aquatic macrophytes on a shaded section of the River Lambourn, England, from 1971 to 1980. Freshwat. Biol. 12: 1—15.

Ham, S. F., Wright, J. F. & Berrie, A. D. (1982b) The effect of cutting on the growth and recession of the freshwater macrophyte *Ranunculus penicillatus* (Dumort.) Bab. var. *calcareus* (R. W. Butcher) C. D. K. Cook, J. Environ. Manage. 15: 263—271.

Howard-Williams C., Davies J. & Pickmere, S. (1982) The dynamics of growth, the effects of changing area and nitrate uptake by watercress *Nasturtium officinale* R. Br. in a New Zealand stream. J. appl. Ecol. 19: 589—601.

Hydraulics Research Ltd. (1985) The hydraulic roughness of vegetation in open channels. Report IT 281 (SR36). Hydraulics Research Ltd., Wallingford, 39 pp.

Hynes, H. B. N. (1970) The Ecology of Running Waters, Liverpool University Press, 555 pp.

Jenkins, J. T. & Proctor, M. C. F. (1985) Water velocity, growth-form and diffusion resistances to photosynthetic CO_2 uptake in aquatic bryophytes. Plant, Cell and Environment 8: 317—323.

Jorga, W. & Weise, G. (1978) Biomasseenwicklung submerser Makrophyten in langsam fliessenden Gewässern in Beziehung zum Sauerstoffhaushalt. Intern. Revue ges. Hydrobiol. 62: 209—234.

Kern-Hansen, U. & Dawson, F. H. (1978) The standing crop of aquatic plants of lowland streams in Denmark and the inter-relationships of nutrients in plant, sediment and water. Proc. 5th Eur. Weed Res. Coun. Symp. Aquat. Weeds, pp. 143—150.

Kofoid, C. A. (1908) The plankton of the Illinois River 1894—1899. Part III constituent organisms and their seasonal distribution. Bull. Illinois State Lab. Nat. Hist. VIII Article 1: 1—361.

Ladle, M. & Casey, H. (1971) Growth and nutrient relationships of *Ranunculus penicillatus* var. *calcareus* in a small chalk stream. Proc. Eur. Weed Res. Counc. 3rd int. Symp. Aquatic Weeds 1971, pp. 53—63.

Le Cren, E. D. & Lowe-McConnell, R. H. (1980) The functioning of freshwater ecosystems IBP 22 Eds. Cambridge Univ. Press, 588 pp.

Lewis, K. (1973a) The effect of suspended coal particles on the life forms of the aquatic moss *Eurhynchium riparioides* (Hedw.) I. The gametophyte plant. Freshw. Biol. 3: 251—257.

Lewis, K. (1973b) The effect of suspended coal particles on the life forms of the aquatic moss *Eurhychium ripariodes* (Hedw.) II. The effect on spore germination and regeneration of apical tips. Freshw. Biol. 3: 391—396.

Lowe, R. L. (1979) Phytobenthic Ecology and Regulated streams. *In*: Ward, J. V. & Stanford, J. A., (eds.), Proceeding First International Symposium on Regulated Streams (Erie, Pennsylvania, 18—20 April 1979), pp. 25—34.

Madsen, J. D. (1986) The production and physiological ecology of the submerged aquatic macrophyte community in Badfish Creek, Wisconsin, Ph.D. Thesis, Dept. of Botany, Univ. of Wisconsin, Madison.

Marshall, E. J. P. and Westlake, D. F. (1978) Recent studies on the role of aquatic macrophytes in their ecosystem. Proc. Eur. Weed Res. Counc. 5th int. Symp. Aquatic Weeds, pp. 43—51.

Michaelis, F. B. (1976) Watercress (*Nasturtium microphyllum* (Boenn.) Rchb and *N. officinale*) in New Zealand cold springs. Aquat. Biol. 2: 317—325.

Mills, C. A. (1981) The spawning of roach *Rutilis rutilis* L. in a chalk stream. Fish. Manage. 12: 49—54.

Mitchell, D. S. (1974) Aquatic Vegetation and its Use and Control. UNESCO, Paris, 135 pp.

Murphy, K. J. & Eaton, J. W. (1982) The management of aquatic plants in a navigable canal system used for amenity and recreation. 6th Int. Symp. Aquatic Weeds (20—25 September 1982). Eur. Weed Res. Soc., pp. 141—151.

Murphy, K. J. & Eaton, J. W. (1983) Effects of pleasure-boat traffic on macrophyte growth in canals. J. applied Ecol. 20: 713—729.

Pieterse, A. H. (1978) The water hyacinth (*Eichhornia crassipes*), a review. Abstracts on Tropical Agriculture 4 (2): 9—42.

Pitlo, R. H. & Dawson, F. H. (in press) Flow resistance by aquatic vegetation. *In*: Pieterse, A. H. & Murphy, K. J. (eds.), Aquatic Weeds. Oxford University Press.

Pond, R. H. 1903 (1905) The Biological relation of aquatic plants to the substratum. U.S. Commission of Fish and Fisheries, Report of the Commissioner 24: 484—526.

Raven, J., Beardall, J. & Griffiths, H. (1982) Inorganic C-sources for *Lemnaea, Cladophora* and *Ranunculus* in a fast-flowing stream: measurements of gas exchange and of Carbon Isotope ratio and their ecological implications. Oecologia (Berl.) 53: 68—78.

Rawlence, D. J. & Whitton, J. S. (1976) An elemental survey of the aquatic macrophytes, water and plankton from the Waikato River, North Island, New Zealand. Mauri ora 4: 121—131.

Rzóska, J. (1976) The Nile, Biology of an ancient river. Monographiae Biologicae 29, Dr. W. Junk, The Hague.

Sastroutomo, S. S., Ikusima, I., Numata, M. & Izumi, S. (1979) The importance of turions in the propagation of pondweed (*Potamogeton crispus* L.). Ecological Review 19: 75—88.

Say, P. J. (1978) Le Riou-Mort, affluent du Lot pollué par métaux lourds. I Etude préliminaire de la chimie et des algues benthiques. Annls Limnol. 14 (1—2): 113—131.

Scott, A. (1985) Distribution, growth and feeding of postemergent grayling *Thymallus thymallus* in an English river. Trans. Amer. Fish. Soc. 114: 525—531.

Sculthorpe, C. D. (1967) The Biology of Aquatic Vascular Plants. E. Arnold.

Ska, B. & Van de Borght, P. (1986) The problem of *Ranunculus* development in the river Semois. Proceedings, EWRS/AAB 7th Symposium on Aquatic Weeds, pp. 307—314.

Slack, K. V. & Feltz, H. R. (1968) Tree leaf control on low-flow water quality in a small Birginia stream. Environm. Sci. Technol. 2: 126—131.

308

Szczpanski, A. (1976) Limiting factors and productivity of macrophytes. Folia geobot. et phytotaxonom. 11: 337—432.

Szemes, G. (1967) Systematisches Verzeichnis der Pflanzenwelt der Donau mit einer zusammenfassenden Erlauterung. *In*: Liepolt, R. (ed.), Limnologie der Donau (V). E. Schweizerbartsche Verlagsbuchhandlung, Stuttgart, pp. 70—131.

Tansley, A. G. (1911) Types of British Vegetation. Cambridge University Press, 416 pp.

Tansley, A. G. (1953) The British Islands and their vegetation. Cambridge University Press, xxxviii + 930 pp.

Taubaev, T. (1970) Flora of Middle Asiatic waterbodies and its value for the national economy. Inst. Botan., Akad. Nauk Uzbek. SSR, FAN, Tashkent, 491 pp.

Thommen, G. H. & Westlake, D. F. (1981) Factors affecting the distribution of populations of *Apium nodiflorum* and *Nasturtium officinale* in small chalk streams. Aquat. Bot. 11: 21—36.

Thyssen, N. (1982) Aspects of the oxygen dynamics of a macrophyte dominated lowland stream. *In*: Symoens, J. J., Hooper, S. S. & Compère, P. (eds.), Studies on Aquatic Vascular Plants. Royal Botanical Society of Belgium, Brussels, pp. 202—213.

Warming, E. (1909) Ecology of Plants. An introduction to the study of plant communities. Engl. transl. by Groom, P. & Balfour, I.B. Oxford Univ. Press, Oxford, 422 pp.

Wehr, J. D. & Whitton, B. B. (1986) Ecological factors relating to the morphological variation in the aquatic moss *Rhynchostegium riparoides* (Hedw.) C. Jens. J. Bryol. 14: 269—280.

Westlake, D. F. (1966) A model for quantitative studies of photosynthesis by higher plants in streams. Air & Wat. Pollut. Int. J. 10: 883—896.

Westlake, D. F. (1966b) The light climate for plants in rivers. *In*: Bainbridge, R., Evans, G. C. & Rackman, O. (eds.), Light as an ecological factor. Brit. Ecol. Soc. Symp. No. 6, pp. 99—120.

Westlake, D. F. (1967) Some effects of low-velocity currents on the metabolism of aquatic macrophytes. J. exp. Biol. 18: 187—205.

Westlake, D. F. (1968) The biology of aquatic weeds in relation to their management. Proc. 9th Brit. Weed Control Conf., pp. 372—379.

Westlake, D. F. (1975) Macrophytes. *In*: Whitton, B. A. (ed.), River Ecology. Blackwell Sci. Publs, Oxford, pp. 106—128.

Westlake, D. F. (1981) The development and structure of aquatic weed populations. Proc. Aquatic Weeds and their Control, pp. 33—47.

Westlake, D. F. (1982) The primary productivity of water plants. *In*: Symoens, J. J., Hooper, S. S. & Compère, P. (eds.), Studies on Aquatic Vascular Plants. Royal Botanical Society of Belgium, Brussels, pp. 165—180.

Westlake, D. F. & Dawson, F. H. (1982) Thirty years of weed cutting on a chalk stream. Proc. Eur. Weed Res. Soc. 6th Symp. Aquat. Weeds, pp. 132—140.

Westlake, D. F. & Dawson, F. H. (1986) The management of *Ranunculus calcareus* by pre-emptive cutting in southern England. European Weed Research Society, Association of Applied Biologists, 7th International Symposium on Aquatic Weeds (September 1986), pp. 395—400.

Westlake, D. F., Casey, H., Dawson, F. H., Ladle, M., Mann, R. H. K. & Marker, A. F. H. (1972) The chalk-stream ecosystem. *In*: Kajak, Z. & Hillbricht-Ilkowska, A. (eds.), Productivity problems of freshwaters. Proc. IBP/UNESCO Symposium (Kazimierz Dolny, Poland, 6—12 May 1970), pp. 615—635.

Whitton, B. A. (1972) Environmental limits of plants in flowing waters. *In*: Edwards, R. W. & Garrod, D. J. (eds.), Conservation and Productivity of Natural Waters. Symposia of the Zoological Society of London 29, pp. 3—19.

Wiegleb, G. (1988) Analysis of flora and vegetation in rivers — concepts and applications. *In*: Symoens, J. J. (ed.), Handbook of Vegetation Science, vol. 15. Vegetation of inland waters. Dr. W. Junk. This volume, pp. 311—340.

Witcomb, D. M. (1968) The fauna of aquatic plants. Proceedings 9th British Weed Control Conference. British Crop Protection Council, pp. 382—391.

Zhadin, V. I. & Gerd, S. V. (1961) Fauna and flora of the rivers lakes and reservoirs of the USSR. (Engl. transl. Israel Program for Scientific Translations, Jerusalem 1963).

ANALYSIS OF FLORA AND VEGETATION IN RIVERS: CONCEPTS AND APPLICATIONS

GERHARD WIEGLEB

INTRODUCTION

The following plant groups are generally considered as macrophytes: macrophytic green, red and brown algae, stone worts, mosses and liverworts, and hydrophytic vascular plants (with the main growth forms of the submerged, floating leaved and free-floating plants) and helophytic vascular plants (incl. creeping and high-growing graminoids as well as dicotyledons). The sections of rivers where physical and chemical conditions allow the growth of macrophytes are called the "macrophyte region" (Roll 1938). They can also be considered as "macrophyte-dominated ecosystems" in the sense of den Hartog (1979). Macrophytes are connected to other compartments of the ecosystem in several ways (see Dahl & Wiegleb 1984):

— Besides phytoplankton and "Aufwuchs" algae, macrophytes are very important primary producers connecting the inorganic environment to the biotic community. The organic substances produced are mainly not used by herbivores but are transferred into the detritus food chain.
— Macrophytes are food and habitat for the macrofauna that shows many functional relations to the plants. Furthermore relations to epiphytic bacteria and algae as well as to the phytoplankton are evident.
— Macrophytes have a strong chemical effect on the water. Nutrient elimination from the water as well as nutrient pumping from the sediment into the water have been discussed. Furthermore, oxygen production and elimination of toxic substances and microbes are important.
— Macrophytes also have physical effects on the environment. Especially sediment stabilization, differentiation of the current velocity and the influence on the microclimate and on quantity and quality of the light consumption shall be mentioned.

311

J. J. Symoens (ed.), Vegetation of inland waters. ISBN 90—6193—196—7.
© *1988, Kluwer Academic Publishers, Dordrecht. Printed in the Netherlands.*

In the field of applied ecology several concepts are based on the importance of macrophytes in rivers. Kohler (1978a) stressed the use of macrophytes as bioindicators for anthropogenic disturbance, mainly pollution. Dahl & Wiegleb (1984) pointed out the role of the macrophytes in the 'rehabilitation of water courses'. Macrophytes share a large part of the function of a river (see Nunnally & Keller 1979) in a whole landscape. Such concepts are of special importance in a landscape that is dominated by the activity of man and in which questions of management, protection and assessment of vegetation gain special attention.

River vegetation like all vegetation may be studied on different geographical scales with different aims. The choice of the scale is not arbitrary but directly depends on the purpose of the study. The appropriateness of the scale for a given purpose may itself be an object of study (see Bouxin & LeBoulenge 1983).

The largest geographical scale is the biogeographical region. Within such a region factors like climate, geology and tectonics may be most influential on the recent distribution of plants. On this scale zonal, zonal-altitudinal and oceanic-continental variation can be shown. A smaller entity is the physiographic region which is mainly characterized by a relatively uniformous geomorphology even though the geology may change on a smaller scale. Within this geographical scale river sections of one river (longitudinal variation) or equivalent sections of adjacent rivers are compared as to their floristic composition and vegetation structure. On a very small scale the distribution of plants within one river section may be studied. There are several characters of rivers leading to some complication in this kind of study, like for example the vertical stratification of the vegetation, including the special problem of the complex geometry of the submerged stratum under the influence of the varying current, and small scale changes in the physical environment causing a spatial heterogeneity, which in most cases does not follow such clear cut zonation patterns as in the vegetation of stagnant waters.

Phenomena on a large geographical scale are relatively stable. However, phenomena on a small scale are highly dynamical and subject to great changes even under natural conditions in short periods of time. Furthermore, small scale phenomena are much more influenced by catastrophic events caused by human activities. Thus more refined methods are needed in small scale studies than in large and medium scale studies. On a medium scale anthropogenic impact tends to standardize the regional conditions especially in lowland areas. The influences of channelization, construction of weirs and reservoirs, eutrophication and pollution are reflected by the vegetation in each physiographic region in a more or less regular way.

The problem of river typology is not only a botanical problem. Many authors have reviewed the problem from different points of view without finding an absolute solution (see Schmassmann 1955, Illies & Botoseanu 1963, Steffan 1965, Pennak 1971, Hawkes 1975, Persoone 1979). Some dialectical contradictions arise. Even though the river is a continuum with the upper region influencing the lower reaches (Vannote *et al.* 1980) the differentiation in river zones may be adequate for descriptive purpose. Despite obvious similarities between rivers, each river shows its distinct features of individuality (Ruess 1954).

Besides the special properties of rivers and the plants colonizing them the purpose of the study should be kept in mind when carrying out a sample. The main purposes are:

1. Inventory of the river flora, mapping of the distribution of species. This purpose is purely descriptive and does not refer to any concept of vegetation.
2. Classification of vegetation. The classificatory approach itself aims to produce a set of concepts (hierarchical or non-hierarchical) that can be used for the exact description of the phenomena observed.
3. Study of the geographical distribution of both species and vegetation types, in most cases using distribution maps. In the first instance this procedure is also descriptive, but it can be used for hypothesis generation, too.
4. Study of the vegetation dynamics. Vegetation dynamics can best be understood as a population process, thus presence and abundance of species is the main interest of study. Often, however, also vegetation types are ordered in succession series. This approach is either descriptive or the results are linked to habitat parameters.
5. Study of the relation between plants and environment. Both the occurrence of single species and of vegetation types are correlated to habitat parameters. This approach seeks to generate hypotheses on the question, why plants occur in a given habitat.

The value of non-classificatory approaches does not need any further justification for the study of distribution, dynamics and habitat relations. The power of the vegetation concept as an analytical tool seems to be limited. Vegetation is a pragmatic concept that should be mainly used for description and should never imply any metaphysical ideas, related to the community and ecosystem concept.

HISTORICAL AND METHODOLOGICAL OUTLINE

For studying river vegetation, the Braun-Blanquet method which fits vegetation units into a hierarchical order, using so-called character or differential species has been widely used. An outline of the results of this approach for Central Europe has been given by Müller (1977), the development of the approach in Eastern Europe is documented in Moravec et al. (1983). Derived from this approach are two other ones. Passarge (1964) and Wiegleb (1978) used sociological species groups, however, no formal concept as to river vegetation has been developed. Mériaux (1978) characterized his Braun-Blanquet associations using area type spectra. This type of geographical analysis is common within the Braun-Blanquet school.

.   The floristical approach in a strict sense does not make any explicit reference to the concept of vegetation as an analytical tool (see above). Only the occurrence and abundance of a given species counts. The results of such analyses are presented in tables, diagrams, or distribution maps. This approach has been used by Holmes & Whitton (1975, 1977), Kohler (1978b), Holmes (1983) and Wiegleb (1984b) for river zones or river sections. Holmes (1980) used floristical lists of total rivers.

In Table I some methods are compared in more detail as to their basic implications. The method of Wiegleb and Herr has been developed in the study of lowland river vegetation in northern Germany (see Wiegleb 1984b for the philosophy of this approach). Cover is usually estimated using the Londo scale. In some permanent plots also the estimation of percent cover has been tried. The river sections usually have a length between 50 and 100 m, they are irregularly distributed along the course of the river and are mainly concentrated around bridges and other sites than can be easily defined geographically. All plants below the actual water level are sampled. The reference to cover gives a clear idea of the vegetation structure when taking into consideration also the growth form. The metric Londo scale is especially suitable for computation. Using this method it is easy to carry out quick surveys of larger areas by choosing for example one sample site every 5 or 10 km of river course. However, this method is unfavourable in areas with small scale inhomogeneity as in highland areas where logitudinal zonation does not follow a uniform pattern. There is also a certain danger of sampling disturbed sections near the bridges. But the choice of the samples is quasirandom and is not carried out according to characters of the vegetation itself.

The method of Kohler (see Kohler et al. 1971, Kutscher 1984) has been developed in the southern German highland area. Estimation usually refers to frequency, but in some of Kohler's definitions (Kohler 1978b) also abundance is mentioned. The river sections have a length of several hundred meters up to 3 km. If possible, the total course of the river is investigated and all plants below the actual water level are recorded. This method leads to a total inventory of a river or river

Table I. Comparison of sampling methods in vegetation survey of rivers.

Method (examples)	Estimation scale	Area of reference	Special characteristics
Wiegleb and Herr (Wiegleb 1984)	Cover: Londo scale, percent in permanent plots	River sections of a length between 50 and 100 m; irregularly distributed along the course; actual water level	— reference to cover gives clear idea of vegetation structure — metrical scale suitable for computation — quick survey of great areas — danger of releveing disturbed sections near the bridges — unfavourable in areas of small scale geographic inhomogeneity
Kohler (Kohler *et al.* 1971, Kutscher 1984)	Frequency: Kohler scale, several versions (3, 5 and 6 step), definitions partly intermingled with abundance	river sections of a length up to 3 km; if possible, total course; actual water level	— total inventarization of a river system, but very work intensive — showing clearly river zonation — levelling seasonal variation, may be of disadvantage when following succession — relatively unsuitable for classification purpose
Holmes and Whitton (Holmes 1983)	Relative abundance: 5 step scale	river sections, length 0.5 to 1 km; regularly distributed along river course; differentiation between river and bank	— compromise between methods above, more objective selection of plots — relatively great chance of total inventory — use of relative abundance appropriate in highland rivers with sparse vegetation
Classical Braun-Blanquet (for example Mériaux 1978, Müller 1973, Weber-Oldecop 1969)	Cover: Braun-Blanquet scale, mostly with sociability	homogeneous plots; between 1 and 25 m²; often several plots in one river section	— inventarization impossible, additional mapping or lists required — size of plots below statistical minimum area, classification almost impossible — because of small sample plots no advantage from reference to cover — great subjectivity in choice of plots

system, however, is very labour-intensive. The results clearly show longitudinal zonation patterns. On the other hand, because of the long river sections vegetation processes in a smaller scale cannot be followed during succession studies. Seasonal variation as well as variation between the years cannot be studied using this method. The results are relatively unsuitable for classification and comparison to ecological data, because the habitat conditions in such long sections cannot be considered as homogenous.

A third method was developed by Holmes & Whitton (1975) and applied to British vegetation by Holmes (1983). The estimation usually refers to relative abundance. The river sections have a length of 500 to 1000 m and are regularly distributed along the river course (every 5 to 10 km). Thus in this case, the structure of the vegetation itself is not used as a criterion for choosing a sample site, either. The sample is differentiated into river and bank vegetation. This method is a compromise between the two above-named. The river sections have an intermediate length. The selection of plots is even more objective than in the Wiegleb and Herr method and the chance of a total inventory is almost as great as in the Kohler method. The reference to relative abundance is appropriate in rivers with sparse vegetation. On the other hand, the values estimated with this method are not directly comparable to other estimations of cover and abundance.

The classical Braun-Blanquet method (Weber-Oldecop 1969, Müller 1977, Mériaux 1978) uses cover for estimation of species. In some cases also sociability is given. The sample plots are usually homogenous (or what the researcher believes to be homogenous), often several plots are chosen within one river section. Especially the reference to very small areas is a main disadvantage of the method. The size of the plots which is between 1 and 25 m^2 is far below the statistical minimum area of aquatic communities. Often only polycormons of vegetatively reproducing plants are sampled. This procedure leads to samples which are difficult to classify and which do not show any characteristic distribution in space. The reference to small areas also makes an inventory almost impossible and additional mapping or listing of species is required (e.g. Weber 1976). The main objection against this method in general is the subjective choice of the sampling plots according to the method of the experienced phytosociologist (Pignatti 1980). In this way already existing prejudice is reproduced again and again.

The methods of Wiegleb and Herr, Kohler, and Holmes and Whitton have several characters in common. They are applicable to surveying macrophyte vegetation and flora on a medium or large geographical scale. Similar methods have been developed by Smith (1978), Novakova & Rydlo (1981), Merry et al. (1981) and Bouxin & LeBoulenge (1983). Classification of such samples inevitably leads to the identity of river typology and vegetation typology. There is no need for pressing the results of such classification into a hierarchical system, and for synonymization of vegetation types. Descriptions of local and regional surveys may be compared, but in a more informal way.

For the study of small scale pattern, and also for succession, several methods have been used, for example the very precise study of Jones (1955) mapping individual plants, or the studies of Ham *et al.* (1981, 1982) and Wiegleb (1983) also referring to small sample plots and using more precise estimation methods.

River vegetation characterized by dominant species has been described by Pietsch (1974) similar to the procedure of Maristo (1941) in lakes. So far this approach has not led to a formal classification, either. It is strongly related to the following approaches that emphasize the structure of vegetation. The formation approach recognizing dominant growth form as most important for vegetation description is implicitly incorporated into the Braun-Blanquet approach when uniting classes related by growth form and not by sociological progression (for example Westhoff & den Held 1975, Ellenberg 1978). An explicit formation approach is uncommon.

Fundamental to all structural approaches to terrestrial vegetation are the life forms according to Raunkiaer. Usually also aquatic plants are included in life form systems (Ellenberg & Mueller-Dombois 1967). The result, however, is very dissatisfactory. The terrestrial life forms are on principle inappropriate for aquatic plants. The diversity of life cycles in aquatic plants needs a more differentiated treatment. Furthermore, much information reported by older authors is simply wrong or inexact leading to wrong assignments, for example in the case of *Najas* and *Potamogeton* (see Brux *et al.* 1987).

Already in the beginning of vegetation research Gams (1918) favoured the synusial approach. For aquatic plants it has been further developed by Vaarama (1938), den Hartog & Segal (1964) and Mäkirinta (1981). The synusial approach can be applied in two different ways, either considering the strata actually present, or considering the strata that may exist under given ecological circumstances. A main difficulty for the synusial approach is a valid definition of growth form (as opposed to the life form concept; see discussion in Weber 1976, Mäkirinta 1978, Schmidt 1981 and Schuyler 1984, to name only the most divergent viewpoints). Many species show great plasticity and are difficult to classify. A combination of synusial approach with floristic (frequency) and dynamic (stability) criteria has been used by Wiegleb (1984b). Using this procedure the running waters vegetation can be divided into a small number of types (major types) that exhibit characteristic geographical distribution.

So far the synusial approach has led to a consistent description of medium scale patterns, because similar habitat types are often colonized by plants of similar growth form although belonging to different species or even genera.

Each type can be divided into several local and regional subtypes. The criteria for classification of aquatic communities according to this approach follow now. If consequently applied it may lead to an operational definition of plant communities and to a kind of identification key of aquatic communities:

(a) *Dominance.* Dominance of a species is rated higher in water courses than presence only. Presence, with low cover or low number of individuals, may be due to immigration from the upper course or a tributary.

(b) *Growth form.* From observations in permanent plots it can be concluded that the predominant growth form is a much more stable element in the community than the species composition.

(c) *Frequency.* If possible, groups or clusters should be produced which are characterized by a certain number of species with high frequency. This aim cannot always be achieved in combination with a and b.

Geographical approaches are often interrelated with structural and dynamic approaches, and many concepts are frequently confused. The biome concept was developed by Clements for large scale description of terrestrial vegetation. Walter & Breckle (1983) proposed a very sophisticated system, distinguishing between true climatical biomes and oro- and pedobiomes. In this sense river vegetation is a special pedobiome showing similar composition through different climatic regions. The concept of 'zonal vegetation' (Walter & Breckle 1983) is historically closely related to the biome concept. Within the context of this concept landscape and vegetation were mainly treated from a terrestrial point of view with the aquatic habitat being more or less 'distortions'. Nevertheless, aquatic vegetation (especially riverine vegetation, Walter & Breckle 1986) shows kinds of variation that coincide with the great vegetation zones of the earth, however, this aspect is insufficiently studied. The biogeoncoen is a smaller entity of a large biome (see Walter & Breckle 1983). The description of such landscape units (river valleys, Decamps *et al.* 1979) is still unfamiliar in vegetation science. A related concept is the sigma association which has been developed within the Braun-Blanquet school.

The study of geographical variation in aquatic plant communities has been badly neglected. The exclusion from scientific interest is based on early phytosociological papers (e.g. Allorge 1922, Koch 1926, Steffen 1931) that postulated the identity of aquatic communities over long geographical ranges. This idea found theoretical support in a misinterpretation of the concept of zonal vegetation (see above).

On the other hand, Gams (1925) and Roll (1942) pointed out that geographical variation of aquatic communities might be worth while studying. In the last years evidence has become available that those early workers were basically right. Weber-Oldecop (1977) proposed a scheme of river classification that is related to geomorphological and pedological variation in his study area. Holmes (1983) found remarkable differences in river floras in England related to geology, geomorphology and oceanity. Haslam (1978) already anticipated many of Holmes' results in a more anecdotal way. Also in stagnant waters, geographical variation becomes more and more obvious. Landolt (1984) gave information on the distribution of Lemnaceae in relation to climate, Felzines (1979) found strong influence of continental and altitudinal gradients in French pond floras, and van Vierssen (1983) stressed the regional differences in European brackish water communities.

Butcher (1933) used geographically defined river types as a basis of study. Within these types the vegetation is compared with respect to its floristic similarity. The approach is related to the floristic one but the direction of investigation is reverse.

A successful classification of rivers requires an evaluation of the importance of factors actually influencing the occurrence of macrophytes. The most important factors or factor group that have an influence on the presence or absence of macrophytes and on the differentiation of their communities are the following (see also Westlake 1973, Dawson 1988):

(a) *Hydrological regime.* Important factors are the constancy of discharge and the amplitude of water level fluctuations which are both regulated by the ratio of surface water and ground water flowing into the river. Upland rivers with strong spates, especially in spring and summer, are usually devoid of macrophytes.

(b) *Current velocity.* This is a very complex factor, being dependent on the hydrological regime, the relief of the landscape, and during the last years, also on river regulation. Closely related to current velocity are factors like sediment characteristics, temperature amplitude and oxygen content. The differentiation between rhitral and potamal water (Weber-Oldecop 1977) has proven to be very useful.

(c) *Calcium hydrogen carbonate complex.* The most important factor is the hydrogen carbonate ion, because it can be used by some plant species, and not by others. Soft water rivers are poor in hydrogen carbonate, hard water rivers are rich. The calcium itself does not play an important role in the differentiation of vegetation. Calcium sulphate waters are similar to soft-water rivers. The special properties of calcium-rich waters are often not caused by the chemical characters of the water, but by the influence of the rocks on the hydrological regime (for example in the chalk streams in England and France).

(d) *Nutrients like N and P.* The direct influence of these nutrients may be less important than formerly believed. The nutrient content of the water and that of the sediment have to be regarded separately. Similar nutrient levels have a different effect on the river community depending on the current velocity and other factors. Undifferentiated labels like 'oligotrophic' or 'eutrophic' should be avoided in describing rivers.

(e) *Anthropogenic influence.* Especially the construction and management of rivers (mechanical influence) as well as input of nutrients and toxic substances (chemical influence) have changed the ecological situation in rivers very drastically during the past years. The anthropogenic influence on river vegetation especially in case of mechanical influence is often underrated and not well studied.

The most elaborated system of river types based on such kind of factors has been used by Holmes (1983) for describing the distribution of aquatic plants in British rivers. His maps of the main vegetation types are well in accordance with river types as defined mainly by physical factors.

It seems well appropriate to distinguish on a geographical base between for example lowland rivers (rivers that have their source below 150 m of altitude), lowland rivers with source or larger parts of their drainage area in highland, upland rivers incl. rivers on high plateaus, and mountainous rivers. The special case are rivers that pass lakes or that spring from sources with high regular water outflow.

In continental phytosociology usually another approach was followed. Here the species composition was analyzed first, and afterwards correlations to the habitat factors were sought. As long as only associations classified with the Braun-Blanquet method were available, it was impossible to find any correlation between vegetation types and river characteristics, because the different concepts were aggregated on different geographical scales. Only Kohler *et al.* (1971) succeeded in correlating some of their river zones as defined by ecological species groups with some chemical habitat parameters and used their results for indication of water quality in general. However, it soon became evident that such kind of general application was impossible because of the individualistic nature of every river system with its own history, its special hydrogeochemical circumstances and its special flora. More intense analysis of ecological species groups and their relation to phytosociological species groups have been carried out by Felzines (1979) and Wiegleb (1978) but only in stagnant waters so far.

The concept of the indicator values (Ellenberg 1979) has also been applied to aquatic plants. The non-appropriateness of these values has been pointed out by Wiegleb (1984a). A formal system of indicator values using a more differentiated set of parameters has been developed by Schmidt (1981). So far, the empirical basis of Schmidt's work is only local and too small for assigning reliable values to each species.

Dynamic approaches are in most cases related to the succession theory of Clements (1916) regarding succession as a sort of ontogeny of a community ruled deterministically by the habitat conditions. The climax concept in a stricter sense has rarely been applied to aquatic vegetation. In many older publications stagnant water vegetation is treated as a stage of development from open water to woodland, as reconstructed by zonation. Running water vegetation is treated as a pioneer community. However, medium term stable communities may occur, for example the very competitive nymphaeid communities in stagnant waters (Wiegleb 1981a) or the *Ranunculus-Rorippa* complex vegetation in rivers (Dawson *et al.* 1978).

'Structure types' as a sort of non-deterministic climax have been introduced by Wiegleb (1981a) for the description of aquatic communities. These structure types are defined dynamically as well as by dominant growth form, thus being related to the synusial approach. The structure types are embedded in a hierarchy of climaxes as described by Schwegler (1981). Because of its flexibility the concept is well appropriate for the description of vegetation change. The concept of potential natural vegetation has been introduced to river vegetation by Dahl & Wiegleb (1984). It is related to the concept of the 'synanthropic' series which were first constructed by Kohler (1975) for south German rivers, followed by Wiegleb (1981b) for north German rivers. It must be admitted that both concepts are based on speculation to a high degree.

DISTRIBUTION OF VEGETATION TYPES IN RIVERS (WITH SPECIAL CONSIDERATION OF EUROPEAN AND JAPANESE FLUVIATILE HABITATS)

It has been stated above that vegetation science regarded aquatic vegetation to be uniform over large geographical areas. In contrast to this point of view, here the hypothesis is stressed that species composition in rivers exhibits all the kinds of geographic variation that is observed in land plants making these aspects worth while studying. The main difficulty for a scientific approach to this problem is the nature of the information available. Different sources can be used, like the general chorological literature, regional floras, the broadly scattered phytosociological literature and limnological literature.

A superficial survey of the accessible facts shows the following:

(a) Many so-called cosmopolitan aquatic species (*Ceratophyllum demersum, Potamogeton pectinatus, Zannichellia palustris, Ruppia maritima*) are taxonomically difficult. Their identity over long ranges is not only unproven but can be doubted with a high degree of certainty.

(b) No species has an overall distribution in Europe. Many species are restricted to relatively small areas. There are many endemic species (Cook 1983). Species that are common to Europe and America, Asia or North Africa in most cases only colonize water in small parts of Europe.

(c) Within their areas species are inhomogeneously distributed. Thus not only the causes of delimitation but also the regularities of non-occurrence within the potential area need to be studied.

(d) In detail many contradictions would become evident.

At least three complexes of geographical variation can be observed in aquatic communities as well as in terrestrial ones.

1. *Climatic variance.* The climatic variation in Europe is mainly characterized by two gradients, the north-south gradient and the oceanic-continental gradient, which are not independent in their effect. Of special interest are the questions
   — whether the summer is long and hot enough to guarantee the reproduction of annual species,
   — whether the winter is cold enough to guarantee the stratification of seed and turions of species adapted to cold conditions,
   — whether the winters are so cold that rivers regularly freeze, and
   — whether the precipitation is unevenly distributed over the year that periods of drought/spate will occur.

Those factors will determine the outer limits of the potential areas of aquatic species in rivers.

In Europe several species groups of similar distribution can be distinguished. There is a group of about 10 boreal species that only occur north of 60° latitude, for example *Sagittaria natans* and *Myriophyllum exalbescens*. Many species (about 30) have a boreal-temperate distribution, rarely crossing the 40th degree of latitude southwards, for example *Potamogeton perfoliatus, P. alpinus* and *Sparganium emersum*. Some species (about 30) are totally restricted to the temperate zone, being either amphiatlantic (*Potamogeton epihydrus*), endemic (microarealophytic, *Oenanthe fluviatilis*), introduced in scattered places with no further spreading so far (*Elodea ernstiae*) or continental species (*Hydrilla verticillata*). Of the more than 30 truly continental species some occur in the temperate as well as in the boreal zone (*Nymphaea candida*) or in the temperate and meridional zone (*Caldesia parnassi-folia*), even continental endemics occur (*Najas tenuissima*). Another group of about 15 species is oceanic-boreal distributed (*Isoetes lacustris, Lobelia dortmanna*), including suboceanic species with a more widespread distribution like *Myriophyllum alterniflorum.*

More than 20 species are restricted to the mediterranean and submediterranean zone, with *Ranunculus saniculifolius* and *Callitriche leniscula* being the most widespread Mediterranean aquatics. Others are small scale endemics like *Potamogeton siculus*. Some more south-ernly distributed subtropic species reach their northern limit in the Mediterranean like *Paspalum paspaloides*. At least the same number of tropical and subtropical species was introduced in the Mediterranean, but have never spread from their respective area of introduction. A few introduced species tend to take over a mediterranean-oceanic distribu-tion (*Vallisneria spiralis, Lagarosiphon major*). The mediterranean-oceanic distribution pattern is the most frequent one in aquatic species (at least 50 species according to Casper and Krausch 1980, 1981). Also a large number of species (appr. 30) show mediterranean-tem-perate distribution never crossing the 60 degree northwards (for example *Potamogeton nodosus, P. trichoides*).

2. *Geological variance.* The geology mainly influences the formation of the substratum and the geochemical conditions of a given drainage area. In some structurally similar vegetation types a significant difference occurs between calcareous (neutral) and non-calcareous (acid) rivers. This has been described many times, especially for small batrachid and potamid communities, with *Callitriche hamulata, Myriophyllum alterniflorum, Ranunculus peltatus,* and *R. penicillatus* growing mainly in acid rivers and *Potamogeton coloratus, Groen-landia densa, Zannichellia palustris* and *R. calcareus* preferring calcareous ones. The distinction is not always as clear as might be expected, hydrological conditions may interfere as well as mosaic occurrence of calcareous and non-calcareous rocks. Geological conditions that lead to sandy or sandy-stony sediments are most favourable for the colonization with macrophytes. These sediment types guarantee the highest sediment stability and the lowest degree of natural turbidity.

3. *Geomorphological variance.* From static point of view the geomor-phology includes height, shape and relief of the landscape, thus including also altitudinal variation. Well known cases of altitudinal variation are *Sagittaria sagittifolia* occurring in the lowlands of several parts of Europe, while other species like *Nasturtium officinale* and *Butomus umbellatus* are restricted to lowland and hilly countries. Other species only occur in mountainous areas like *Ranunculus*

lutulentus. The actual morphology of the landscape is determined by dynamical processes that are related to climatic conditions (weathering, glaciation, wind, erosion etc.). Geomorphological processes lead to zonal phenomena like the "climate morphological" zones resulting in special types of valley formation (Büdel 1971) and also to the formation of medium scale morphological landscape types or physiographic regions. The morphology of the physiographic region influences the morphology of the river itself, which influences the physical habitat factors such as gradient, current velocity, hydrological regime, and substrate condition that directly determine the composition of the river flora (see above).

No other vegetation type is more directly influenced by geomorphological processes than vegetation of rivers and river banks. On a smaller scale a different pattern can be observed in each individual stream with regard to upper, middle and lower course (Fahy 1974). Those geomorphological circumstances will be different again in physiographic regions of different kind, thus creating a pattern of geomorphic stages, that will surely be reflected by the riverine vegetation.

For a synoptic treatment, Europe is subdivided into three main flora zones (see Meusel *et al.* 1965):

1. The boreal zone, which is roughly equivalent to the boreal zonobiome VIII (cold temperate zonobiome with cold summers) according to Walter & Breckle (1983).
2. The temperate zone which is equivalent to the nemoral zonobiome VI (typical temperate zonobiome with short winter period), including the adjacent zono-ecotones (Walter & Breckle 1983). Flora zones 1 and 2 are comprised within the ectotropical zone of retarded valley formation according to Büdel (1971).
3. The meridional zone incl. the submeridional zone. These zones correspond to the mediterranean zonobiome IV (Winter-humid zonobiome with summer drought; Walter & Breckle 1983). They belong to subtropical zone of mixed relief formation according to Büdel (1971).

A more sophisticated division of Europe into approximately 20 physiographic regions similar to the zoogeographic subdivisions of Illies (1978) is desirable, however, because of lack of empirical data from most parts of Europe such procedure is impossible.

The following treatment mainly refers to rivers of a width between 5 and 50 m. Larger rivers show special characters tending partly towards stagnant waters. It is in parts purely hypothetical, however, there is strong evidence that the basic assumptions are right. The ideas were developed based on intensive survey of river vegetation in Northern

Germany (1978—1983). In Denmark, Britain, France, Belgium, Spain and Italy more superficial surveys were carried out in selected areas (1981—1984). A comparative survey was carried out in Japan (1985).

As the river vegetation of the boreal zone of Europe has not been studied so far it is not considered here. The information available in the literature is very sparse. There is some evidence for the widespread occurrence of *Sparganium* rich communities and communities with isoetid species that do not occur in rivers in southern parts of Europe (see Stake 1967, 1968, Rautava 1964, Quennersted 1965).

In the following the major types of both the temperate and the meridional zone are outlined.

Type 1: *Callitriche*-rich communities

Structurally the *Callitriche* species are intermediate between the parvopotamids and the batrachiids. Most *Callitriche* species do not show any specific preference for hydrochemical parameters, but show a distinct geographical distribution. For example *C. platycarpa* is widely distributed throughout the northwest European lowlands, while it is replaced by *C. cophocarpa* in similar rivers in the German highland area. *C. obtusangula* occurs abundant both in the Atlantic part of the temperate zone and the West Mediterranean area. *Callitriche*-dominated communities make up a significant proportion of the sites investigated throughout Western Europe.

At least two structurally different subtypes can furthermore be distinguished:

(1.1) Species poor dominance stands, only accompanied by mosses and floating grasses. Such kind of communities can also be found in ditches and ponds.
(1.2) Species rich stands with various species of various growth forms. Such kind of communities (with *C. hamulata*) are especially frequent in soft water rivers where species like *Myriophyllum alterniflorum, Ranunculus peltatus*, and *R. penicillatus* occur. In more acid and nutrient poor rivers species like *Juncus bulbosus, Isolepis fluitans* and others that are traditionally included as character species of the stagnant waters, communities of "*Littorelletea*" may reach high abundances in the community.

Type 2: *Batrachium* communities

Though being similar in growth form to the preceding type on a large scale these communities show a different distribution. Many *Batrachium* species are restricted to the western parts of Europe (Cook 1983).

Three structurally different subtypes can be distinguished according to the actual growth form of the dominant *Batrachium* species:

(2.1) Floating leaved *Batrachium* communities. They are formed by medium sized *Batrachium* species (*Ranunculus peltatus, R. penicillatus* and their hybrids) which share ecological and structural similarities with the *Callitriche* species. Because of the taxonomic difficulties within this group of plants the exact distribution of their communities is still insufficiently known.

(2.2) Small *Batrachium* communities without floating leaves. These communities are structurally similar to the preceding, however, do not include floating leaved plants. Characteristic are *Ranunculus trichophyllus* and *R. circinatus*. These communities mainly occur in hard water rivers of the highlands. They colonize small upper courses where many other aquatic plants cannot grow.

(2.3) Large *Batrachium* communities without floating leaves. The large *Batrachium* communities are formed by *Ranunculus fluitans, R. calcareus* and *R. fluitans* hybrids without floating leaves. These plants may reach a length of up to 8 m. Because of the wide distribution of this type a strong differentiation in regional subtypes can be recognized. There are:
— very species poor variants, mainly in soft water rivers in highlands;
— a variant with *Myriophyllum spicatum* and *Potamogeton pectinatus* that is widespread all over Europe;
— a variant in slow flowing river sections of the highlands with *Sparganium emersum*;
— a southern variant with *Potamogeton nodosus*, etc.

It becomes visible that the subdivision is influenced by hydrochemical, physical and geographical factors as well. *Batrachium* communities are more frequent in Atlantic regions, for example Britain and Denmark, than in continental regions. They are mainly distributed in rhitral waters. In Southern Europe they are mainly represented by the second subtype.

Type 3: Parvopotamid communities incl. myriophyllid and elodeid communities

This type is often associated with *Callitriche* and small *Batrachium* communities to which also structural similarities can be found. It is a floristically very inhomogeneous group which exhibits a stronger differentiation according to hydrochemistry than can be observed in the other growthform groups. A very important community is the

Potamogeton pectinatus community which is favoured in different types of rivers by anthropogenic distortion of various kind. One group of plants (*Zannichellia palustris, Groenlandia densa, Hippuris vulgaris*) is restricted to hard water rivers, and there often intermingled with *R. fluitans*. The communities formed by the neophytic *Elodea* species (*E. canadensis* and *E. nuttallii*) may be regarded as an independent type. Because of their different growth characteristics they tend to fill the whole space of a water thus forming dense dominant stands. Elodeids are usually very irregularly distributed being very frequent in one river system or physiographic region and almost absent in another.

Parvopotamid communities are rare in the northern part of the temperate zone (e.g. Denmark). They occur with a certain frequency over large parts of Germany, France and Britain. They are more frequent in the Mediterranean area, mainly with *Zannichellia palustris, Potamogeton pectinatus* and *Myriophyllum spicatum* as dominants.

Type 4: Magnopotamid communities

These communities formed by broad leaved *Potamogeton* species are often associated with *R. fluitans* or *Sparganium emersum* but also form stands of their own, mainly in slowly flowing waters of the lowlands.

The typical species of this group in the temperate region are *P. perfoliatus, P. lucens* and *P. alpinus*.

Quite peculiar is the position of some communities with floating leaf producing species. The community with *P. gramineus* has a northern distribution. The *Potamogeton coloratus* community is somewhat isolated within the hydrophyte communities. In Central Europe, the species colonizes extreme habitats. It has its centre of distribution in the Mediterranean area. The community is found in hard water rivers that are nutrient poor (especially poor in phosphate and ammonia). Because of the low competitive strength of *P. coloratus*, regular distortion is necessary for its maintenance. The floristical structure of the stands is usually simple with some other hard water species and Characeae.

Another special case is the *Potamogeton nodosus* community. *Potamogeton nodosus* is much more frequent in Southern Europe where it colonizes both running and stagnant water. In nothern areas it only colonizes rivers that do not freeze in winter. *P. nodosus* is almost a nymphaeid in its growth form. The *P. nodosus* community can be regarded as a southern vicariant of the *P. natans*-rich *Sparganium emersum* community. It is the most characteristic vegetation type of the Mediterranean area (Litav & Agami 1976, DeMarchi *et al.* 1979, Vrhovsek *et al.* 1981, and Penuelas & Catalan 1983).

Type 5: *Sparganium emersum* communities

The characteristic species of the floating leaved communities of rivers is *Sparganium emersum*. *Sparganium*-dominated communities are distributed in lowland rivers and also slow flowing sections of upland rivers all over the northern hemisphere (Klotz & Köck 1984, Baskhiria; Hilbig & Schamsran 1977, Mongolia; Wiegleb & Kadono 1988, Japan; Smith 1978, Wisconsin; Haslam 1978, Eastern North America). Because of their wide distribution and also their wide tolerance against several hydrochemical factors many sub-divisions can be recognized. Only the two most important shall be mentioned here. There is one species and growth form rich variant that is characterized by the co-occurrence of *Sparganium* with *Potamogeton natans, Nuphar lutea* and *Sagittaria sagittifolia*. It is mainly distributed in lowland rivers of northern Germany with current velocities between 0.1 and 0.3 m/sec and can be furthermore subdivided into several local types.

A second subtype is usually only formed by *Sparganium* and *P. natans*. It also colonizes faster flowing sections (0.2 to 0.7 m/sec) including narrow upper courses in case of lack of other competitive species. This type is also frequent in upland areas in mosaic with rhitral *Ranunculus* communities.

The characteristic species of this type are very variable in their growth form ranging from vallisneriid over nymphaeid to helophytic. *Sparganium* dominated sites are rare in the true Mediterranean area. There true vallisnerids (*Vallisneria spiralis*) can be frequently found which are absent in the cold temperate zone.

Type 6: Pleustophyte communities

Pleustophyte communities naturally do not belong to the dominant vegetation types of running water. Pure *Lemna* stands are only formed in case of total stagnation. One pleustophytic community can be found regularly in waters up to 0.2 m/sec current velocity all over Europe. This is the community dominated by *Ceratophyllum demersum*. In Southern Europe also *Ceratophyllum submersum* and *Salvinia natans* may occur under such circumstances.

Type 7: Helophyte communities

Helophyte communities are usually an integral part of the river community and together with true helophytes complex vegetation types are formed. There are four main types of helophytes that take part in the formation of river communities:

— Submersed perennial grasses like *Glyceria, Agrostis* and *Phalaris* which are associated everywhere but can become dominant in narrow waters of the highlands. In the Mediterranean, these northern grasses are replaced by southern ones of the genera *Paspalum* and *Echinochloa*.

— Low perennial reed swamp plants (mainly dicotyledons like *Berula, Sium, Apium, Veronica, Nasturtium, Oenanthe, Rorippa, Myosotis*). According to hydrological circumstances and river morphological characters like shape of the bank these species occur only sparsely or in dense stands. Often complex communities like "*Ranunculo-Sietum*" have been described, however, the occurrence of *Ranunculus* and *Berula* is regulated by totally different ecological factors.

— Low annual plants, especially of the genus *Polygonum*. They are mainly restricted to parts of the river bed that are only occasionally flooded, however, species like *P. hydropiper* frequently grow submersed.

— High reed swamp plants, mainly graminoids like *Phragmites, Typha, Sparganium erectum, Schoenoplectus lacustris* and *Glyceria maxima*. Stands of these plants are less frequent in case of increased anthropogenic influence on the rivers. The tendency of these plants to form floating forms seems to be quite different in various regions. For example *Schoenoplectus* and *Sparganium* are much more aquatic in Italy than in Northern Germany.

A comparison of the situation in Europe with the one in Japan (Wiegleb & Kadono 1988) shows several similarities but also some characteristic differences. The climatic zones in Japan are much more clear-cut than in Europe. There is no distinct oceanic-continental gradient. The fact that the geological circumstances are more homogenous is quite important. In Japan, the *Sparganium emersum* community also has a northern distribution. It is the most frequent community in Hokkaido, which belongs to a boreal-temperate zono-ectotone VI—VIII (Walter & Breckle 1983). The species composition is quite similar to the one in Central Europe. The community is also frequent in the northern and central parts of Honshu which represent the temperate zonobiome. A similar distribution is shown by the *Batrachium* communities, however, the diversity in growth forms of *Batrachium* is much smaller in Japan than in Europe. *Callitriche* communities occur all over the country formed by one species only, *C. palustris*.

Similar to Europe, parvopotamid communities are rare in the northern part (Hokkaido). They are most frequent and abundant in the northern and central part of Honshu, however, get less frequent in the south. The most important species are *Potamogeton octandrus, P. oxyphyllus* and their hybrids. *P. pectinatus* is rare in Japan. The situation in the magnopotamid communities is also different from Europe. The floating leaved *P. distinctus* occurs as a dominant scattered all over the country. In the southern and western parts *P. malaianus* is almost as important as *P. nodosus* in the comparable Mediterranean area. However, *P. malaianus* rarely forms floating leaves, thus showing a quite different prevailing growth form. Besides the *P. malaianus* community the southern part (which belongs to the warm-temperate zonobiome V) is furthermore characterized by *Trapa*, *Vallisneria* and most of all elodeid communities (with the neophytic species *Egeria densa* and *Elodea nuttallii*, and the indigenous *Hydrilla verticillata*).

In the grasses there is a clear distinction between the northern genera like *Glyceria*, *Agrostis*, *Phalaris*, *Alopecurus* and *Beckmannia*, and the southern ones like *Paspalum*, *Echinochloa* and *Leersia*. Characteristic plants that have no equivalent in Europe are *Phragmites japonica* (which is able to form creeping shoots) and *Polygonum thunbergii* (which is perennial in contrast to other *Polygonum* species). In Hokkaido many helophytes can be frequently found floating, especially *Sparganium erectum*, *Cicuta virosa* and *Equisetum fluviatile*. *Oenanthe javanica* is distributed all over the country. Mosses are less frequent in Japanese rivers than in European ones. This is true even in mountainous areas.

DYNAMICS OF RIVER VEGETATION

For more than 70 years, descriptive life history studies of aquatic species and descriptive studies of the dynamics of aquatic communities have been carried out independently. Only during the last years, a branch of succession theory has evolved, that focusses its interest on life history attributes (Noble & Slatyer 1980, Van der Valk 1981, Lepš *et al.* 1982). But so far no explicit application of these ideas to the study of river vegetation is known.

The use of descriptive information on aquatic plants and their communities for predicting succession is related to many difficulties.

For a given type of community at least the following information is required:

(a) Geographical variation of the community type;
(b) change of the community structure under given circumstances;
(c) exact data on habitat parameters; and
(d) exact data on life history attributes of at least the dominant species.

The literature concerning vegetation change in aquatic communities has become too numerous to be listed. The approaches to the problem can be grouped in the following way (see also Dawson 1988):

(a) Intensive studies that follow the variation of vegetation within a year and from year to year (Wiegleb 1983, Ham *et al.* 1981, Ham *et al.* 1982);

(b) restudy of sites after several years (e.g. Kohler, Pensel & Zeltner 1980, Kohler & Zeltner 1981); and

(c) evaluation of changes after long periods of time, so far only in stagnant waters (e.g. Stuckey 1971, Volker & Smith 1965).

General principles of change, which were supposed by Segal (1971) based on his observation in Dutch waters, cannot be realized. Exact causes for the observed changes can only be given in case of intensive study. In studies of type b and c it remains more or less open what has happened in-between the observation times. Such results cannot be used for predicting vegetation change in the future.

Here only one example of vegetation change is discussed. Within a study on the phytosociological behaviour of *P. alpinus* (Brux *et al.* 1987) 55 sites were sampled repeatedly over several years. The most complete record (10 years) exists in the case of the River Lethe (for details see Wiegleb 1983, 1984a. It is shown in Table II. The site is heavily influenced by man. In the beginning of 1979 the river was dredged. The vegetation is regularly cut in August, and the water is highly eutrophicated with nutrients from the surrounding agricultural area. Nevertheless, the vegetation composition remained almost constant over the years. Only minor floristical changes occurred. Every year the vegetation is made up by changing proportions of *P. alpinus, P. natans, Ranunculus peltatus, Myriophyllum alterniflorum, Sparganium emersum* and *Callitriche platycarpa.* Very remarkable is the cyclical change in dominance between *Myriophyllum alterniflorum* and *Sparganium emersum.* In 1986 the vegetation development was quite different. Following a high cover of *Sp. emersum* in 1985, *Potamogeton natans* became dominant with approx. 60% cover. There was no obvious change in the hydrochemical conditions, and no catastrophic event could be observed.

The study of the vegetation dynamics in all 55 sites in total gave a very good example of the independence of species behaviour. There was no relation of the behaviour of *Potamogeton alpinus* to the observed behaviour of other species nor to the vegetation as a whole. This is strong backing of the individualistic concept, as has been formulated by Gleason (1926). No sociological relations of the species exist. Stands with obviously similar habitat conditions (see Wiegleb & Todeskino 1983) do not show identical development of vegetation. Similar kind of disturbance does not produce an ideal convergence of vegetation succession. This makes it impossible to speculate on the exact way of vegetation process in every site under study.

It must be added that the judgement of stability or non-stability is

Table II. Vegetation change in site Lethe 7, a lowland river in Lower Saxony (1 = list of hydrophytes only, p = present, cover values are expressed with the Londo scale).

Year	77[1]	78	79	80	81	82	83	84	85	86
Total cover %	—	50	30	55	55	60	50	80	70	95
Potamogeton alpinus	p	.2	.1	+	.2	.1	.2	.1	.1	.7
Ranunculus peltatus	p	.2	+	.4	.1	+	.4	1.2	.1	1.2
Myriophyllum alterniflorum	p	3.	+	.7	3.	1.2	1.2	3.	.4	1.2
Sparganium emersum	p	1.	2.	3.	1.2	3.	2.	1.2	5.	1.2
Potamogeton natans	p	.2	.4	.4	.4	.7	2.	2.	.7	6.
Callitriche platycarpa	p	.2	.1	.4	.4	.4	.2	.4	+	.2
Lemna minor	p	+	+	+	+	+	.2	+	·	+
Potamogeton pusillus	p	+	·	·	·	·	·	·	·	+
Nitella flexilis	·	.2	·	·	+	·	·	·	·	·
Callitriche cf. *obtusangula*	·	·	·	·	·	.1	+	+	·	·
Glyceria fluitans	·	+	.2	.4	.2	.2	.2	.4	.1	.4
Myosotis palustris	·	+	.1	.1	+	+	.1	+	+	+
Phalaris arundinacea	·	+	.1	.1	+	.1	.1	.1	+	.4
Polygonum hydropiper	·	+	·	+	+	.1	.1	+	.1	+
Nasturtium officinale	·	+	·	·	+	·	·	·	·	·
Sparganium erectum	·	+	+	·	+	·	·	·	·	·
Agrostis stolonifera	·	+	+	+	+	+	+	·	·	+
Bidens tripartita	·	+	+	·	·	·	·	·	·	·
Hydrocotyle vulgaris	·	+	·	·	·	·	·	·	·	·
Alisma plantago-aquatica	·	·	+	·	·	·	·	·	·	·
Eleocharis palustris	·	·	+	·	·	·	·	·	·	·
Juncus effusus	·	·	·	+	+	+	+	+	+	+
Bidens cernua	·	·	·	+	·	+	·	+	+	·
Rorippa amphibia	·	·	·	+	·	·	·	·	·	·
Filamentous algae	·	·	·	·	p	·	p	·	p	·
Ranunculus repens	·	·	·	·	·	·	·	·	+	·
Poa palustris	·	·	·	·	·	·	+	·	·	·
Lotus uliginosus	·	·	·	·	·	·	·	·	·	+
Calamagrostis canescens	·	·	·	·	·	·	·	·	·	+

dependent on the geographical scale, too. On a medium geographical scale there is some regularity again. A scheme for the development of river communities under anthropogenic influence is presented in Table III. So far it is only valid for the northern part of Germany. 3 different stages are distinguished: (from left to right) quasi-natural, moderately disturbed and heavily disturbed. In the upper half of the table the lowland rivers are listed. In coastal rivers and other sluggish flowing waters, species and growth form rich floating leaved communities are transferred into lemnid or algal communities under heavy disturbance. Under moderate disturbance species poor nymphaeid or parvopotamid communities are found.

Table III. Reaction of macrophyte communities in Lower Saxonian water courses to anthropogenic disturbance.

Water types	quasi-natural	moderately disturbed	heavily disturbed
Lowland			
coastal waters and other sluggish flowing waters	species and growth form rich floating leaved communities	species poor floating leaved communities, increasing portion of parvopotamids and lemnids	lemnid and algal communities
potamal brooks	species rich *Sparganium emersum* dominated communities with magnopotamids	*Sparganium emersum* community, various variants	monodominant stands of nymphaeids and parvopotamids, filamentous algae
rhitral influenced brooks	*Callitriche* and *Batrachium* communities, or other structurally similar stands	*Elodea* and *Sparganium* rich stands with *R. fluitans* or *R. peltatus*	no macrophytes, filamentous algae
Highlands			
calcareous brooks	species rich stands with hard water species like *Groenlandia densa*, *Hippuris vulgaris*, *Myriophyllum spicatum*	species poor stands with *Zannichellia, R. trichophyllus* or *R. fluitans*	helophyte stands (*Berula, Phalaris* etc.), filamentous algae, sometimes *P. pectinatus*
non-calcareous brooks	species poor stands dominated by *R. fluitans* or *C. hamulata*	*Callitriche*-dominated stands	helophyte stands
larger upland rivers	growth form rich stands with magnopotamids, *Sparganium* and *R. fluitans*	*Ranunculus fluitans* community with parvopotamids	*P. pectinatus* community or devastation

In slow flowing potamal brooks and rivers the species rich *Sparganium emersum* community occurs under natural circumstances. It is transferred via species poor variants into stands dominated by single nymphaeids, parvopotamids or filamentous algae.

In faster flowing rhitral water *Callitriche* and *Batrachium* rich stands are first replaced by *Elodea* and *Sparganium*. In case of heavy disturbance either no plants or only filamentous algae remain. The larger rivers in the lowlands are already that much disturbed that no reconstruction of such kind of series is possible.

In the highlands calcareous and non-calcareous brooks and rivers must be distinguished. In calcareous brooks species rich stands of the hard water flora often occur in complex with floating helophytes and grasses. They are transformed into species poor stands of *Zannichellia* or small *Batrachium* species, ending in stands that are devoid of true hydrophytes except *P. pectinatus*. Often only helophytes and filamentous algae remain.

In non-calcareous brooks the sites are poorer in species under natural circumstances with *Batrachium* and *Callitriche* species dominating. First the *Batrachium* species get extinct. Later only helophytes remain. In the case of the larger upland rivers the difference between calcareous and non-calcareous cannot be detected because the drainage areas usually cover both types of rocks. Species rich vegetation with *Sparganium, Batrachium* and broad-leaved *Potamogeton* occur. In case of minor disturbance *R. fluitans* and parvopotamids get dominant. In case of heavy disturbance *P. pectinatus* is left or the rivers are totally devastated.

Observations leading to such kind of information are of special importance for nature conservancy purpose. They lead to a rational judgement about the question which kind of river vegetation is typical for a given physiographic region.

PERSPECTIVES FOR FUTURE RESEARCH

In the preceding an outline on approaches and methods in river vegetation studies has been given and also some results considering the distribution of aquatic plants and their communities in space and time are presented.

At present vegetation science in rivers has hardly reached the descriptive stage. For river vegetation there is no such information available as for example for forest, grassland or coastal vegetation which until now were favourite study objects in vegetation science. There is an urgent need for large scale survey to gather some basic information on the distribution of river communities in space, and there is also an urgent need for long term investigation in selected places to gather some information on the stability of the species compositions observed in the large scale survey.

To reach these aims use of modern technology like big computers is necessary to store and process the data gathered in the field. Nation-wide survey programmes using valid methods have to be carried out in all European countries similar to the work of Holmes (1983).

In the future it will be necessary to develop a science that is at least partly predictive. For all geographical scales hypotheses should be generated and tested. This procedure seems to be most difficult on a small scale level. When referring to mosaic habitats the development of vegetation does not follow regular patterns and is not in direct accordance with the present measurable habitat parameters. However, if life history of the species involved is sufficiently known and the constraints of habitat parameters are exactly defined also testing of hypotheses should be possible. On a medium and large scale hypothesis generating is much easier.

For example the following hypotheses can be tested:

(1) The aquatic vegetation changes according to the climatic zones similar to the terrestrial vegetation;
(2) The aquatic vegetation shows a distinct oceanic-continental gradient;
(3) Lowland rivers differ significantly in species composition and vegetation structure from upland rivers;
(4) Within homogeneous physiographic regions sand rivers differ in species composition from clay rivers;
(5) Substratum stability is one of the main requirements for macro-phyte growth in general exceeding the importance of hydrochemical conditions by far;

Following these guidelines will certainly lead to a better under-standing of the nature of aquatic vegetation.

ACKNOWLEDGEMENTS

Parts of the manuscript are based on an unpublished literature study on the relation between geomorphology and river vegetation which was carried out together with W. Herr (Oldenburg). F. H. Dawson and D. Westlake (Wareham, Great Britain) critically read the manuscript.

REFERENCES

Allorge, P. (1922) Les associations végétales du Vexin fançais. Diss. Paris.

336

Bouxin, G. & LeBoulenge, E. (1983) A phytosociological system based on a multi-scaled pattern analysis: a first example. Vegetatio 54: 3—16.

Brux, H., Todeskino, D. & Wiegleb, G. (1987) Growth and reproduction of *Potamogeton alpinus* Balbis growing in disturbed habitats. Arch. Hydrobiol. Beih. 27: 115—127.

Büdel, J. (1971) Das natürliche System der Geomorphologie, mit kritischen Gängen zum Formenschatz der Tropen. Würzburger Geogr. Arbeiten 34.

Butcher, R. W. (1933) On the distribution of macrophytic vegetation in the rivers of Britain. J. Ecol. 21: 58—91.

Casper, S. J. & Krausch, H. D. (1980) Pteridophyta and Anthophyta, 1. Teil. *In*: Ettl, H., Gerloff, J. & Heynig, H. (eds.), Süßwasserflora von Mitteleuropa, Bd. 23. Stuttgart.

Casper, S. J. & Krausch, H. D. (1981) Pteridophyta und Anthophyta, 2. Teil. *In*: Ettl, H., Gerloff, J. & Heynig, H. (eds.), Süßwasserflora von Mitteleuropa, Bd. 24. Stuttgart.

Clements, F. E. (1916) Plant succession: an analysis of the development of vegetation. Washington.

Cook, C. D. K. (1983) Aquatic plants endemic to Europe and the Mediterranean. Bot. Jahrb. Syst. 103: 539—582.

Dahl, H. J. & Wiegleb, G. (1984) Gewässerschutz und Wasserwirtschaft der Zukunft — Grundlagen eines zukünftigen Fließgewässerschutzes. Jahrb. Naturschutz Landschaftspflege 36: 26—65.

Dawson, F. H. (1988) Water flow and the vegetation of running waters. *In*: Symoens, J. J. (ed.), Handbook of vegetation science, vol. 15. Vegetation of inland waters. Dr W. Junk. This volume, pp. 283—309.

Dawson, F. H., Castellano, E. & Ladle, M. (1978) Concept of species succession in relation to river vegetation and management. Verh. int. Ver. theor. angew. Limnol. 20: 1429—1434.

Decamps, H., Capblanq, J., Casanova, H. & Tourenq, J. N. (1979) Hydrobiology of some regulated rivers in the southwest of France. *In*: Ward, J. V. & Stanford, J. A. (eds.), The ecology of regulated streams, pp. 273—288. New York and London.

De Marchi, A., Zanotti Censoni, A., Corbetta, F. & Ghetti, P. F. (1979) Cenosi macrofitiche alveali del torrente Parma in rapporto a morfologia e tipologia dei sedimenti. Ateneo Parmense, Acta Nat. 15: 221—240.

den Hartog, C. (1979) Seagrasses and seagrass ecosystems, an appraisal of the research approach. Aquat. Bot. 7: 105—117.

den Hartog, C. & Segal, S. (1964) A new classification of the waterplant communities. Acta Bot. Neerl. 13: 367—393.

Ellenberg, H. (1978) Vegetation Mitteleuropas mit den Alpen. 2. Aufl. Stuttgart.

Ellenberg, H. (1979) Zeigerwerte der Gefäßpflanzen Mitteleuropas. 2. Aufl. Scripta Geobotanica 9. Göttingen.

Ellenberg, H. & Müller-Dombois, D. W. (1967) A key to Raunkiaer plant life forms with revised subdivisions. Ber. Geobot. Inst. ETH Rübel 37: 56—73.

Fahy, G. (1974) Geomorphology. Dublin.

Felzines, J. C. (1979) L'analyse factorielle des correspondances et l'information mutuelle entre les espèces et les facteurs du milieu: application à l'écologie des macrophytes aquatiques et palustres. Bull. Soc. bot. N. France 32: 39—63.

Gams, H. (1918) Prinzipienfragen der Vegetationsforschung. Ein Beitrag zur Begriffsklärung und Methodik der Biocoenologie. Vierteljahrsschr. Naturf. Ges. Zürich 63: 293—493.

Gams, H. (1925) Wasserpflanzen als Indikatoren. Fischerei-Zeitung 28: 914—918.

Gleason, H. A. (1926) The individualistic concept of the plant association. Bull. Torr. Bot. Club 53: 7—26.

Ham, S. F., Cooling, D. A., Hiley, F. D., McLeish, P. R., Scorgie, H. R. A. & Berrie, A. D. (1982) Growth and recession of aquatic macrophytes on a shaded section of the River Lambourn, England, from 1971 to 1980. Freshwater Biol. 12: 1—15.

Ham, S. F., Wright, J. F. & Berrie, A. D. (1981) Growth and recession of aquatic macrophytes on an unshaded section of River Lambourn, England, from 1971 to 1976. Freshwater Biol. 11: 381—390.

Haslam, S. M. (1978) River plants. The macrophytic vegetation of water courses. Cambridge.

Hawkes, H. A. (1975) River zonation and classification. In: Whitton, B. A. (ed.), River ecology Oxford, pp. 372—374.

Hilbig, W. & Schamsran, Z. (1977) Beitrag zur Kenntnis der Vegetation im Choval Aimak (Mongolische Volksrepublik). Arch. Naturschutz Landschaftsforschg. 17: 35—82.

Holmes, N. T. H. (1980) Preliminary results from river macrophyte survey and implications for conservation. Nature Conservancy Council, Chief Scientists Team Notes 24. London.

Holmes, N. (1983) Typing British rivers according to their flora. Focus on Nature Conservation 4. Huntingdon.

Holmes, N. T. H. & Whitton, B. A. (1975) Macrophytes of the River Tweed. Trans. bot. Soc. Edinburgh 42: 369—381.

Holmes, N. T. H. & Whitton, B. A. (1977) Macrophytic vegetation of the River Swale, Yorkshire. Freshwater Biol. 7: 545—558.

Illies, J. (1978) Limnofauna europaea. Stuttgart.

Illies, J. & Botoseanu, L. (1963) Problèmes et méthodes de la classification et de la zonation écologique des eaux courantes, considérées surtout du point de vue faunistique. Mitt. int. Ver. theor. angew. Limnol. 12: 1—57.

Jones, H. (1955) Studies on the ecology of the River Rheidol. I. Plant colonization and permanent quadrat records in the main stream of the lower Rheidol. J. Ecol. 43: 462—476.

Klotz, S. & Köck, U. V. (1984) Vergleichende geobotanische Untersuchungen in der Baschkirischen ASSR. 3. Teil: Wasserpflanzen-, Flußufer- und Halophytenvegetation. Feddes Repert. 95: 381—408.

Koch, W. (1926) Die Vegetation der Linthebene unter besonderer Berücksichtigung der Verhältnisse in der Nordostschweiz. Jahrb. St. Gallener Naturw. Ges. 61: 1—146.

Kohler, A. (1975) Veränderungen natürlicher submerser Fließgewässervegetation durch organische Belastung. Daten Dok. Umweltschutz 14: 59—66.

Kohler, A. (1978a) Wasserpflanzen als Bioindikatoren. Beih. Veröff. Naturschutz Landschaftspflege Bad.-Württ. 11: 259—281.

Kohler, A. (1978b) Methoden der Kartierung von Flora und Vegetation von Süßwasserbiotopen. Landschaft + Stadt 10: 73—85.

Kohler, A., Pensel, T. & Zeltner, G. H. (1980) Veränderungen von Flora und Vegetation in den Fließgewässern der Friedberger Au (bei Augsburg) zwischen 1972 und 1978. Verh. Ges. Ökol. 8: 343—350.

Kohler, A., Vollrath, H. & Beisl, E. (1971) Zur Verbreitung, Vergesellschaftung und Ökologie der Gefäßmakrophyten im Fließwassersystem Moosach. Arch. Hydrobiol. 69: 33—365.

Kohler, A. & Zeltner, G. H. (1981) Der Einfluß von Be- und Entlastung auf die Vegetation von Fließgewässern. Daten Dok. Umweltschutz, Sonderr. Umwelttagung 31: 127—139.

Kutscher, G. (1984) Verbreitung und Ökologie höherer Wasserpflanzen in Fließgewässern der Schwäbischen Alb. Diss. Univ. Hohenheim.

Landolt, E. (1984) Verbreitungsmuster in der Familie Lemnaceae und ihre ökologische Bedeutung. Verh. Ges. Ökol. 12: 241—254.

338

Lepš, J., Osbornova-Kosinova, J. & Rejmanek, M. (1982) Community stability, complexity and species life history strategies. Vegetatio 50: 53—63.

Litav, M. & Agami, M. (1976) Relationship between water pollution and the flora of two coastal rivers of Israel. Aquat. Bot. 2: 23—41.

Mäkirinta, U. (1978) Ein neues ökomorphologisches Lebensformensystem der aquatischen Makrophyten. Phytocoenologia 4: 445—470.

Mäkirinta, U. (1981) Die Berücksichtigung der Synusien bei der Beschreibung der Wasservegetation mit der floristisch-soziologischen Methode. In: Dierschke, H. (ed.): Syntaxonomie, p. 169—179. Vaduz.

Maristo, L. (1941) Die Seetypen Finnlands auf floristischer und vegetationsphysiognomischer Grundlage. Ann. Bot. Soc. Zool. -Bot. Fenn. Vanamo 15: 1—312.

Mériaux, J. L. (1978) Etude analytique et comparative de la végétation aquatique d'étangs et marais du Nord de la France (Vallée de la Sensée et Bassin Houillier du Nord-Pas de Calais). Documents Phytosociologiques N.S. 3: 244 pp.

Merry, D. G., Slater, F. M. & Randerson, P. F. (1981) The riparian and aquatic vegetation of the River Wye. J. Biogeogr. 8: 313—327.

Meusel, H., Jäger, E. & Weinert, E. (1965) Vergleichende Chorologie der zentraleuropäischen Flora. Bd. 1. Jena.

Moravec, J. et al. (1983) Rostlinna spolecenstva ceske socialisticke rupubliky a jejich ohrozeni. Serreoceskou prirodou.

Müller, T. (1977) Verband: Ranunculion fluitantis Neuhäusl 59. In: Oberdorfer, E. (ed.), Süddeutsche Pflanzengesellschaften, Teil I, 2. Aufl., Stuttgart, pp. 89—99.

Noble, I. R. & Slatyer, R. D. (1980) The use of vital attributes to predict successional changes in plant communities subject to recurrent disturbances. Vegetatio 43: 5—21.

Novakova, H. & Rydlo, J. (1981) Kvétena opatovického kanálu. Prace a studie — prir. Parbubice 12: 35—44.

Nunnally, N. R. & Keller, E. (1979) Use of fluvial processes to minimize adverse effects of stream channelization. Report No. 144. Water Resour. Res. Inst., Univ. N. Carolina. Charlottes, N.C.

Passarge, H. (1964) Pflanzengesellschaften des nordostdeutschen Flachlandes I. Jena.

Pennak, R. W. (1971) Toward a classification of lotic habitats. Hydrobiologia 38: 321—334.

Penuelas, J. & Catalan, J. (1983) Distribution longitudinale des bryophytes d'un fleuve méditerranéen du N.E. de l'Espagne: Le Fluvia. Ann. Limnol. 19: 179—185.

Persoone, G. (1979) Proposal for a biotypological classification of water courses of the European community. In: James, A. & Evison, L. (eds.), Biological indicators of water quality, chapter 7. 32 pp. Chichester.

Pietsch, W. (1974) Ökologische Untersuchung und Bewertung von Fließgewässern mit Hilfe höherer Wasserpflanzen — ein Beitrag zur Belastung aquatischer Ökosysteme. Mitt. Sekt. Geobot. Phytotax. Biol. Ges. DDR 1974: 13—29.

Pignatti, S. (1980) Reflections on the phytosociological approach and the epistemological basis of vegetation science. Vegetatio 42: 181—186.

Quennerstedt, N. (1965) The major rivers of northern Sweden. Acta phytogeogr. Suec. 50: 198—204.

Rautava, E. (1964) Über die Wasservegetation des Flusses Vaskojoki im nördlichsten Finnland. Ann. Univ. Turku A II 32: 69—93.

Roll, H. (1938) Allgemein wichtige Ergebnisse für die Pflanzensoziologie bei der Untersuchung von Fließgewässern in Holstein. Feddes Repert. Beih. 101: 108—112.

Roll, H. (1942) Pflanzensoziologie und Gewässerkunde. Ber. Dtsch. Bot. Ges. 60: 135—145.

Ruess, K. (1954) Die Makrophyten-Vegetation des fließenden Wassers im Gebiet um Seeon (Chiemgau). Diss. München.

Schmassmann, H. (1955) Die Stoffhaushaltstypen der Fließgewässer. Arch. Hydrobiol. Suppl. 22: 504—509.

Schmidt, D. (1981) Pflanzensoziologische und ökologische Untersuchungen der Gewässer um Güstrow. Natur u. Naturschutz Mecklenburg 17: 1—130.

Schuyler, A. E. (1984) Classification of life forms and growth forms of aquatic macrophytes. Bartonia 50: 8—11.

Schwegler, H. (1981) Stabilitätsbegriffe für biologische Systeme. Angew. Bot. 55: 129—137.

Segal, S. (1971) Principles on structure, zonation and succession of aquatic macrophytes. Hydrobiologia 12: 89—95.

Smith, S. G. (1978) Aquatic macrophytes of the Pine and Popple River system, Florence and Forest Counties, Wisconsin. Trans. Wisc. Acad. Sci. Arts Lett. 66: 148—185.

Stake, E. (1967) Higher vegetation and nitrogen in a rivulet in Central Sweden. Schweiz. Z. Hydrol. 29: 107—125.

Stake, E. (1968) Higher vegetation and phosphorus in a small stream in Central Sweden. Schweiz. Z. Hydrol. 30: 353—373.

Steffan, A. W. (1965) Zur Statik und Dynamik im Ökosystem der Fließgewässer und zu Möglichkeiten ihrer Klassifizierung. In: Tüxen, R. (ed.), Biosoziologie Den Haag, pp. 65—110.

Steffen, H. (1931) Die Pflanzengesellschaften Ostpreußens. Jena.

Stuckey, R. L. (1971) Changes of vascular aquatic flowering plants during 70 years in Put-In-Bay harbour, Lake Erie, Ohio. Ohio. J. Sci. 71: 321—342.

Vaarama, A. (1938) Wasservegetationsstudien am Großsee Kallavesi. Ann. Bot. Soc. Vanamo 13(1): 1—314.

Van der Valk, A. G. (1981) Succession in wetlands: a Gleasonian approach. Ecology 62: 688—696.

van Vierssen, W. (1983) The influence of human activities on the functioning of macrophyte dominated aquatic ecosystems in the coastal area of western Europe. Proc. Int. Symp. Aquat. Macrophytes (Nijmegen, 18—23 September 1983), pp. 273—281.

Vannote, R. L., Minshall, G. W., Cummins, K. W., Sedell, J. R. & Cushing, C. E. (1980) The river continuum concept. Can. J. Fish. aquat. Sci. 37: 130—137.

Volker, R. & Smith, S. G. (1965) Changes in the aquatic vascular flora of Lake East Okoboji in historic times. Proc. Iowa Acad. Sci. 72: 65—72.

Vrhovsek, D., Martincic, A. & Kralj, M. (1981) Evaluation of the polluted River Savinja with the help of macrophytes. Hydrobiologia 80: 97—110.

Walter, H. & Breckle, S. W. (1983) Ökologie der Erde, Bd. 1: Ökologische Grundlagen in globaler Sicht. Stuttgart.

Walter, H. & Breckle, S. W. (1986) Ökologie der Erde, Bd. 3: Spezielle Ökologie der gemäßigten und arktischen Zonen Euro-Nordasiens. Stuttgart.

Weber, H. E. (1976) Die Vegetation der Hase von der Quelle bis Quakenbrück. Osnabrücker naturwiss. Mitt. 4: 131—190.

Weber-Oldecop, D. W. (1969) Wasserpflanzengesellschaften im östlichen Niedersachsen. Diss. TU Hannover.

Weber-Oldecop, D. W. (1977) Fließgewässertypologie in Niedersachsen auf floristisch-soziologischer Grundlage. Göttinger Florist. Rundbr. 10: 73—80.

Westhoff, V. & den Held, A. J. (1975) Plantengemeenschappen in Nederland. 2. ed. Zutphen.

Westlake, D. F. (1973) Aquatic macrophytes in rivers: a review. Pol. Arch. Hydrobiol. 20: 31—40.

Wiegleb, G. (1978) Vergleich ökologischer und soziologischer Artengruppen von Makrophyten des Süßwassers. Verh. Ges. Ökol. Kiel 1977: 243—249.

340

Wiegleb, G. (1981a) Probleme der syntaxonomischen Gliederung der Potametea. *In*: Dierschke, H. (ed.), Syntaxonomie, p. 207—249. Vaduz.

Wiegleb, G. (1981b) Struktur, Verbreitung und Bewertung von Makrophytengesellschaften niedersächsischer Fließgewässer. Limnologica (Berlin) 13: 427—448.

Wiegleb, G. (1983) Recherches méthodologiques sur les groupement végétaux des eaux courantes. Coll. Phytosociol. 10: Végétations aquatiques et amphibies (Lille, 1981): 69—83.

Wiegleb, G. (1984a) A study of habitat conditions of the macrophytic vegetation in selected river systems in Western Lower Saxony (FRG). Aquat. Bot. 18: 313—352.

Wiegleb, G. (1984b) Makrophytenkartierung in Niedersachsen — Methoden, Ziele und erste Ergebnisse. Inf. Naturschutz Landschaftspflege 4: 109—136.

Wiegleb, G. & Herr, W. (1984) Zur Entwicklung vegetationskundlicher Begriffsbildung am Beispiel der Fließwasservegetation Mitteleuropas. Tüxenia 4: 303—325.

Wiegleb, G. & Kadono, Y. (1988) Composition, structure and distribution of macrophytic communities in Japanese rivers. Bot. Jahrb. Syst. (in press).

Wiegleb, G. & Todeskino, D. (1983) Habitat conditions of *Potamogeton alpinus* Balbis stands and relations to the plants biological characters. Proc. Int. Symp. Aquat. Macrophytes (Nijmegen, 18—23 September 1983), pp. 311—316.

AQUATIC PLANTS IN EXTREME ENVIRONMENTS

JOHN M. MELACK

1. INTRODUCTION

The intent of this chapter is to review the ecology of aquatic plants that inhabit physically and chemically extreme environments of inland waters. Such plants are fascinating in their own right, and their special characteristics have attracted the attention of a wide variety of disciplines. For instance, cyanobacterial mats in hot, saline habitats are modern analogs of Precambrian communities and therefore offer information about the origin of life (Schopf 1983). Furthermore, because microbial mats may have been the precursors of petroleum deposits such as the enormous oil shale reserves associated with the Green River Formation, U.S.A. (Parker & Leo 1965, Eugster & Hardie 1975), investigation of the diagenesis of the modern mats may help explain petroleum formation. Algae of highly saline waters are also a commercially exploited resource because of their high protein, glycerol or β-carotene content (Smithsonian Science Information Exchange 1980, Shelef & Soeder 1980). Other examples where the study of plants in extreme environments aids modern science and technology include the search for extraterrestrial life (Billingham 1981, Ponnamperuma 1976) and the design of life-support systems for space crafts (Taub 1974). The application of new genetic engineering techniques now make it possible that salt tolerant crops can be developed with contributions of halophyte genomes (Hollaender 1979, Szalay & MacDonald 1980). Likewise, elucidation of photosynethtic mechanisms is being augmented by studies that use organisms from extreme habitats (e.g. Cohen *et al.* 1975, Robinson *et al.* 1982).

The material presented here will be circumscribed by criteria that delineate the taxa, the habitats and the environmental conditions. The pool of organisms represents seven divisions of plants inclusive of the Cyanophyta, Chlorophyta, Chrysophyta, Euglenophyta, Pyrrophyta, Bryophyta and Tracheophyta. Only plants inhabiting inland waters are considered; marine and brackish (*sensu* den Hartog 1967) habitats are

341

J. J. Symoens (ed.), Vegetation of inland waters. ISBN 90—6193—196—7.
© *1988, Kluwer Academic Publishers, Dordrecht. Printed in the Netherlands.*

excluded. The plants must be submerged in water; this criterion excludes many tracheophytes that may be rooted underwater but extend into the air. The environmental bounds selected are near those tolerated by actively growing and reproducing plants and sufficiently severe to restrict the flora to very few species. In addition, examples were chosen from inland waters which had not experienced human interference. In most cases habitats were selected that maintained the conditions as delineated without large fluctuations. Four sets of environmental extremes will be considered:

1. Salinities greater than 50 grams of total dissolved solids per liter;
2. pHs greater than 10 and salinities exceeding 30 grams of total dissolved solids per liter;
3. Temperatures exceeding 50 °C.
4. Temperatures near 0 °C that do become subfreezing.

Organisms living in low pH waters (pH < 3) are included in the section on high temperatures because of the coincidence of high temperature and acidity in some springs. Furthermore, references to acid-mine drainage and acidification of dilute waters by atmospheric deposition contaminated with strong acids are listed below in the context of stresses caused by man-made pollution.

Several environmental conditions that could appear extreme in some circumstances were excluded for the following reasons:

(a) While prolonged anoxia is a severe problem for strict aerobes, life is abundant in numerous locales with low or no oxygen. Furthermore, temporary anaerobiosis is common in aerobes (Hook & Crawford 1978). Moreover, oxygen toxicity exists for a wide variety of anaerobes and aerobes (Kuenen 1979).
(b) High or low insolation and intermittent darkness are widespread conditions experienced by many plants (Harris 1978, Wright & Burton 1981). Brief mention of responses to high light are included in sections on snow algae and benthic mats in saline lakes.
(c) Pollution from industrial, urban and agricultural activities is altering many environments and is a severe stress for the biota in some situations. The influences of oil spills (Baker 1970, 1979, McCauley 1966), ionizing radiation from radioactive material (Nasin & James 1978, Ichikawa 1981), toxic metals (Stokes et al. 1973, Ehrlich 1978, Rushforth et al. 1981, Woolhouse 1981, Whitton et al. 1981, Lampkin & Sommerfeld 1982), pesticides and polychlorinated biphenyls (O'Connors et al. 1978, Bednarz 1981, Södergren & Gelin 1983), acidification (Drabløs & Tollan 1980), excess nutrients (Barica & Mur 1980), and thermal effluents (Patrick 1969, Squires et al. 1979) are well documented and remain the

focus of active study. In contrast to the environmental conditions treated in this review, pollution is often an acute, exotic stress.

Some bacteria can live under all the extreme conditions considered here and under others even more severe (Vallentyne 1963, Gould & Corry 1980, Shilo 1979, Heinrich 1976). An interested reader will find recent reviews of microbes living at high temperatures (Amelunxen & Murdock 1978, Brock 1978, Tansey & Brock 1978), low temperatures (Morita 1975, Baross & Morita 1978, Inniss & Ingraham 1978), in hypersaline waters (Larsen 1967, Kushner 1978, Brown 1978, 1983, Caplan & Ginzburg 1978), at extreme pHs (Langworthy 1978, Horikoshi & Akiba 1982), in contact with man-made toxins (Alexander 1976) and under high pressures (ZoBell 1970, Wirsen & Jannasch 1975, Jannasch 1979, Jannasch & Wirsen 1983).

2. HYPERSALINE WATERS

The designation of waters with concentrations of total dissolved solids (TDS) in excess of 50 g l^{-1} as hypersaline follows the classification by Beadle (1959) which was derived largely from biological tolerances of salinity. Hypersaline waters occur worldwide (McCarraher 1972) and include a great variety of chemical compositions and concentration ranges; the upper limits are set by the solubility of the most soluble minerals present. For example, there are soda lakes precipitating trona, Mg-Na-SO_4 bitterns depositing epsomite and bloedite and Na-SO_4-Cl waters precipitating mirabilite and halite (Eugster & Hardie 1978). However, the relevance of the geochemistry of the major solutes to the biology of hypersaline waters is not straightforward. For instance, while the Dead Sea's chemical composition is principally Na, Mg and Cl, an inhibitory effect of high Ca and the relatively high bromine may reduce the species' diversity (Nemenz 1970, Nissenbaum 1975). Some cyano-bacteria are sensitive to Ca and Mg at high but not low salinities (Baas Becking 1931). Furthermore, in shallow saline lakes fluctuations in salinity are common and high insolation and extreme temperatures occur, and these factors impart additional complexity in evaluating the responses of the biota (e.g. Melack 1976, Borowitzka 1981, Brock & Lane 1983).

Phytoplankton

The number of species of phytoplankton is low in saline lakes and decreases further as salinity increases until phytoplankton disappear in concentrated brines (Iltis 1974, Hammer *et al.* 1983; Table I). Factors

Table I. Some records of the number of species of phytoplankton in saline lakes. Taxonomic and sampling uncertainties make the number of species per lake only approximate.

Lake Location	Salinity g l^{-1}	Major Salts	Number of Species	References
Corangamite Australia	25—36	NaCl	11	Hammer 1981 Bayly and Williams 1966
Werowrap Australia	23—56	NaHCO$_3$-NaCl	8	Walker 1973
Gallocanta Spain	32—105	NaCl	20	Comin *et al.* 1983
Bodou Chad	34—48	Na$_2$CO$_3$	19	Iltis 1969
Big Quill Canada	43—53	MgSO$_4$-Na$_2$SO$_4$	18	Hammer *et al.* 1983 Hammer 1978
Mono USA	73—90	Na$_2$CO$_3$-NaCl	15	Mason 1967, Melack 1983, Dana *et al.* 1977
Great Salt USA				Felix and Rushforth 1979
Farmington Bay	55	NaCl	21	
South Arm	113—121	NaCl	9	
North Arm	342	NaCl	1	
Hunazoko Antartica	213	NaCl	4	Tominaga and Fukui 1981
Dead Sea Israel	320	MgCl$_2$-NaCl	1	Nissenbaum 1975
Don Juan Pond Antarctica	470—540	CaCl$_2$	0	Meyer *et al.* 1962 Wright and Burton 1981

other than the concentration of total dissolved solids surely contribute to this low diversity, and what little is known of the ecological conditions pertinent to phytoplankton in hypersaline waters will be summarized or referenced here. Soda lakes are included in Table I but will be treated separately (see Section 3) because they combine high pH with high salinity and have a distinctive assemblage of species.

Few floristic studies with adequate sampling and careful taxonomy are available for hypersaline lakes. Much of the distributional evidence is provided by collections from expeditions or surveys (e.g. Hutchinson 1937, Beadle 1943, Rawson & Moore 1944) or as ancillary data from limnological studies which often include only numerically dominant species (e.g. Wetzel 1964, Mason 1967, Melack & Kilham 1974, Hammer 1981). One of the more thorough regional analyses of algal distribution was done for the saline lakes of Saskatchewan, Canada, by Hammer *et al.* (1983). Collections were made in 41 lakes ranging in salinity from 3.2 to 428 g TDS 1^{-1}. These lakes are predominately solutions of sodium, magnesium and sulfate (Hammer 1978). Abundant, planktonic algae with broad salinity tolerances that extended above 50 g TDS 1^{-1} include the chlorophytes, *Dunaliella salina* (Dunal.) Teodoresco and *Rhizoclonium hieroglyphicum* (Ag.) Kuetz, the cyanophytes, *Lyngbya birgei* G. M. Smith, *Microcystis aeruginosa* Kuetz. emend. Elenkin, *Oscillatoria tenuis* Ag., *O. utermoehli* (Uter.) J. de Toni and *Nodularia spumigena* Mertens, and the bacillariophytes, *Melosira granulata* (Ehr.) Ralfs, *Stephanodiscus niagarae* Ehr. and *Chaetocerus elmorei* Boyer. Fourteen species were restricted to hypersaline lakes; eleven of these were diatoms, and the others were a dinoflagellate and euglenoids. In lakes with salinities above 100 g TDS 1^{-1}, chlorophytes usually dominated although diatoms remained important.

Of the phytoplankton that may be considered characteristic of hypersaline lakes (cf. Hammer *et al.* 1983), the chlorophyte, *Dunaliella* (Volvocales), is known to grow at the highest salinities; it is euryhaline (Borowitzka 1981) and inhabits chloride, sulfate and carbonate waters (Massjuk 1973). In contrast to other Volvocales *Dunaliella* lacks a rigid cell wall, and this, in combination with a high protein content, makes it a promising candidate for mass cultivation for food (Gibbs & Duffus 1976). Furthermore, *Dunaliella* produces such large amounts of glycerol to aid its osmoregulation (Brown & Borowitzka 1979) that it probably can be commercially exploited (Ben-Amotz & Avron 1980). Therefore, *Dunaliella* has received considerable attention by physiologists and biochemists interested in adaptation to high salinity and the regulation of glycerol production (Brown 1976, Kauss 1978, Caplan & Ginzburg 1978, Brown & Borowitzka 1978, Ben-Amotz & Avron 1980 and other references cited below). Although earlier workers suggested that *Dunaliella* accumulated NaCl internally to balance the salts in the external medium (e.g. Ginzburg 1969), most of its *in vitro* enzymes are not salt resistant (Johnson *et al.* 1968, Ben-Amotz & Avron 1972, Heimer 1973). Actually, *Dunaliella* produces intracellular concentrations of glycerol comparable to the osmolality of the external solution (Craigie & McLachlan 1964, Ben-Amotz & Avron 1973, Borowitzka & Brown 1974, Ben-Amotz 1975, Borowitzka *et al.* 1977). Glycerol is a compatible solute, i.e. it is a low molecular weight, neutral solute that does not inhibit enzymes even at concentrations as high as 4.4 molal (Borowitza & Brown 1974). However, Gimmler & Schirling (1978) provide evidence, perhaps questionable on technical grounds (Brown & Borowitzka 1979), that Na and K can contribute to osmoregulation in *Dunaliella parva*, and point out that a contribution from these solutes is energetically advantageous to the cell. In contrast to *Dunaliella*, the halophilic heterotrophic bacteria *Halobacterium* and *Halococcus* balance external salinity with high concentrations of Na, K and Cl in the cytoplasm (Christian & Waltho 1962, Ginzburg *et al.* 1971, Brown 1976, Caplan & Ginzburg 1978, Borowitzka 1981, Yancey *et al.* 1982)), and most of their enzymes require high salt concentrations to function properly.

The ecology of *Dunaliella* is less well known than its osmoregulatory physiology. Massjuk's (1973) comprehensive review of the biology of *Dunaliella* is available in translations with only limited circulation. More up-to-date and succinct information on the alga's ecology is provided by Brock (1975), Brown & Borowitzka (1979), Borowitzka (1981) and Post (1981). Of the two halophilic species originally described by Teodoresco (1905), *D. salina* (Dunal) Teod. accumulates carotenoids at high salinities and appears red while *D. viridis* Teod. remains green. Additional green species known from salt lakes include *D. bioculata* Butcher, *D. minuta* Lerche, *D. parva* Lerche and *D. euchlora* Lerche (Butcher 1959, Borowitzka 1981). *Dunaliella*'s growth forms include vegetative, palmelloid and encysted cells, and aplanospores, and are determined by salinity and other environmental conditions. For example, *D. salina* usually exists as motile vegetative cells that reproduce asexually at salinities from 40 to *ca.* 350 g TDS 1^{-1} and when nutrients are adequate. Benthic mats of palmelloid forms occur in dilute regions of the Great Salt Lake (Brock 1976); haploid aplanospores also form at lower salinities (< 40 g TDS 1^{-1}, Loeblich 1972). Sexual reproduction by the formation and fusion of isogametes, apparently induced by reduced salinity, was reported by Lerche (1937), but has yet to be confirmed by others (A. Gibor, personal communication).

Dunaliella grows actively over a very wide range of salinities. For example, Ben-Amotz & Avron (1980) reported an optimal growth of *D. salina* at *c.* 1 *M* NaCl but a change in exponential growth rate by a factor of only 1.5 over the range from 1.0 to 3.0 *M* NaCl. *D. viridis* grows over a range from 0.15 to 5.0 *M* NaCl (i.e., 9 to 304 g TDS 1^{-1}, Brock 1975) with an optimum between 1.0 and 1.5 *M* (Borowitzka *et al.* 1977). Apparently, *Dunaliella*'s dominance at salinities above 200 g TDS 1^{-1} indicates its competitive superiority even at suboptimal growth rates. *Dunaliella* grows near optimally between *c.* 20 °C and 35 °C (Gibor 1956, Van Acken & McNulty 1973, Massjuk 1973), and survives subzero temperatures (Massjuk 1973, Post 1977). Photosynthetic optima between 35 and 40 °C are reported by Borowitza (1981) and Massjuk (1973). The high intracellular concentration of glycerol may reduce the effects of heat and cold stress on enzymes (Borowitzka 1981). *Dunaliella* is often exposed to high insolation in the lakes where it flourishes and accumulates large amounts of carotenoids, mainly β-carotene (Aasen *et al.* 1969), that may have a photoprotective role (Loeblich 1972). Motile cells do not survive continuous darkness for periods exceeding 2 weeks (Oren & Shilo 1982) and heterotrophy appears poorly developed (Gibor 1956). The results summarized here are based largely on studies that attempted to optimize conditions for growth and examined one factor at a time. In natural environments some of the laboratory conditions (e.g. continuous light, elevated CO_2) do not occur. Moreover, the major factors affecting cell growth interact and a single optimum is unlikely. For example, Ginzburg & Ginzburg (1981) reported that the temperature optimum was higher at high light intensities than at lower intensities.

Dunaliella parva is the only eukaryotic alga in the Dead Sea, a hypersaline (*c.* 320 g TDS 1^{-1}) magnesium-sodium-chloride water (Nissenbaum 1975). In February 1979 the Dead Sea underwent complete overturn for the first time in many years (Steinhorn *et al.* 1979). Subsequent to this event Oren and Shilo (1982) followed the population dynamics of *D. parva* and correlated the alga's changes in abundance with physical and chemical factors in the lake; no grazers were present. *D. parva* increased to a maximum density of 8800 cells ml^{-1} in the summer of 1980 and then declined until finally collapsing in January—February 1981 when the cells were distributed well below the euphotic zone. The summer bloom was confined to the thin layer (5—10 m) of less saline water in the epilimnion that formed after the winter floods. Laboratory experiments by Oren & Shilo (1982) indicated that dilution and phosphate enrichment are required for augmented growth of *D. parva* in the Dead Sea. Light limitation was probably the main cause of the decline.

Among the many genera of cyanobacteria that grow in hypersaline waters, the unicellular species *Aphanothece halophytica* is the most salt-tolerant (Borowitzka 1981). Synonyms for this coccoid cyano-bacterium probably include *Aphanocapsa halophytica, Aphanothece packardii, Polycystis packardii, Microcystis packardii, Aphanothece utahensis* and *Coccochloris elabens* (Brock 1976, Bauld 1981). This organism occurs both in the plankton and attached to substrata (see below). *A. halophytica* grows in solutions of NaCl containing 30 to 350 g TDS 1^{-1} (Borowitzka 1981). However, when cultured in Great Salt Lake water enriched with nutrients *A. halophytica* grew best at salinities from 160 to 230 g 1^{-1} perhaps in part because of displace-ment by competitors at higher and lower salinities (Brock 1975, 1976). Isolates from a solar evaporating pond tested by Kao *et al.* (1973) grew optimally at 50 to 100 g NaCl 1^{-1}, thus indicating the considerable range possible among populations from different habitats. Osmotic regulation appears to be maintained by changes in intracellular potassium (Miller *et al.* 1976, Yopp *et al.* 1978) and the organic solutes, betaine and glutamate (Borowitzka 1981). Total protein content also varies as a function of salinity, and the protein content of 76% of dry weight at 1 *M* NaCl reported by Tindall *et al.* (1977) is among the highest known for algae.

Primary productivity by phytoplankton in hypersaline lakes ranges widely (Table II, Hammer 1981). Some of the highest rates of photosynthesis can occur in soda lakes (Table II; see below, Section 3), but moderate to low rates are common in hypersaline waters. The limited information on the factors that determine primary productivity in saline waters indicates that low nutrient levels (Stephens & Gillespie 1976, Melack *et al.* 1982), self shading (Talling *et al.* 1973), low temperature (Campbell 1978) and grazing (Lenz 1982) can be important. In meromictic, saline lakes major seasonal shifts in the contributions to primary productivity by phytoplankton and bacteria can be observed (e.g. Cohen *et al.* 1977b, 1977c; Cloern *et al.* 1983). The low species diversity at all trophic levels in hypersaline waters provides an excellent opportunity for concurrent analysis of ecosystem processes and population dynamics; few such studies exist (e.g. Walker 1973; Post 1981; Vareschi 1978, 1979, 1982, Vareschi & Vareschi 1984 and Vareschi & Jacobs 1984; Melack *et al.* 1985).

Attached algae

Cyanobacterial mats are conspicuous and common features of shallow, hypersaline (Bauld 1981) and thermal (Brock 1978, see Section 4) waters. In such extreme environments the cyanobacteria and associated algae and bacteria are not grazed or disturbed by micro- or macrofauna and can form thick laminated mats. Their similarity to Precambrian fossils has sparked special interest among paleobiologists (Schopf 1983). Bauld (1981) provides a comprehensive review of microbial mats in salt lakes, and describes the mats as "microbial communities which colonize benthic surfaces and form cohesive, prostrate, often laminated, structures of varying preservation potential. The fossil remains of analogous structures, known as stromatolites (Walter 1976), preserve a record of the mat and its activities."

In Antarctica benthic mats are known from a variety of sites. A well-documented occurrence is the thick mat that extends to at least 20 m in Lake Bonney (Taylor Valley, South Victoria Land) in waters that increase in salinity from 1.4 (5 m) to 310 (20 m) g TDS 1^{-1} (Angino *et al.* 1964, Parker *et al.* 1977). The dominant species are *Schizothrix calcicola*, *S. antarctica* and *Phormidum augustimum*. Among the few reports of mats in Australia the stromatolitic mats in 164 g TDS 1^{-1} water on Rottnest Is, Western Australia, are noteworthy because

Table II. Primary productivity by phytoplankton in hypersaline waters and soda lakes. Only lakes with measurements that span a full year are included except where indicated. Note different units for primary productivity.

Lake Location	Salinity g 1^{-1}	Major Solutes	Primary Productivity	References
Soap USA	32	Na, CO_3	650—2920 mg C $m^{-2}d^{-1}$	Walker 1975
Elmenteita Kenya	15—40	Na, CO_3	110—1740 mg O_2 m^{-2} h^{-1}	Melack 1981
Werowrap Australia	23—56	Na, CO_3, Cl	30—2900 mg C m^{-2} d^{-1}	Walker 1975
Borax USA	25—61	Na, CO_3, Cl	10—525 mg C m^{-2} d^{-1}	Wetzel 1964
Big Soda USA	26 (mixolimnion) 80 (monimolimnion)	Na, HCO_3, Cl, SO_4 Na, HCO_3, Cl	90—2800 mg C m^{-2} d^{-1}	Cloern et al. 1983
Bogoria Kenya	70	Na, CO_3	230—3000 mg O_2 m^{-2} h^{-1} (two seasons)	Melack 1981
Mono USA	90	Na, CO_3, Cl	90—310 mg C m^{-2} h^{-1} (summer)	Dana et al. 1977
Little Manitou Canada	70—100	Na, Mg, SO_4, Cl	200—1200 mg C m^{-2} d^{-1}	Hammer 1978 Hayes and Hammer 1978
Sakskoe USSR	58—140	—	100—5500 mg C m^{-2} d^{-1}	Dobrynin 1978
Sasyk USSR	77—203	—	80—2300 mg C m^{-2} d^{-1}	Dobrynin 1978
Great Salt USA South Arm	259	Na, Cl	6—7660 mg C m^{-2} d^{-1}	Stephens and Gillespie 1976
Deep Antarctica	270	Na, Cl	< 9 mg C m^{-2} d^{-1}	Campbell 1978
Pink Australia	163—340	Na, Cl	0—184 mg m^{-2} d^{-1}	Hammer 1981

Botryococcus is the principal component (Bauld 1981). Stratiform, globulari and crenulate stromatolitic, cyanobacterial mats occur in two ephemeral hydromagnesite-aragonite lakes near the southeastern end of the Coorong Lagoon, South Australia (Walter *et al.* 1973). North American examples are surprisingly limited which surely reflects the predilections of the scientists more than the paucity of algae mats. The better documented sites of note here are dominated by the palmelloid stage of *Dunaliella salina*, but include *Aphanothece halophytica* and probably a *Phormidium* (northern arm, Great Salt Lake, Brock 1976; see above) and the chlorophyte *Ctenocladus circinnatus* (see below). Post (1979) observed *Dunaliella salina* and *D. viridis* living in oxygen-rich domes in a salt crust (99.9% NaCl) that formed in the Great Salt Lake. In Laguna Mormona, a closed lagoon on the Pacific coast of Baja California, Mexico, the most prominent cyanobacterial mats are laminated formations made by *Microcoleus chthonoplastes* (Horodyski *et al.* 1977). In Africa documentation of benthic mats is very limited (Bauld 1981). As part of a study of flamingo feeding, Tuite (1981) reported chlorophyll values ranging from 20 to 60 mg m^{-2} in saline lakes in eastern Africa. Melack (1976) measured benthic photosynthetic rates up to 700 mg O_2 m^{-2}h^{-1} in Lake Elmenteita, a Kenyan soda lake. Imhoff *et al.* (1979) encountered mats of filamentous and unicellular cyanobacteria in salt lakes with 160—375 g TDS l^{-1} located in the Wadi Natrun, Egypt.

Cyanobacterial mats growing under especially unusual conditions were discovered by Anderson (1958) in Hot Lake, Washington (USA). This small, meromictic lake had an average salinity of 100 g TDS l^{-1} at the surface that increased to 400 g TDS l^{-1} at the bottom. Solar heating raised the temperature in the anoxic, H_2S-rich monimolimnion to about 50 °C during the summer. The mat was composed of *Plectonema nostocorum* Gom., *Oscillatoria chlorina* Gom., *Anacystis thermalis* (Menegh.) Drouet and Daily and *Gomphosphaeria aponia* Kuetz. A similar situation was discovered near Elat, Israel, by Por (1968), and the, so-called, Solar Lake has been the object of intensive study since then (Eckstein 1970, Cohen *et al.* 1975a, 1975b; Cohen *et al.* 1977a, 1977b, 1977c; Krumbein *et al.* 1977; Jørgensen & Cohen 1977, Jørgensen *et al.* 1979, Jørgensen *et al.* 1983, Revsbech *et al.* 1983). The physicochemical and biological aspects of this small, hypersaline lake are clearly summarized by Bauld (1981) and only those features of the cyanobacterial mats that are of general botanical interest are discussed here. The first major discovery made at the Solar Lake was the ability of the benthic cyanobacterium, *Oscillatoria limnetica*, to perform both oxygenic and anoxygenic photosynthesis. In the latter process, photosystem I alone is operative, H_2S is the electron donor, elemental sulfur is deposited externally and no oxygen is evolved. Besides the ecological importance of anoxygenic photosynthesis to cyanobacteria (cf. Padan 1979), it also has evolutionary implications linking anoxygenic (typical of bacteria) and oxygenic (typical of higher plants) photosynthesis.

Of the four mat types that carpet the entire bottom of the Solar

Lake, the so-called flat mat in the littoral is the best studied. During the summer this finely laminated mat is composed of a layer of diatoms and coccoid cyanobacteria (i.e. *Aphanothece halophytica* and *Aphanocapsa littoralis*) that overlie a layer of filamentous cyanobacteria (i.e. *Phormidium* sp. and *Microcoleus* sp.) and gliding bacteria (*Chloroflexus* sp.). The top layer produces carotenoids and appears brownish-orange, while the lower layer is dark green and is protected from photo-oxidative stress. In the winter a dark green mat composed largely of *Oscillatoria limnetica, O. salina* and other filamentous cyanobacteria is present. The flocculose mat of *O. limnetica* that forms at depths of 4.5—5.0 m during winter stratification experiences high H_2S concentrations (up to $39 \, mg \, l^{-1}$) and very low light intensities (c. 0.5% of incident light). Under these conditions *O. limnetica* utilizes H_2S as an electron donor and photosynthesizes anoxygenically. After holomixis and oxygenation of the whole water column in the summer, *O. limnetica* in the flocculose mat switches to oxygenic photosynthesis.

Distinct vertical zonations with very steep microgradients develop in cyanobacterial mats. Earlier work by Jørgensen *et al.* (1979) and Krumbein *et al.* (1977) demonstrated maximum photosynthetic rates at depths of 1—2 mm in the mat and a diel cycle of NH_4 and H_2S increase throughout the upper 2—4 mm during the night and depletion during the day. Recent application of microelectrode techniques have revealed very steep gradients of sulfide and oxygen, coexistence of the two compounds in a layer only 0.25 mm thick during the day and a biologically-mediated turnover time of sulfide of only 21 seconds (Revsbech *et al.* 1983). Vertical profiles of photosynthetic activity measured with a new oxygen microelectrode method indicated a photic zone only 0.8 mm thick in one mat and photosynthesis to 2.5, 4.5 and > 10 mm in three other mats (Jørgensen *et al.* 1983). A maximum photosynthetic rate of $770 \, mg \, O_2 \, m^{-2}h^{-1}$ was recorded at midday under clear skies (c. 1660 $\mu E \, m^{-2} \, s^{-1}$) for a flat mat.

Cyanobacterial mats are often well-developed in shallow, saline lakes which are usually susceptible to large and frequent variations in salinity and water level. Therefore, the organisms that constitute the mats must be capable of osmoregulation and be resistant to desiccation. Little is known about the osmoregulatory ability of the major, mat-forming cyanobacteria (cf. Borowitzka 1981). Resumption of metabolic activity after wetting of dried mats of *Microcoleus vaginatus* and other cyanobacteria has been observed (Durrell & Shields 1961, Drouet 1982, Bauld 1981).

Benthic mats of cyanobacteria are an important component of the salina system used to produce industrial and domestic NaCl by solar evaporation of seawater (Nixon 1974, Davis 1978, Jones *et al.* 1984, Bauld 1981). The darkly pigmented mats increase evaporation by enhancing the absorption of solar radiation. More importantly, the mats greatly reduce seepage loss through the bottom. In addition, the mats may reduce contamination of the salt by organic sediments.

Macroalgae are plentiful in some hypersaline lakes and are represented by several families of the chlorophyta, i.e. Cladophoraceae, Ulvaceae, Charophyceae and Chlorophyceae (Beadle 1943, Ungar 1974, McCarraher 1977, Burne *et al.* 1980, Brock 1981, Brock & Lane 1983, Hammer *et al.* 1983). For example, in Saskatchewan lakes *Enteromorpha prolifera* (Fl. Dan.) J. G. Ag. occurred up to salinities of 182 g TDS 1^{-1} and *Cladophora crispata* (Ruth) Kuetz. was abundant from 3 to 100 g TDS 1^{-1} (Hammer *et al.* 1983).

The charophyte, *Lamprothamnium papulosum* (Wallr.) J. Gr., occurs over the widest range of salinities among the Australian macroalgae (Burne *et al.* 1980, Brock 1981). Brock & Lane (1983) found plants growing in waters from 9 to 125 g TDS 1^{-1}. Burne *et al.* (1980) reported substantial photosynthetic ^{14}C uptake at salinities about three times those of the lakes from which plants were collected (i.e. 70 g TDS 1^{-1}). However, they also found that zygotes recovered from halite deposits in a salt lake germinated in sea water (35 g TDS 1^{-1}) but not at salinities of 53 or 70 g TDS 1^{-1}. These results and other field observations lead Burne *et al.* (1980) to conclude that the persistence of charophytes such as *L. papulosum* in saline waters depends upon occasional supply of fresh water. *L. papulosum* maintains its turgor approximately constant over a range of salinities from nearly fresh to twice seawater principally by varying internal osmotic pressure by altering concentrations of K^+ and Cl^- (Bisson & Kirst 1980a, b). Its vacuolar sucrose concentration is unusually high for an alga but is not correlated with salinity (Bisson and Kirst 1983).

Ctenocladus Borzi (Chlorophyceae) is a filamentous alga that lives in inland saline waters where sodium is the dominant cation; most collections are from North America and others are from Peru, Sicily and Siberia (Blinn & Stein 1970). It usually grows epilithically or as periphyton but can appear in the plankton. The nomenclature was confused by the introduction of a second generic name, *Lochmiopsis*, but the experimental work of Blinn and Stein (1970) indicates that the morphological differences occur in response to changes in environmental conditions and that *Ctenocladus* is a monotypic genus.

Ctenocladus circinnatus is abundant in eight Saskatchewan saline lakes that range from 17 to 214 g TDS 1^{-1} (Hammer *et al.* 1983), in Mono Lake, California, at a salinity of 90 g TDS 1^{-1} that varies little each year (Melack 1983, D. Herbst, personal communication) and in ponds in British Columbia that vary seasonally from *c.* 10 to greater than 100 g TDS 1^{-1} (Blinn & Stein 1970). In Borax Lake, California, the periphyton composed largely of *Cterocladus* can account for about

70% of the annual primary productivity (Wetzel 1964). Laboratory studies of *Ctenocladus circinnatus* by Blinn (1970, 1971) demonstrate that normal vegetative growth occurs when Na/Mg ratios are above 1.3 and pHs are above 8.8. However, Hammer *et al.* (1983) reported abundant *Ctenocladus* in a lake with a Na/Mg ratio of 0.8. *Ctenocladus* forms akinetes in response to unfavorable conditions, and these thick-walled cells germinate best at salinities below 1600 mOsm, between 9 and 26 °C, between pHs of 8.4 to 11.0 and in moderate light (Blinn 1971). The akinetes survive encrusted in salt and buried in snow and ice.

Submerged vascular plants

Very few taxa of angiosperms and no bryophytes grow and reproduce submerged in saline, inland waters (Ungar 1974, Brock 1981a, Brock & Lane 1983). Among those genera that do, i.e. *Zannichellia, Lepilaena, Althenia, Potamogeton* and *Ruppia* (all representatives of the Potamogetonaceae; den Hartog 1970, 1981), little is known of their ecology and physiology with the exception of *Ruppia*. While none can be definitely classified as obligate halophytes (cf. Barbour 1970), several species can tolerate wide ranges of salinity with sudden and large fluctuations and waters where the dominant salts are sulfates, carbonates or chlorides. Although the extensive literature on the biology of halophytes (e.g. Waisel 1972, Chapman 1960, Dainty 1979, Stewart *et al.* 1979, Jefferies *et al.* 1979, Winter 1979, Flowers *et al.* 1977, McMillan & Moseley 1967, Epstein 1969, Uphof 1941, Poljakoff-Mayber & Gale 1975, Haines & Dunn 1985, Sen & Rajpurohit 1982, Reimold & Queen 1974, Munns *et al.* 1983) seldom concerns submerged species of inland waters, much can be learned from it to help guide further study of these taxa.

The angiosperms that live submerged in lakes and have the greatest tolerance of high salinities are members of the genus *Ruppia*. According to Jacobs & Brock (1982) the genus belongs in the family Potamogetonaceae; Hutchinson (1959) had recognized the family Ruppiaceae as distinct from the Potamogetonaceae, and further confusion occurs at the species level. Especially troublesome is the taxon called *R. maritima* L. which subsumes much of the variability known in the genus (Verhoeven 1979).

Ruppia has a cosmopolitan but discontinuous distribution in primarily brackish, marine and inland saline waters (Verhoeven 1979, Brock 1981a). The genus is found on all the continents, on isolated islands and ranges from Tierra del Fuego to 68 °N. The large literature devoted to *Ruppia* can be accessed via the thorough review of western European species by Verhoeven (1979, 1980a, 1980b) and via the works of

Conover (1964), McMillan & Moseley (1967) and Brock (1982a, 1982b). Most of these studies deal with species living in coastal, marine environments. Much less is known about species of inland, hypersaline waters; a summary of this work follows.

Records of *Ruppia* living in waters with salinities that can exceed 50 g TDS 1^{-1} are listed in Table III. This compilation illustrates the

Table III. Some records of *Ruppia* in hypersaline, inland waters.

Species	Salinity g 1^{-1}	Major Salts	Location	Reference
R. maritima	saturated solution (ca. 390)	MgSO$_4$	Hot L. (= Epsom L.) Okanogan Co., Washington USA	St. John and Courtney 1924 Anderson 1958
	25—61	NaCl-Na$_2$CO$_3$ (high B)	Borax L. Lake Co., California USA	Wetzel 1964
R. cirrhosa	3—110	NaCl	Salinas, Carmargue France	Verhoeven 1979
R. tuberosa	13—230	NaCl	Lakes of southeastern South Australia	Brock 1982a
R. polycarpa	1—125	NaCl	Lakes of Western Australia	Brock and Lane 1983
R. megacarpa	10—150	NaCl	Lakes of Western Australia	Brock and Lane 1983

range of chemistries in which *Ruppia* occurs; the list of sites is not exhaustive. Experiments with concentrated seawater corroborate the tolerance of hypersaline conditions by *R. cirrhosa* (Verhoeven 1979) and *R. maritima* (McMillan & Moseley 1967; Bourn 1935). However, the *R. maritima* used by Bourn did not thrive or produce seeds at salinities of 52.5 g TDS 1^{-1}. In fact, the likelihood that some populations of the same species have different salinity tolerances confounds detailed comparison of these data; examination of ecotypic variation is definitely needed. Furthermore, salinities in the habitats where *Ruppia* occurs often vary, and it is possible that persistence of the plants depends on intermittent periods at the low end of the salinity ranges.

Few data are available to further characterize the habitats of *Ruppia*. Nutrient concentrations range from low to high, and pHs are alkaline and can exceed 10 (Verhoeven 1979). Waters with healthy stands of *Ruppia* are clear and shallow (Verhoeven 1979, Congdon & McComb

1979); experimental shading reduced *Ruppia cirrhosa* growth (Verhoeven 1980b).

Ruppia's tolerance of variable and high salinities must depend on its ability to osmoregulate. Brock (1981b) found that the fluid within the tissue of three Australian species (*R. megacarpa* Mason, *R. polycarpa* Mason, and *R. tuberosa* Davis) was hypertonic to the surrounding water. The organic solute, proline, usually increased as the ambient salinity increased. The proline content of *R. maritima* L. is also known to rise as a function of salinity (Stewart & Lee 1974). Brock (1981b) found 47.7 mg proline/g dry weight in *R. tuberosa* growing at 60 g TDS 1^{-1}; this finding is consistent with those of others for a variety of halophytes. Enzyme activities were not inhibited by such accumulations of proline in those halophytes tested by Stewart and Lee (1974). If Brock's assumptions that proline is confined to the protoplasm and that the cytoplasm occupies 10% of the cell volume are correct, the proline can account for an equivalent of 50% of the salinity of the external water. Proline would, therefore, make an important contribution to the osmoregulation by *Ruppia*.

Some of the inland lakes where *Ruppia* occurs are ephemeral, and the genus has a life history adapted to such conditions (Brock 1982a, 1982b, Brock & Lane 1983). *Ruppia* requires as few as three months of inundation to germinate, mature and reproduce. Growth of its rhizomes enables rapid colonization of neighboring wetland. Although the resistance to desiccation of the vegetative parts of *Ruppia* is low, the seeds do persist in dry and hypersaline locales (Verhoeven 1979) and some species produce resistent, asexual turions on the rhizomes. In addition, *Ruppia* disperses via several means. Vegetative fragments float well and root rapidly once they encounter a suitable substratum. *Ruppia* produces large numbers of seeds, and these are carried by currents, wind and waterfowl. The seeds remain viable after passage through birds' guts (Ridley 1930).

Primary productivity by *Ruppia maritima* in Borax Lake (California) was measured by Wetzel (1964) and compared with the contributions of the phytoplankton and periphyton to the lake's annual productivity. Wetzel enclosed plants growing in the lake within plexiglas chambers and introduced Na_2 $^{14}CO_3$; he assayed for incorporation of the ^{14}C as CO_2 after combustion of the aboveground plant at the end of each incubation. Carbon uptake rates ranged from 0.0 to 982 mg C m^{-2} d^{-1}. The *Ruppia* contributed only 7% of the total annual primary productivity in the lake. However, *Ruppia*'s growing season was only 75 days, and its maximal standing crop was only 64 grams ash-free dry weight m^{-2}. This standing crop is low compared to those summarized by Verhoeven (1980b) for largely marine habitats.

3. ALKALINE, SALINE WATERS

Alkaline (pH $>$ 10) and saline (TDS $>$ 30 g 1^{-1}) waters contain large quantities of sodium and carbonate (Eugster & Hardie 1978) and are often called soda lakes. Such solutions are usually derived from the weathering of igneous rocks followed by evaporative concentration (Garrells & MacKenzie 1967). In rare circumstances carbonatite lavas contribute sodium and carbonate (von Knorring & duBois 1961, Dawson 1962). The pH of soda lakes is usually between 10 and 11 and is very well buffered; such pHs are at the upper extreme of the pHs of natural waters (Baas Becking et al. 1960). Soda lakes are distributed worldwide (McCarraher 1972) and are especially well represented and studied in eastern Africa (Melack 1981, Vareschi et al. 1981, Livingstone & Melack, 1984). Many of these lakes are shallow and fluctuate widely in salinity. Alkalinities in Lake Nakuru (Kenya), for example, have ranged from 122 meq 1^{-1} in 1969 (Mclack & Kilham 1974) to 1440 meq 1^{-1} in 1961 (Talling & Talling 1965).

The flora of alkaline, saline lakes is more depauperate than other waters of comparable salinity. No submerged angiosperms or macroalgae occur in soda lakes. The phytoplankton and benthic mats (see Section 2) are primarily cyanobacteria, diatoms and chlorophytes (Jenkin 1936, Hecky & Kilham 1973, Iltis 1974, Melack 1976). An especially striking feature of soda lakes throughout Africa is persistent, almost unialgal populations of the cyanobacterium *Spirulina platensis* (Norst.) Geitl. (Iltis 1969, 1971, Melack 1979a, 1979b, Vareschi 1982). *S. platensis* is known from lakes ranging in salinity from 8.5 to 170 g TDS 1^{-1} and attains maximal development between 22 and 62 g TDS 1^{-1} (Iltis 1968).

The nearly unialgal blooms of *Spirulina platensis* can approach the highest photosynthetic rates known for natural communities (Talling et al. 1973, Melack & Kilham 1974, Melack 1979, 1981). Talling et al. (1973) attributed these exceptional photosynthetic rates to high algal biomass in the euphotic zone and a high photosynthetic capacity (i.e. mg O_2/mg chlorophyll *a* h). These productive populations of *S. platensis* can support huge flocks of Lesser Flamingoes (Vareschi 1978) and large quantities of zooplankton (Vareschi & Vareschi 1984) and fish (Vareschi 1979). The combination of consumption by grazers and algal respiration often nearly balances daily production and leads to surprisingly constant algal abundance for months and, sometimes years (Vareschi 1982, Vareschi & Jacobs 1984). Such constancy can, however, be abruptly interrupted for reasons yet to be understood (Melack 1976, 1979b, Vareschi 1982).

The recent discovery that *S. platensis* has a long history of use as a human food (Leonard & Compère 1967) and an extraordinary protein content (Clément *et al.* 1967) has sparked research on its mass culture, physiology, ultrastructure and biochemistry (e.g. Zarrouk 1966, Ogawa & Terui 1971, Remy *et al.* 1976, Arai *et al.* 1976, Chernyad'ev *et al.* 1978, Ogawa *et al.* 1979, Aiba *et al.* 1977, Eykelenburg 1977, 1978, 1979, Matty & Smith 1978, Ogawa *et al.* 1978, Owers-Narhi *et al.* 1979, Masaki *et al.* 1979, Titu *et al.* 1980, Vincenzini *et al.* 1980, Pande *et al.* 1981, Devi *et al.* 1981, Sharma & Singh 1981, Nigam *et al.* 1981, Rao & Argos 1981, Gatesoupe & Robin 1981, Riccardi *et al.* 1981). Some of this biochemical and physiological work is motivated by the availability of large amounts of *S. platensis* and some by the alga's exceptional photosynthetic activity (e.g. Robinson *et al.*1982). Mass culture has progressed beyond the laboratory and now is represented by several commercial, outdoor operations (e.g. Durand-Chastel 1980, Venkataraman *et al.* 1980, Soong 1980, Tel-Or 1980). Although relevant to an understanding of the biology of *S. platensis*, much of this work does not pertain to the ecological conditions experienced by natural populations.

Diatoms, chlorophytes, and other cyanobacteria can be important in temperate and tropical soda lakes. In the alkaline, saline waters of eastern Africa *Nitzschia frustulum* (Kütz) Grun., *Nitzschia sigma* (Kütz) W. Sm., *Rhopalodia gibberula* (Ehr.) O. Müll., and *Anomoeneis spaerophora* (Ehr.) Pfitz are numerically dominant in lakes that exceed $30 \text{ g TDS } 1^{-1}$ (Hecky & Kilham 1973). The association of particular diatom assemblages with particular ranges of salinity is a valuable aid to tracing the history of a lake with diatoms preserved in the lake's sediments (Hecky 1971, Holdship 1976). The nitrogen fixing cyanobacterium *Anabaenopsis arnoldii* Aptekari can be an especially important component in African soda lakes (Melack 1976, Rhodes 1981). In Mono Lake (California) the chlorophyte *Coccomyxa* sp. numerically dominates the phytoplankton during all seasons and at all depths (Melack 1983). An unusual feature of the annual cycle is the persistence of an abundant population of *Coccomyxa* in the cold, dimly lighted hypolimnion throughout the summer and its bloom in the winter throughout the lake.

4. HIGH TEMPERATURE WATERS

Temperatures exceeding 50 °C are sufficiently high to greatly reduce the number of species of aquatic plants (Brock 1978). Only prokaryotes live above 60—62 °C, and no photosynthetic cyanobacteria live above *c.* 73 °C. The most thermophilic nitrogen fixer grows up to *c.* 64 °C. With the exception of the peculiar organism *Cyanidium caldarium* (see below), no other eukaryotes grow above 50 °C other than fungi (Tansey

& Brock 1972). One species of chlorophyte, *Mougeotia*, has been found living up to 47 °C (Stockner 1967) and mosses reach similar maxima.

Natural high temperature, aquatic habitats are primarily geothermally heated (Waring 1965). In some circumstances solar heating can raise temperatures above 50 °C in chemically stratified or shallow waters (Kalecsinsky 1901, Eckstein 1970). Coincident with high temperatures can be other environmental extremes. The temperatures themselves are often remarkably constant but can vary considerably (Brock 1978). The pH of thermal springs varies from about 1 to 10.5 and many springs are acidic with pHs from 2—4 (Brock 1971). Potentially toxic elements such as fluoride, arsenic, aluminum, copper and sulfide can reach high concentrations and do restrict the distribution of some species (Castenholz 1976, 1977, Castenholz & Wickstrom 1975); however, hypersaline springs are often not hot (Brock 1970). Phosphate can be unusually high while inorganic nitrogen varies widely and can be very low (Brock 1970).

Recent reviews of the biology of organisms living at high temperatures (e.g. Brock 1967a, 1970, 1978, Castenholz 1969, 1973, Castenholz & Wickstrom 1975, Tansey & Brock 1978, Arango 1981) provide excellent perspectives on the ecology of aquatic plants in these extreme environments. Therefore, only selected highlights will be presented here.

Large differences in algal standing crop are associated with the thermal gradients of hot springs. For example, in Yellowstone National Park (USA) Brock & Brock (1966) and Brock (1967b) observed a peak algal biomass between 55 ° and 60 °C with lower abundances near the upper limit for cyanobacteria (73—75 °C) and a sharp decline below 40 °C where grazers become important (Brock *et al.* 1969). In contrast, primary productivity, which is often very high, is greater in regions with temperatures below and above 55 to 60 °C than in the zones with the maximum biomass (Stockner 1968, Brock 1970, Wiegert & Fraleigh 1972).

The two genera of cyanobacteria with highest thermal limits are *Mastigocladus* and *Synechococcus* (Brock 1978). *Mastigocladus laminosus* (Ag.) Cohn (Nostocales) occurs world-wide in thermal waters as warm as 64 °C (Castenholz 1969). It is the most thermophilic nitrogen fixing cyanobacterium (Stewart 1970). *Synechococcus lividus* Copeland lives in hot springs with temperatures as high as 73 °C (see below) and is found in New Zealand (Castenholz 1976), North America, and Japan, but seems to be absent from Iceland (Castenholz 1969). Neither the ultrastructure (Edwards *et al.* 1968, Holt & Edward 1972), nor the phycobilisomes (Edwards & Gantt 1971) differ from non-thermal cyanobacteria. Peary & Castenholz (1964) reported four distinct, genetically determined, temperature strains of *S. lividus* with growth optima at about 45, 50, 55 and 65 °C. Growth rates vary considerably among the strains; the lowest temperature strain grew about five times faster than the highest temperature strain. Meeks & Castenholz (1971) found that the highest temperature strain had minimum and maximum temperatures for growth of 54 °C and 72 °C. Furthermore, field measurements of photosynthesis (Brock 1967b) and growth (Brock & Brock 1968) also delineate an upper bound near 72 °C. However, Brock (1978) reports that growth rates were greater at 68 °C than at 57 °C and attributes this difference from the laboratory studies of Peary & Castenholz (1964) and Meeks & Castenholz (1971) to the lack of light limitation at the higher temperature because of a thinner mat.

A variety of stromatolitic structures growing in hot springs may help interpret Precambrian microfossils (Brock 1978). In Yellowstone National Park (USA) recent research has concentrated on two types of stromatilites: (1) flat, laminated mats produced by the photosynthetic bacterium *Chloroflexus* and the cyanobacterium *Synechococcus*, and (2) conical formations produced largely by species of *Phormidium* (e.g. Walter *et al.* 1972, Doemel & Brock 1974, 1977, Weller *et al.* 1975, Brock 1978). The *Synechococcus* is restricted to the upper 0.2 to 1 mm of the mats, and the growth optimum of the mats is between 52 and 56 °C. The major biological and geological implication of the formation of laminated structures primarily by the photosynthetic activity of the bacterium *Chloroflexus* is that microfossils such as those in the Precambrian Gunflint cherts need not have been formed by oxygenic cyanobacteria. The conical structures form by actively growing *Phormidium* in waters between 32 °C and 59 °C with pHs of 7 to 9.

Cyanidium caldarium is the only photosynthetic organism living in habitats with pHs less than 5 and temperatures above 40 °C, Brock (1978) devotes a whole chapter to *Cyanidium* and reviews its taxonomy, biochemistry, ultrastructure, and ecology. The taxonomic status of *Cyanidium* is confused by its peculiar cellular characteristics. The current consensus places the genus in Rhodophyta (Chapman 1974, DeLuca & Moretti 1983), and the division of the genus into several species has been proposed (Merola *et al.* 1982). *Cyanidium* has true chloroplasts, nucleus and mitochondria; it contains chlorophyll *a*, *c*-phycocyanin, β-carotene and several other carotenoids; its storage product is polyglucoside, and it has a cell wall (Allen 1959, Brock 1978). The temperature optimum of *ca.* 45 °C for growth and photosynthesis and a temperature maximum for growth and existence in nature of *c.* 57 °C was established by Doemel & Brock (1970, 1971) from field collections and measurements and from experiments with pure cultures in the laboratory. The pH optimum is fairly broad (2—4), but the organism can live in $1 N$ H_2SO_4 (Allen 1959). No growth occurs at pH 5 or above but photosynthesis can occur at pH 7 (Doemel & Brock 1971). *Cyanidium* can grow under bright sunlight, adapts to dim light, and can grow heterotrophically (Allen 1959, Brock 1978). It can assimilate ammonia, nitrate, urea, casein hydrolysate and casein as nitrogen sources. *Cyanidium* is distributed world-wide in appropriate habitats and has been collected and confirmed by culture in the United States, Italy, Iceland, New Zealand, Japan, and confirmed only microscopically in Indonesia, El Salvador and Dominica (Brock 1978). It is common in warm acid soils as well as acid springs. *Cyanidium* forms dense mats that are sometimes associated with bacteria but can be almost purely *Cyanidium*.

Aquatic habitats with pH less than 3 include both thermal and non-thermal waters, and recent interest in acidification of natural waters (see citations in Introduction) and in acidophilic organisms (Langworthy 1978) has increased investigation of naturally acidic habitats. Brock (1973, 1978) presents convincing evidence derived from field collections, cultures and literature citations that cyanobacteria do not live at pH values below 4. Eukaryotic algae and mosses (*Sphagnum*) can be abundant and fairly diverse in such acidic waters (Brock 1978). Four major divisions of algae are represented: Rhodophyta (*Cyanidium caldarium*), Bacillariophyta (e.g. *Pinnularia, Eunotia*), Euglenophyta (*Euglena mutabilis*) and Chlorophyta (Volvocales, *Chlamydomonas*; Chlorococcales, *Chlorella*; Ulothrichales, *Ulothrix* and Zygnematales, *Zygogonium*). *Zygogonium* mats are common in acidic waters (pH 2 to 3) of Yellowstone National Park (Lynn & Brock 1969). Photosynthesis by this alga has a temperature optimum of 25 °C and a broad pH optimum from 1 to 5. Further information about lakes with naturally low pH is available in Ueno (1958), Brock & Brock (1970), Satake & Saijo (1974).

5. LOW TEMPERATURE WATERS

Among the aquatic plants known from cold water (cf. Hutchinson 1967, 1975, Schulthorpe 1967, Kalff 1970, Tilzer 1972) the algae living in snow and ice deserve consideration here because of their exceptional ability to grow and reproduce near 0 °C and their tolerance of repeated freezing and thawing while in a vegetative stage. These cryophilic algae can impart striking blue, green and especially red color to snow fields and glaciers in mountains and polar regions and are the subject of a venerable literature dating from Aristotle (cf. Kol 1968, Fott 1970). This diverse flora is largely represented by chlorophytes but does include chrysophytes, euglenoids, pyrrhophytes, cryptomonads, xanthophytes, diatoms and cyanobacteria (Hoham 1980). Kol (1968) describes 110 taxa of algae from snow and ice, but the taxonomy of many of these species is uncertain. Recent examination of the developmental stages and morphological variation in zygotes has reduced the number of previously described species (e.g. Hoham 1975a; Hoham & Mullet 1977; Hoham *et al.* 1979) and further elucidation of life histories is needed. For example, the volvocalean *Chloromonas nivalis* (Chod.) Hoh. & Mull., formerly *Chloromonas cryophila*, includes stages that resemble *Carteria* Dies. and *Scotiella nivalis* (Chod.) Fritsch, *S. cryophila* Chod. and *S. polyptera* Fritsch (Hoham & Mullet 1977, 1978). Fukushima's (1963) and Kol's (1968) comprehensive treatments of geographical and ecological aspects of cryophilic algae provide background for the modern studies of ultrastructure, physiology, and ecology that will be highlighted here.

The Chlamydomonadaceae (Chlorophyta) are the major, cryophilic algae in many places; Hoham (1980) has recently reviewed this group. Chlamydomonads are distributed worldwide from near sea level in polar regions to over 5000 m in mountainous areas. Algal blooms occur during spring and early summer as a combination of meltwater and increased availability of nutrients and light (e.g. Curl *et al.* 1972, Hoham 1975a, Hoham & Mullet 1977) initiate germination of overwintering zygospores at the snow-soil boundary or at the new snow-old snow interface in persistent snowfields. The zoospores then swim toward the surface of the snow and appear as conspicuous blooms only a few days after germination. These vegetative cells can tolerate hours of subfreezing temperatures each day and continue to grow during intermittent periods with meltwater. The zoospores divide to form asexual or sexual daughter cells, and syngamy can occur within less than one week. The factors causing syngamy are not known. The new zygotes are metabolically active (see below) at least initially but do pass the year in the snow or soil until germinating the following spring. Meiosis can be expedited by freezing the zygotes after several months of dormancy. Melting snowbanks must persist for sufficient time to allow completion of the life cycle each year if these algae are to remain in the same locality.

Chlamydomonas nivalis (Bau.) Wille is often the dominant alga in red snow (Garric 1965, Kol 1968, Thomas 1972). The red color is caused by the xanthophyll, astaxanthin (Viala 1966) and other ketocarotenoids (Czygan 1970). After nutrient enrichment and under low light, the red carotenoids in the resting spores decrease while chlorophyll and other carotenes increase (Czygan 1970). The relative influence of light and nutrients on the pigment composition of *C. nivalis* and other snow algae needs further analysis.

Zygotes of *C. nivalis* are roughly spherical, from 10—50 μ in diameter and may have a mucilaginous covering with bacteria and debris attached on the outside of the thick cell wall (Weiss 1983). This algal-bacterial association was found in all samples examined from Yellowstone Park, Wyoming, and throughout the Sierra Nevada, California. The bacteria were lacking in samples from nearby snowbanks without *C. nivalis*.

While cultures of *C. nivalis* grew best at 4 °C (Czygan 1970), field measurements of ^{14}C uptake indicated substantial rates at temperatures as low as −3 °C but maximal rates at 10 °C or 20 °C (Mosser *et al.* 1977). The few measurements of photosynthetic activity ($\mu g\,C\,mm^{-3}\,h^{-1}$) by natural populations at about 0 °C that included *C. nivalis* are low and range from 0.05—0.97 (Fogg 1967; Mosser *et al.* 1977) to 5.7—34.2 (Thomas 1972). Thomas's values may be high because he overestimated the CO_2 content of the snow. In melted snow and subzero snow (*c.* −4.7 °C), photosynthetic rates were substantially reduced. Mosser *et al.* (1977) observed no reduction in ^{14}C uptake under full sunlight in mid-summer at 2900—3350 m above sea level. Under similar conditions Thomas (1972) found slightly less uptake at the surface than at a depth of 2—6 cm in the snow (i.e. *c.* 20—50% of full sunlight). Cryophilic algae such as *C. nivalis* are clearly well adapted to cold, high light environments, and further investigation of their photosynthetic activity should be rewarding.

The optimum temperature for growth of *Chloromonas pichinchae* (Lagerh.) Wille (Chlorophyta, Volvocales) is 1 °C, the coldest optimum yet discovered for a cryophilic alga (Hoham 1975b). *C. pichinchae* continues growing well at 5 °C but at 10 °C cells divide slowly and do not separate. If placed in culture medium from 1 to 5 °C after exposure to 10 °C medium, this palmelloid condition is replaced by single biflagellate cells. If rapidly growing cultures are subjected to alternating periods of 24 hours of −5 °C and then 5 °C, large lipid bodies are present in the vegetative cells after one week.

In the Stuart Range of Washington (U.S.A.), *C. pichinchae* formed green patches near the snow's surface in shaded areas and distinct bands to depths of 25 cm but was not found in exposed areas above timberline (Hoham 1975a). Although lacking eyespots, *C. pichinchae* expresses phototaxis; it moves toward the surface at dawn and dusk and to depths of 10—15 cm at midday (Hoham 1975a). The role of different wavelengths or intensities of light in these movements is not known.

Dissolved solutes, dust, and forest litter are potential sources of nutrients for snow algae (Hoham 1976, Fjerdingstad *et al.* 1978). Growth of axenic cultures of *C. pichinchae* enriched with extracts of coniferous leaves, bark, and pollen is enhanced (Hoham 1978). *C. pinchinchae* also grew better in snowmelt with 18 μg PO_4-P 1^{-1} collected under a coniferous canopy than in meltwaters from open areas at higher elevations with less phosphorus. Axenic, nutrient-depleted cultures of *C. pichinchae* showed enhanced growth as a function of increasing concentrations of NH_4Cl, L-glutamic acid, K_2HPO_4, and $Na_2C_3H_5[OH]_2PO_4$ (Hoham 1980).

Cryophytic algae are the primary producers in a fascinating eco-system in melting snowbanks. Bacteria and a variety of grazers such as protozoans, rotifers, and insects (Hoham 1980) also occur in the snow. Examination of the supply of nutrients and light to the algae, the trophic interactions and responses of the biota to atmospheric deposi-tion of trace metals and strong acids are likely to be fruitful endeavors.

6. CONCLUSION

Extreme aquatic environments provide the ecologist with an excellent opportunity to study populations, communities and ecosystems in an integrated manner. As is evident from this review, considerable infor-mation about the biology of the aquatic plants that live under extreme conditions is available; much remains to be discovered. A few inte-grated analyses of the ecology of hypersaline, sodic or hot waters have been conducted (see pertinent sections above). More effort along similar lines would be very rewarding.

Examination of the adaptation and persistence of organisms under extreme conditions has a special relevance in today's industrialized world. Human activities are drastically altering habitats and introducing exotic chemicals to many places. The ultimate, man-caused disturbance of the earth, a nuclear war, is not an impossibility. The aftermath of such a catastrophe would include major reduction in sunlight, sub-freezing temperatures, very acidic and toxic precipitation and greatly elevated ultraviolet and ionizing radiation (Turco *et al.* 1983, Ehrlich *et al.* 1983). The depauperate biological communities of extant, extreme environments provide some perspective on the kinds of organisms that may withstand nuclear war and severe industrial pollution.

The extreme conditions in which aquatic plants live are truly amazing, and one is reminded of Charles Darwin's comment in *Voyage of H.M.S. Beagle*, "Well may we affirm, that every part of the world is habitable." In fact, bounds do exist (Baas Becking *et al.* 1960, Brock 1967; see above), and scholarly debate has centered around the biological definition of extreme environments (e.g. Henderson 1913, Vallentyne 1963, Kushner 1981, Alexander 1976). In a provocative essay Vallentyne (1963) argues that "The fact that most living species conform physiologically and ecologically to average Earth conditions should not be taken to indicate any inherent environmentally based physicochemical conservatism of living matter. Adaptation has taken

364

place." Brock (1968) takes issue with this premise and states that essential constituents of cells such as DNA and ATP are stable only within certain physicochemical limits and that biological evolution is therefore constrained. The recent, albeit controversial, evidence for bacteria living at 250 °C requires reevaluation of this argument (Baross & Deming 1983, Trent et al. 1984, White 1984). A global homeostasis maintained in part by the biosphere, i.e., the 'Gaia' hypothesis (Lovelock & Margulis 1974), is consistent with Brock's perspective. Yet, one is still intrigued by the exotic, and the adaptability of species does excite one's imagination.

ACKNOWLEDGEMENTS

I thank L. Meeker for bibliographic assistance, D. Mustard and M. Fujii for typing the text, and S. MacIntyre, J. Sickman, A. Gibor, and B. Prezelin for critical reading of the manuscript. Financial support was provided by NSF grant DEB 81—11398.

REFERENCES

Aasen, A. J., Eimhjellen, K. E. & Liaaen-Jensen, S. (1969) An extreme source of β-carotene. Acta chem. Scand. 23: 2544—2545.

Aiba, S. & Ogawa, T. (1977) Assessment of growth yield of a blue-green alga, Spirulina platensis, in axenic and continuous culture. J. Gen. Microbiol. 102: 179—182.

Alexander, M. (1976) Natural selection and the ecology of microbial adaptation in a biosphere. In: Heinrich, M. R. (ed.), Extreme environments. Academic Press, New York, pp. 3—25.

Allen, M. B. (1959) Studies with Cyanidium caldarium, an anomalously pigmented chlorophyte. Arch. Mikrobiol. 32: 270—277.

Amelunxen, R. E. & Murdock, A. L. (1978) Microbial life at high temperature: mechanisms and molecular aspects. In: Kushner, D. J. (ed.), Microbial life in extreme environments. Academic Press, London, pp. 217—278.

Anderson, G. C. (1958) Some limnological features of a shallow, saline meromictic lake. Limnol. Oceanogr. 3: 259—270.

Angino, E. E., Armitage, K. B. & Tash, J. L. (1964) Physicochemical limnology of Lake Bonney, Antarctica. Limnol. Oceanogr. 9: 207—217.

Arai, S., Yamashita, M. & Fujimaki, M. (1976) Enzymatic modification for improving nutritional qualities and acceptability of proteins extracted from photosynthetic microorganisms Spirulina maxima and Rhodopseudomonas capsulatus. J. Nutr. Sci. Vitaminol. 22: 447—456.

Arango, M. (1981) Responses of microorganisms to temperature. In: Lange, O. L., Nobel, P. S., Osmond, C. B. & Ziegler, H. (eds.), Physiological plant ecology. I. Responses to the physical environment. Springer-Verlag, Berlin, pp. 339—369.

Baas Becking, L. G. M., Kaplan, I. R. & Moore, D. (1960) Limits of the natural environment in terms of pH and oxidation-reduction potentials. J. Geol. 68: 243—284.

Baker, J. M. (1970) The effects of oils on plants. Environ. Pollut. 1: 27—44.

Baker, J. M. (1979) Responses of salt marsh vegetation to oil spills and refinery effluents. *In*: Jefferies, R. L. & Davy, A. J. (eds.), Ecological processes in coastal environments. Blackwell Scientific Publications, Oxford, pp. 529—542.

Barbour, M. G. (1970) Is any angiosperm an obligate halophyte? Am. Midl. Nat. 84: 105—120.

Barica, J. & Mur, L. (eds.) (1980) Hypertrophic ecosystems. Dr. W. Junk Publ., The Hague, 348 pp.

Baross, J. A. & Morita, R. Y. (1978) Microbial life at low temperatures: ecological aspects. *In*: Kushner, D. J. (ed.), Microbial life in extreme environments. Academic Press, London, pp. 9—71.

Baross, J. A. & Deming, J. W. (1983) Growth of 'black smoker' bacteria at temperatures of at least 250 °C. Nature 303: 423—426.

Bauld, J. 1981. Occurrence of benthic microbial mats in saline lakes. Hydrobiologia 81: 87—111.

Bayly, I. A. E. & Williams, W. D. (1966) Chemical and biological studies on some saline lakes of south-east Australia. Aust. J. mar. Freshwat. Res. 17: 177—228.

Beadle, L. C. (1943) An ecological survey of some inland saline waters of Algeria. J. Linn. Soc. (Zool.) 43: 218—242.

Beadle, L. C. (1959) Osmotic and ionic regulation in relation to the classification of brackish and inland saline waters. Arch. Oceanogr. Limnol. (Roma), Suppl. 11: 143—151.

Bednarz, T. (1981) The effect of pesticides on the growth of green and blue-green algae cultures. Acta Hydrobiol. 23: 155—172.

Ben-Amotz, A. (1975) Adaptation of the unicellular alga *Dunaliella parva* to a saline environment. J. Phycol. 11: 50—54.

Ben-Amotz, A. & Avron, M. (1972) Photosynthetic activities of the halophilic alga *Dunaliella parva*. Plant. Physiol. 49: 240—243.

Ben-Amotz, A. & Avron, M. (1973) The role of glycerol in the osmotic regulation of the halophilic alga *Dunaliella parva*. Plant Physiol. 51: 875—878.

Ben-Amotz, A. & Avron, M. (1980) Osmoregulation in the halophilic algae *Dunaliella* and *Asteromonas*. *In*: Rains, D. W. and Valentine, R. C. (eds.), Genetic engineering of osmoregulation. Plenum Press, New York, pp. 91—99.

Ben-Amotz, A. & Avron, M. (1980) Glycerol, β-carotene and dry algal meal production by commercial cultivation of *Dunaliella*. *In*: Shelef, G. & Soeder, C. J. (eds.), Algal biomass. Elsevier/North Holland Biomedical Press, Amsterdam, pp. 603—610.

Billingham, J. (ed.) (1981) Life in the universe. MIT Press, Cambridge, Mass, 461 pp.

Bisson, M. A. & Kirst, G. O. (1980a) *Lamprothamnium*, a euryhaline charophyte. I. Osmotic relations and membrane potential at steady state. J. exp. Bot. 31: 1223—1235.

Bisson, M. A. & Kirst, G. O. (1980b) *Lamprothamnium*, a euryhaline charophyte. II. Time course of turgor regulation. J. exp. Bot. 31: 1237—1244.

Bisson, M. A. & Kirst, G. O. (1983) Osmotic adaptations of charophyte algae in the Coorong, South Australia and other Australian lakes. Hydrobiologia 105: 45—51.

Blinn, D. W. & Stein, J. R. (1970) Distribution and taxonomic reappraisal of *Ctenocladus* (Chlorophyceae: Chaetophorales). J. Phycol. 6: 101—105.

Blinn, D. W. (1970) The influence of sodium on the development of *Ctenocladus circinnatus* Borzi (Chlorophyceae). Phycologia 9: 49—54.

Blinn, D. W. (1971) Autecology of a filamentous alga, *Ctenocladus circinnatus* (Chlorophyceae), in saline environments. Can. J. Bot. 49: 735—743.

Borowitzka, L. J. (1981) The microflora: adaptations to life in extreme environments. Hydrobiologia 81: 33—46.

Borowitza, L. J. & Brown, A. D. (1974) The salt relations of marine and halophilic species of the unicellular green alga, *Dunaliella*. Arch. Microbiol. 96: 37—52.

Borowitza, L. J., Kessly, D. S. & Brown, A. D. (1977) The salt relations of *Dunaliella*. Further observations on glycerol production and its regulation. Arch. Microbiol. 113: 131—138.

Bourn, W. S. (1935) Sea-water tolerance of *Ruppia maritima* L. Contr. Boyce Thompson Inst. 7: 240—255.

Brock, M. A. (1981a) The ecology of halophytes in the south-east of South Australia. Hydrobiologia 81: 23—32.

Brock, M. A. (1981b) Accumulation of proline in a submerged aquatic halophyte, *Ruppia* L. Oecologia 51: 217—219.

Brock, M. A. (1982a) Biology of the salinity tolerant genus *Ruppia* L. in saline lakes in South Australia. I. Morphological variation within and between species and eco-physiology. Aquat. Bot. 13: 219—248.

Brock, M. A. (1982b) Biology of the salinity tolerant genus *Ruppia* L. in saline lakes in South Australia. II. Population ecology and reproductive biology. Aquat. Bot. 13: 249—268.

Brock, M. A. & Lane, J. A. K. (1983) The aquatic flora of saline wetlands in Western Australia in relation to salinity and permanence. Hydrobiologia 105: 63—76.

Brock, M. L., Wiegert, R. G. & Brock, T. D. (1969) Feeding by *Paracoenia* and *Ephydra* (Diptera: Ephydridae) on microorganisms of hot springs. Ecology 50: 192—200.

Brock, T. D. (1967a) Life at high temperature. Science 158: 1012—1019.

Brock, T. D. (1967b) Relationship between standing crop and primary productivity along a hot spring thermal gradient. Ecology 48: 566—571.

Brock, T. D. (1967c) Microorganisms adapted to high temperatures. Nature 214: 882—885.

Brock, T. D. (1970) High temperature systems. Ann. Rev. Ecol. Syst. 1: 191—220.

Brock, T. D. (1971) Bimodal distribution of pH values of thermal springs of the world. Bull. Geol. Soc. Amer. 82: 1393—1394.

Brock, T. D. (1973) Lower pH limit for the existence of blue-green algae: evolutionary and ecological implications. Science 179: 480—483.

Brock, T. D. (1975) Salinity and the ecology of *Dunaliella* from the Great Salt Lake. J. gen. Microbiol. 89: 285—292.

Brock T. D. (1976) Halophilic-blue-green algae. Arch. Microbiol. 107: 109—111.

Brock, T. D. (1978) Thermophilic microorganisms and life at high temperature. Springer-Verlag, New York, 465 pp.

Brock, T. D. & Brock, M. L. (1966) Temperature optima for algal development in Yellowstone and Iceland hot springs. Nature 209: 733—734.

Brock, T. D. & Brock, M. L. (1968) Measurement of steady-state growth rates of a thermophilic alga directly in nature. J. Bact. 95: 811—815.

Brock, T. D. & Brock, M. L. (1970) The algae of Waimangu Cauldron (New Zealand): distribution in relation to pH. J. Phycol. 6: 371—375.

Brown, A. B. (1976) Microbial water stress. Bact. Rev. 40: 803—846.

Brown, A. D. (1978) Compatible solutes and extreme water stress in eukaryotic microorganisms. Adv. Microb. Phys. 17: 181—242.

Brown, A. D. (1983) Halophilic procaryotes. *In*: Lange, O. L., Nobel, P. S., Osmond, C. B. & Ziegler, H. (eds.), Physiological plant ecology. III. Responses to the chemical and biological environments. Springer-Verlag, Berlin, pp. 137—162.

Brown, A. D. & Borowitzka, L. J. (1979) Halotolerance of *Dunaliella*. *In*: Levandowsky, M. & Hunter, S. H. (eds.), Physiology and biochemistry of protozoa. Vol. 1. Academic Press, New York, pp. 139—190.

Burne, R. V., Bauld, J. & DeDecker, P. (1980) Saline lake charophytes and their geological significance. J. Sed. Pet. 50: 281—293.

Butcher, R. W. (1959) An introductory account of the smaller algae of British coastal waters. Fish. Invest. Minist. Agric. Fish. Food (Gr. Brit.) Series IV, Part I: 21—24.

Campbell, P. J. (1978) Primary productivity of a hypersaline Antarctic lake. Aust. J. mar. Freshwat. Res. 29: 717—724.

Caplan, S. R. & Ginzburq, M. (1978) Energetics and structure of halophilic microorganisms. Elsevier North-Holland Biomedical Press, Amsterdam, 672 pp.

Castenholz, R. W. (1969a) The thermohilic cyanophytes of Iceland and the upper temperature limit. J. Phycol. 5: 360—368.

Castenholz, R. W. (1969b) Thermophilic blue-green algae and the thermal environment. Bact. Rev. 33: 476—504.

Castenholz, R. W. (1973) Ecology of blue-green algae in hot springs. In: Carr, N. G. & Whitton, B. A. (eds.), The biology of blue-green algae. Blackwell Sci. Publ., Oxford, pp. 379—414.

Castenholz, R. W. (1976) The effect of sulfide on the blue-green algae of hot springs. I. New Zealand and Iceland. J. Phycol. 12: 54—68.

Castenholz, R. W. (1977) The effect of sulfide on the blue-green algae of hot springs. II. Yellowstone National Park. Microb. Ecol. 3: 79—105.

Castenholz, R. W. & Wickstrom, C. E. (1975) Thermal streams. In: Whitton, B. A. (ed.), River ecology. Univ. of California Press, Berkeley, pp. 264—285.

Chapman, D. J. (1974) Taxonomic position of Cyanidium caldarium. The Porphyridiales and Goniotrichales. Nova Hedwig. 25: 673—682.

Chapman, V. J. (1960) Salt marshes and salt deserts of the world. Interscience Pub., New York, 392 pp.

Chernyad'ev, I. I., Terekhova, I. V. Al'bitskaya, O. N. Goronkova, O. I. & Doman, N. G. (1978) Illumination as a factor in the dynamic regulation of photosynthetic metabolism of carbon in Spirulina. Fiz. Rast 25: 815—820.

Christian, J. H. B. & Waltko, J. A. (1962) Solute concentrations within cells of halophilic and non-halophilic bacteria. Biochim. biophys. Acta 65: 506—508.

Clement, G., Giddey, C. and Menzi, R. (1967) Amino acid composition and nutritive value of the alga Spirulina maxima. J. Sci. Fd. Agric. 18: 497—501.

Cloern, J. E., Cole, B. E. & Oremland, R. S. (1983) Autotrophic processes in meromictic Big Soda Lake, Nevada. Limnol. Oceanogr. 28: 1049—1061.

Cohen, Y., Jørgensen, B. B. Padan, E. & Shilo, M. (1975a) Sulfide dependent anoxygenic photosynthesis in the cyanobacterium Oscillatoria limnetica. Nature 257: 489—492.

Cohen, Y., Padan, E. & Shilo, M. (1975b) Facultative anoxygenic photosynthesis in the cyanobacterium Oscillatoria limnetica. J. Bacteriol. 123: 855—861.

Cohen, Y., Krumbein, W. E. Goldberg, M. & Shilo, M. (1977a) Solar Lake (Sinai) 1. Physical and chemical limnology. Limnol. Oceanogr. 22: 597—608.

Cohen, Y., Krumbein, W. E. & Shilo, M. (1977b) Solar Lake (Sinai) 2. Distribution of photosynthetic microorganisms and primary production. Limnol. Oceanogr. 22: 609—620.

Cohen, Y., Krumbein, W. E. & Shilo, M. (1977c) Solar Lake (Sinai) 3. Bacterial distribution and production. Limnol. Oceanogr. 22: 621—634.

Comin, F. A., Alonso, M., Lopez, P. & Comelles, M. (1983) Limnology of Gallocanta Lake, Aragon, northeastern Spain. Hydrobiologia 105: 207—221.

Congdon, R. A. & McComb, A. J. (1979) Productivity of Ruppia : seasonal changes and dependence on light in an Australian estuary. Aquat. Bot. 6: 121—132.

Conover, J. T. (1964) The ecology, seasonal periodicity, and distribution of benthic plants in some Texas lagoons. Bot. Marina 7: 4—41.

Craigie, J. S. & McLachlan, J. (1964) Glycerol as a photosynthetic product in *Dunaliella tertiolecta* Butcher. Can. J. Bot. 47: 777—778.

Curl. H. Jr., Hardy, J. T. & Ellermeier, R. (1972) Spectral absorption of solar radiation in alpine snowfields. Ecology 53: 1189—1194.

Czygan, F.-C. (1970) Blutregen and Blutschnee: Stickstoffmangelzellen von *Haematococcus pluvialis* und *Chlamydomonas nivalis.* Arch. Mikrobiol. 74: 69—76.

Dainty, J. (1979) The ionic and water relations of plants which adjust to a fluctuating saline environment. *In*: Jefferies, R. L. & Davy, A. J. (eds.), Ecological processes in coastal environments. Blackwell Scientific Publ., Oxford, pp. 201—209.

Dana, G., Herbst, D. B., Lovejoy, C., Loeffler, B. & Otsuki, K. (1977) Limnology. *In*: Winkler, D. W. (ed.), An ecological study of Mono Lake, California. Inst. Ecol. Publ. 12. Univ. Calif., Davis, pp. 39—69.

Davis, J. S. (1978) Biological communities of a nutrient enriched salina. Aquat. Bot. 4: 23—42.

Dawson, J. B. (1962) Sodium carbonate lavas from Oldoinyo Lengai, Tanganyika. Nature 195: 1075—1076.

DeLuca, P. & Moretti, A. (1983) Floridosides in *Cyanidium caldarium, Cyanidioschyzon merolae* and *Galdieria sulphuraria* (Rhodophyta, Cyanidiophyceae). J. Phycol. 19: 368—369.

Devi, A. M., Subbulakshmi, G., Devi, K. M. & Venkataraman, L. V. (1981) Studies of the proteins of mass cultivated, blue-green alga (*Spirulina platensis*). J. Agric. Food Chem. 29: 522—525.

Dobrynin, E. G. (1978) [The intensity of photosynthesis in salt lakes of Crimea]. Inf. Byull. Biol. vnutr. vod. 37: 26—29 (in Russian; FBA translation (new series) No. 131).

Doemel, W. N. & Brock, T. D. (1970) The upper temperature limit of *Cyanidium caldarium.* Arch. Mikrobiol. 72: 326—332.

Doemel, W. N. & Brock, T. D. (1971) The physiological ecology of *Cyanidium caldarium.* J. Gen. Microbiol. 67: 17—32.

Doemel, W. N. & Brock, T. D. (1974) Bacterial stromatolites: origin of laminations. Science 184: 1083—1085.

Doemel, W. N. & Brock, T. D. (1977) Structure, growth and decomposition of laminated algal-bacterial mats in alkaline hot springs. Appl. Environ. Microbiol. 34: 433—452.

Drabløs, D. & Tollan, A. (eds.) (1980) Ecological impact of acid precipitation. SNSF, Norway, 383 pp.

Durand-Chastel, H. (1980) Production and use of *Spirulina* in Mexico. *In*: Shelef, G. & Soeder, C. J. (eds.), Algae Biomass. Elsevier/North Holland Biomedical Press, Amsterdam, pp. 51—64.

Eckstein, Y. (1970) Physicochemical limnology and geology of a meromictic pond on the Red Sea shore. Limnol. Oceanogr. 15: 363—372.

Edwards, M. R., Berns, D. S., Ghiorse, W. C. & Holt, S. C. (1968) Ultrastructure of the thermophilic blue-green alga, *Synechococcus lividus* Copeland. J. Phycol. 4: 283—298.

Edwards, M. R. & Gantt (1971) Phycobilisomes of the thermophilic blue-green alga *Synechococcus lividus.* J. Cell. Biol. 50: 896—900.

Ehrlich, H. L. (1978) How microbes cope with heavy metals, arsenic and antimony in their environment. *In*: Kushner, D. J. (ed.), Microbial life in extreme environments. Academic Press, London, pp. 381—408.

Ehrlich, P. R. *et al.* (1983) Long-term biological consequences of nuclear war. Science 222: 1293—1300.

Epstein, E. (1969) Mineral metabolism of halophytes. *In*: Robinson, I. H. (ed.), Ecological aspects of the mineral nutrition of plants. Blackwell Scientific Publ., Oxford, pp. 345—355.

Eugster, H. P. & Hardie, L. A. (1975) Sedimentation in an ancient playa-lake complex: the Wilkins Peak Member of the Green River Formation of Wyoming. Bull. Geol. Soc. Am. 86: 319—334.

Eugster, H. P. & Hardie, L. A. (1978) Saline lakes. *In*: Lerman, A. (ed.), Lakes. Springer Verlag, New York, pp. 237—293.

Eykelenburg, C. van. (1977) On the morphology and ultrastructure of the cell wall of *Spirulina platensis*. Antonie van Leeuwenhoek 43: 89—99.

Eykelenburg, C. van. (1978) A glucan from the cell wall of the cyanobacterium *Spirulina platensis*. Antonie van Leeuwenhoek 44: 321—327.

Eykelenburg, C. van. (1980) Ecophysiological studies on *Spirulina platensis*. Effect of temperature, light intensity and nitrate concentration on growth and ultrastructure. Antonie van Leeuwenhoek 46: 113—127.

Felix, E. A. & Rushforth, S. R. (1979) The algal flora of the Great Salt Lake, Utah, U.S.A. Nova Hedwig. 31: 163—194.

Fjerdingstad, Ein., Vanggaard, L., Kemp, K. & Fjerdingstad, Er. (1978) Trace elements of red snow from Spitsbergen with a comparison with red snow from East-Greenland (Hudson Land). Arch. Hydrobiol. 84: 120—134.

Flowers, T. J., Troke, P. F. & Yeo, A. R. (1977) The mechanism of salt tolerance in halophytes. Ann. Rev. Plant. Physiol. 28: 89—121.

Fogg, G. E. (1967) Observations on the snow algae of the South Orkney Islands. Phil. Trans. Roy. Soc. London, Ser. B 252: 279—287.

Fott, B. (1970) Review of E. Kol. 1968. Kryobiologie. Limnol. Oceanogr. 15: 660—661.

Fukushima, H. (1963) Studies on cryophytes in Japan. J. Yokohama Munic. Univ., Ser. C, Nat. Sci. 43: 1—146.

Garells, R. M. & MacKenzie, F. T. (1967) Origin of the chemical composition of some springs and lakes. *In*: Stumm, W. (ed.), Equilibrium concepts in natural waters systems. Adv. Chem. Ser. no. 67, pp. 222—242.

Garric, R. K. (1965) The cryoflora of the Pacific Northwest. Am. J. Bot. 52: 1—8.

Gatesoupe, F.-J. & Robin, J. H. (1981) Commercial single-cell proteins either as sole food source or in formulated diets for intensive and continuous production of rotifers (*Brachionus plicatilis*). Aquaculture 25: 1—15.

Gibbs, N. & Duffus, C. M. (1976) Natural protoplast *Dunaliella* as a source of protein. Appl. Environ. Microbiol. 31: 602—604.

Gibor, A. (1956) The culture of brine algae. Biol. Bull. 111: 223—229.

Gimmler, H. & Schirling, R. (1978) Cation permeability of the plasmalemma of the halotolerant alga *Dunaliella parva*. II. Cation content and glycerol concentration of the cells as dependent upon external NaCl concentration. Z. Pflanzenphysiol. Bd. 87: 435—444.

Ginzburg, M. (1969) The unusual membrane permeability of two halophilic unicellular organisms. Biochim. biophys. Acta. 173: 370—376.

Ginzburg, M., Sachs, L. & Ginzburg, B. Z. (1971) Ion metabolism in a Halobacterium. II. Ion concentrations in cells at different levels of metabolism. J. Membrane Biol. 5: 78—101.

Ginzburg, M. & Ginzburg, B. Z. (1981) Interrelationships of light, temperature, sodium chloride and carbon source in growth of halotolerant and halophilic strains of *Dunaliella*. Br. phycol. J. 16: 313—324.

Gould, G. W. & Corry, J. E. L. (eds.) (1980) Microbial growth and survival in extremes of environment. Academic Press, New York, 244 pp.

Haines, B. L. & Dunn, E. L. (1985) Coastal marshes. *In*: Chabot, B. F. & Mooney, H.

370

A. (eds.), Physiological ecology of North American plant communities. Chapman and Hall Ltd, London. pp. 323—347.

Hammer, U. T. (1978) The saline lakes of Saskatchewan. III. Chemical characterization. Int. Revue ges. Hydrobiol. 63: 311—335.

Hammer, U. T. (1981) A comparative study of primary production and related factors in four saline lakes in Victoria, Australia. Int. Revue ges. Hydrobiol. 66: 701—743.

Hammer, U. T. (1981) Primary production in saline lakes. Hydrobiologia 81: 47—57.

Hammer, U. T., Shamess, J. & Haynes, R. C. (1983) The distribution and abundance of algae in saline lakes of Saskatchewan, Canada. Hydrobiologia 105: 1—26.

Harris, G. P. (1978) Photosynthesis, productivity and growth: the physiological ecology of phytoplankton. Ergebn. Limnologie 10: 1—171.

Hartog, C. den (1967) Brackish water as an environment for algae. Blumea 15: 31—43.

Hartog, C. den (1970) The sea-grasses of the world. North-Holland Publ. Co., Amsterdam, 275 pp.

Hartog, C. den (1981) Aquatic plant communities of poikilosaline waters. Hydrobiologia 81: 15—22.

Haynes, R. C. & Hammer, U. T. (1978) The saline lakes of Saskatchewan. IV. Primary production of phytoplankton in selected saline ecosystems. Int. Revue ges. Hydrobiol. 63: 337—351.

Hecky, R. E. (1971) The paleolimnology of the alkaline, saline lakes on the Mt. Meru lahar. Ph.D. thesis, Duke Univ., Durham, 121 pp.

Hecky, R. E. & Kilham, P. (1973) Diatoms in alkaline, saline lakes: ecology and geochemical implications. Limnol. Oceanogr. 18: 53—71.

Heimer, Y. M. (1973) The effects of sodium chloride, potassium chloride and glycerol on the activity of nitrate reductase of a salt-tolerant and two non-tolerant plants. Planta 113: 279—281.

Heinrich, M. R. (1976) Extreme environments, mechanisms of microbial adaptation. Academic Press, New York, 362 pp.

Henderson, L. J. (1913) The fitness of the environment. MacMillan Co., New York, 317 pp.

Hoham, R. W. (1975a) The life history and ecology of the snow alga *Chloromonas pichinchae* (Chlorophyta, Volvocales). Phycologia 14: 213—226.

Hoham, R. W. (1975b) Optimum temperatures and temperature ranges for growth of snow algae. Arct. Alp. Res. 7: 13—24.

Hoham, R. W. (1976) The effect of coniferous litter and different snow melt-waters upon the growth of two species of snow algae in axenic culture. Arct. Alp. Res. 8: 377—386.

Hoham, R. W. (1980) Unicellular chlorophytes — snow algae. *In*: Cox, E. R. (ed), Phytoflagellates. Elsevier North-Holland, New York, pp. 61—84.

Hoham, R. W. & Mullet, J. E. (1977) The life history and ecology of the snow alga *Chloromonas cryophila* sp. nov. (Chlorophyta, Volvocales). Phycologia 16: 53—68.

Hoham, R. W. & Mullet, J. E. (1978) *Chloromonas nivalis* (Chod.) Hoh. & Mull. comb. nov., and additional comments on the snow alga, *Scotiella*. Phycologia 17: 106—107.

Hoham, R. W., Roemer, S. C. & Mullet, J. E. (1979) The life history and ecology of the snow alga *Chloromonas brevispina* comb. nov. (Chlorophyta, Volvocales). Phycologia 18: 55—70.

Holdship, S. A. (1976) The paleolimnology of Lake Manyara: a diatom analysis of a 56 m core. Ph.D. thesis, Duke Univ., Durham, 121 pp.

Hollaender, A. (ed.) (1979) The biosaline concept. Plenum Press, New York, 391 pp.

Holt, S. C. & Edwards, M. R. (1972) Fine structure of the thermophilic blue-green alga *Synechococcus lividis* Copeland. A study of frozen-fractured-etched cells. Can. J. Microbiol. 18: 175—181.

Hook, D. D. & Crawford R. M. M. (eds.) (1978) Plant life in anaerobic environments. Ann Arbor Science Publ., Ann Arbor, Mich, 564 pp.

Horikoshi, K. & Akiba, T. (1982) Alkalophilic microorganisms. Springer-Verlag, Berlin, 213 pp.

Horodyski, R. J., Bloeser, B. & Vonder Haas, S. (1977) Laminated algal mats from a coastal lagoon, Laguna Mormona, Baja California, Mexico. J. Sed. Petrol. 47: 680—696.

Hutchinson, G. E. (1937) A contribution to the limnology of arid regions. Trans. Connecticut Acad. Sci. 33: 47—132.

Hutchinson, G. E. (1967) A treatise on limnology, vol. II, Introduction to lake biology and limnoplankton. John Wiley & Sons, New York, 1115 pp.

Hutchinson, G. E. (1975) A treatise on limnology, vol. III, Limnological botany. John Wiley & Sons, New York, 660 pp.

Hutchinson, J. (1959) The families of flowering plants. II. Monocotyledons. Clarendon Press, Oxford, 792 pp.

Ichikawa, S. (1981) Responses to ionizing radiation. In: Lange, O. L., Nobel, P. S., Osmond. C. B. & Ziegler, H. (eds.), Physiological plant ecology. I. Responses to the physical environment. Springer-Verlag, Berlin, pp. 199—228.

Iltis, A. (1968) Tolérance de salinité de Spirulina platensis (Gom.) Geitl. (Cyanophyta) dans les mares natronées du Kanem (Tchad). Cah. O.R.S.T.O.M., Sér. Hydrobiol. 2: 119—125.

Iltis, A. (1969) Phytoplancton des eaux natronées du Kanem (Tchad) 1. Les lacs permanents à Spirulines. Cah. O.R.S.T.O.M., Sér. Hydrobiol. 3: 29—43.

Iltis, A. (1971) Note sur Oscillatoria (sous-genre Spirulina platensis (Nordst.) Bourrelly (Cyanophyta) an Tchad. Cah. O.R.S.T.O.M., Sér. Hydrobiol. 5: 53—72.

Iltis, A. (1974) Phytoplancton des eaux natronées du Kanem (Tchad). VII. Structure des peuplements. Cah. O.R.S.T.O.M. Sér. Hydrobiol. 8: 51—76.

Imhoff, J. F., Sahl, H. G., Soliman, G. S. H. & Truper, H. G. (1979) The Wadi Natrun: chemical composition and microbial mass development in alkaline brines of eutrophic desert lakes. Geomicrobiol. J. 1: 219—234.

Inniss, W. E. & Ingraham, J. L. (1978) Microbial life at low temperatures: mechanisms and molecular aspects. In: Kushner, D. J. (ed.), Microbial life in extreme environments. Academic Press, London, pp. 73—104.

Jacobs, S. W. L. & Brock, M. A. (1982) A revision of the genus Ruppia (Potamogetonaceae) in Australia. Aquat. Bot. 14: 325—337.

Jannasch, H. W. (1979) Microbial turnover of organic matter in the deep sea. BioScience 29: 228—232.

Jannasch, H. W. & Wirson, C. O. (1983) Microbiology of the deep sea. In: Rowe, G. T. (ed.), Deep-sea biology. The Sea, vol. 8. John Wiley and Sons, New York, pp. 231—259.

Jefferies, R. L., Davy, A. J. & Rudmik, T. (1979) The growth strategies of coastal halophytes. In: Jefferies, R. L. & Davy, A. J. (eds.), Ecological processes in coastal environments. Blackwell Scientific Publ., Oxford, pp. 243—268.

Jenkin, P. M. (1936) Reports on the Percy Sladen Expedition to some Rift Valley lakes in Kenya in 1929. VII. Summary of the ecological results with special reference to the alkaline lakes. Ann. Mag. nat. Hist., Ser 10.18: 133—181.

Johnson, M. K., Johnson, E. J., MacElroy, R. D., Speer, H. L. & Bruff, B. S. (1968) Effects of salts on the halophilic alga Dunaliella viridis. J. Bact. 95: 1461—1468.

Jones, A. G., Ewing, C. M. & Melvin, M. V. (1981) Biotechnology of solar saltfields. Hydrobiologia 82: 391—406.

Jørgensen, B. B. & Cohen, Y. (1977) Solar Lake (Sinai) 5. The sulfur cycle of the benthic cyanobacterial mats. Limnol. Oceanogr. 22: 657—666.

Jørgensen, B. B., Revsbech, N. P., Blackburn, T. H. & Cohen, Y. (1979) Diurnal cycle

372

of oxygen and sulfide microgradients and microbial photosynthesis in a cyano-bacterial mat sediment. Appl. Environ. Microbiol. 38: 46—58.

Jørgensen, B. B., Revsbech, N. P. & Cohen, Y. (1983) Photosynthesis and structure of benthic microbial mats: microelectrode and SEM studies of four cyanobacterial communities. Limnol. Oceanogr. 28: 1075—1093.

Kalecsinsky, A. V. (1901) Uber die Ungarischen Warmen und Heissen Kochsalzseen als Naturliche Warmeaccumulatoren, sowie uber die Herstellung von Warmen Salzseen und Warmeaccumulatoren. Z. Gewaessert 4: 226—248.

Kalff, J. (1970) Arctic lake ecosystems. In: Holdgate, M. W. (ed.), Antarctic ecology. Academic Press, New York, pp. 651—663.

Kao, O. H. W., Berns, D. S. & Town, W. R. (1973) The characterization of c-phycocyanin from an extremely halo-tolerant blue-green alga, Coccohloris elabens. Biochem J. 131: 39—50.

Kauss, H. (1978) Osmotic regulation in algae. In: Reinhold, L., Harborne, J. B. & Swain, T. (eds.), Progress in phytochemistry. vol. 5. Pergamon Press, Oxford, pp. 1—27.

Kol, E. (1968) Kryobiologie. Biologie und Limnologie des Schnees und Eises I. Kryovegetation. In: Elster, H.-J. & Ohle, W. (eds.), Die Binnengewässer, Vol. 24, E. Schweizerbart'sche Verlagsbuchhandlung, Stuttgart, 216 pp.

Krumbein, W. E., Cohen, Y. & Shilo, M. (1977) Solar Lake (Sinai) 4. Stromatolitic cyanobacterial mats. Limnol. Oceanogr. 22: 635—656.

Kuenen, J. G. (rapporteur). (1979) Oxygen toxicity group report. In: Shilo, M. (ed.), Strategies of microbial life in extreme environments. Dr. J. Bernhard, Dahlem Konferenzen, Berlin, pp. 223—241.

Kushner, D. J. (1971) Life in extreme environments. In: Buvet, R. & Ponnamperuma, C. (eds.), Chemical evolution and the origin of life. North-Holland Publ. Co, pp. 485—491.

Lampkin, A. J. III & Sommerfeld, M. R. (1982) Algae distribution in a small, intermittent stream receiving acid mine-drainage. J. Phycol. 18: 196—199.

Langworthy, T. A. (1978) Microbial life in extreme pH values. In: Kushner, D. J. (ed.), Microbial life in extreme environments. Academic Press, London, pp. 279—315.

Larsen, H. (1967) Biochemical aspects of extreme halophilism. Adv. Microb. Physiol. 1: 97—132.

Léonard, J. & Compère, P. (1967) Spirulina platensis (Gom.) Geitl., algue bleue de grande valeur alimentaire par sa richesse en protéines. Bull. Jard. Bot. natl. Belg. 37 (Suppl.), 23 pp.

Lenz, P. H. (1982) Population studies of Artemia in Mono Lake, California. Ph.D. thesis, Univ. of California, Santa Barbara, 230 pp.

Lerche, W. (1937) Untersuchungen über Entwicklung und Fortpflanzung in der Gattung Dunaliella. Arch. Protistenk. 88: 236—268.

Livingstone, D. A. & Melack, J. M. (1984) Some lakes of Subsaharan Africa. In: Taub, F. B. (ed.), Lake and reservoir ecosystems. Elsevier, Amsterdam, pp. 467—497.

Loeblich, L. A. (1972) Studies on the brine flagellate, Dunaliella salina. Ph.D. thesis, Univ. of California, San Diego.

Lovelock, J. E. & Margulis, L. (1974) Atmospheric homeostasis by and for the biosphere: the gaia hypothesis. Tellus 26: 1—9.

Lynn, R. & Brock, T. D. (1969) Notes on the ecology of a species of Zygogonium (Kütz) in Yellowstone National Park. J. Phycol. 5: 181—185.

Masaki, R., Wada, K. & Matsubara, H. (1979) Isolation and characterization of two ferredoxin-NADP+ reductases from Spirulina platensis. J. Biochem. 86: 951—962.

Mason, D. T. (1967) Limnology of Mono Lake, California. Univ. Calif. Publ. Zool. 83: 1—102.

Massjuk, N. P. (1973) [Morphology, taxonomy, ecology and geographical distribution of genus *Dunaliella* Téod.] Kiev. (in Russian).

Matty, A. J. & Smith, P. (1978) Evaluation of a yeast, a bacterium and an alga as a protein source for rainbow trout, effect of protein level on growth, gross conversion efficiency and protein conversion efficiency. Aquaculture 14: 235—246.

McCarraher, D. B. (1972) A preliminary bibliography and lake index of the inland mineral waters of the world. FAO Fisheries Circular No. 146. FAO, Rome, 33 pp.

McCarraher, D. B. (1977) Nebraska's sandhill lakes. Nebraska Game and Parks Commission, Lincoln, Nebraska, 67 pp.

McCauley, R. N. (1966) The biological effects of oil pollution in a river. Limnol. Oceanogr. 11: 475—486.

McMillan, C & Moseley, F. N. (1967) Salinity tolerances of five marine spermatophytes of Redfish Bay, Texas. Ecology 48: 503—506.

Meeks, J. C. & Castenholz, R. W. (1971) Growth and photosynthesis in an extreme thermophile *Synechococcus lividus* (Cyanophyta). Arch. Mikrobiol. 78: 25—41.

Melack, J. M. (1976) Limnology and dynamics of phytoplankton in equatorial African lakes. Ph.D. thesis, Duke University, Durham, N. C., 453 pp.

Melack, J. M. (1979a) Photosynthesis and growth of *Spirulina platensis* (Cyanophyta) in an equatorial lake (Lake Simbi, Kenya). Limnol. Oceanogr. 24: 753—760.

Melack, J. M. (1979b) Temporal variability of phytoplankton in tropical lakes. Oecologia 44: 1—7.

Melack, J. M. (1981) Photosynthetic activity of phytoplankton in tropical African soda lakes. Hydrobiologia 81: 71—85.

Melack, J. M. (1983) Large, deep salt lakes: a comparative limnological analysis. Hydrobiologia 105: 223—230.

Melack, J. M. & Kilham, P. (1974) Photosynthetic rates of phytoplankton in East African alkaline, saline lakes. Limnol. Oceanogr. 19: 743—755.

Melack, J. M., Kilham, P. & Fisher, T. R. (1982) Responses of phytoplankton to experimental fertilization with ammonium and phosphate in an African soda lake. Oecologia 52: 321—326.

Melack, J. M., Lenz, P. H. & Cooper, S. D. (1985) The ecology of Mono Lake. Nat. Geogr. Soc. Res. Rep. 20: 461—470.

Merola, A., Castaldo, R., DeLuca, P., Gambardella, R., Musachio, A. & Taddei, R. (1982) Revision of *Cyanidium caldarium*. Three species of acidophilic algae. G. Bot. Ital. 116: 189—195.

Meyer, G. H., Morrow, M. B., Wyss, O., Berg, T. E. & Littlepage, J. L. (1962) Antarctica: the microbiology of an unfrozen saline pond. Science 138: 1103—1104.

Miller, D. M., Jones, J. H., Yopp, J. H., Tindall, D. R. & Schmid, W. E. (1976) Ion metabolism in a halophilic blue-green alga, *Aphanothece halophytica*. Arch. Microbiol. 111: 145—149.

Morita, R. Y. (1975) Psychrophilic bacteria. Bact. Rev. 39: 144—167.

Mosser, J. L., Mosser, A. G. & Brock, T. D. (1977) Photosynthesis in the snow: the alga *Chlamydomonas nivalis* (Chlorophyceae). J. Phycol., 13: 22—27.

Munns, R., Greenway, H. & Kirst, G. O. (1983) Halotolerant eukaryotes. *In*: Lange, O. L. Nobel, P. S. Osmond, C. B. & Ziegler, H. (eds.), Physiological plant ecology III. Responses to the chemical and biological environment. Springer-Verlag, Berlin, pp. 59—135.

Nasim, A. & James, A. P. (1978) Life under conditions of high irradiation. *In*: Kushner, D. J. (ed.), Microbial life in extreme environments. Academic Press, London, pp. 409—439.

Nemenz, H. (1970) Ionenverhältnisse und die Besidlung hyperhaliner Gewässer, besonders durch Insekten. Acta biotheoretica. 19: 148—170.

374

Nigam, B. P., Ramanathan, P. K. & Venkataraman, L. V. (1981) Simplified production technology of blue-green alga *Spirulina platensis* for feed applications in India. Biotech. Lett. 3: 619—622.

Nissenbaum, A. (1975) The microbiology and biogeochemistry of the Dead Sea. Microbial Ecol. 2: 139—161.

Nixon, S. W. (1974) Salina systems. *In*: Odum, H. T., Copeland, B. J. & McMahan, E. A. (eds.), Coastal ecological systems of the United States, vol. 3. The Conservation Foundation, Washington, D.C., pp. 318—341.

O'Connors, H. B. Jr., Wurster, C. F., Powers, C. D., Biggs, D. C. & Rowland, R. G. (1978) Polychlorinated biphenyls may alter marine trophic pathways by reducing phytoplankton size and production. Science 201: 737—739.

Ogawa, T. & Terui, G (1971) Studies on the growth of *Spirulina platensis*. (2) Growth kinetics of an autotrophic culture. J. Ferment. Technol. 50: 143—149.

Ogawa, T., Fujii, T. & Aiba, S. (1978) Growth yield of microalgae: reassessment of Y_{kcal}. Biotech. Bioeng. 20: 1493—1500.

Ogawa, K. *et al.* (1979) Location of the iron-sulfur cluster in *Spirulina platensis* ferredoxin by X-ray analysis. J. Biochem. 81: 529—531.

Oren, A. & Shilo, M. (1982) Population dynamics of *Dunaliella parva* in the Dead Sea. Limnol. Oceanogr. 27: 201—211.

Owers-Narhi, L., Robinson, S. J., DeRoo, C. S. & Yocum, C. F. (1979) Reconstitution of cyanobacterial photophosphorylation by a latent Ca^{+2}-ATP-ase. Biochem. Biophys. Res. Comm. 90: 1025—1031.

Padan, E. (1979) Impact of facultatively anaerobic photoautotrophic metabolism on the ecology of cyanobacteria (blue-green algae). Adv. Microbiol. Ecol. 3: 1—48.

Pande, A. S., Sarkar, R. & Krishnamoorthi, K. D. (1981) Toxicity of copper sulphate to the alga *Spirulina platensis* and the ciliate *Tetrahymena pyriformis*. Indian J. Exp. Biol. 19: 500—502.

Parker, B. C., Hoehn, R. C., Paterson, R. A., Craft, J. A., Lane, L. S., Stavros, R. W., Sugg, H. G., Whitehurst, J. T., Fortner, R. D. & Weand, B. L. (1977) Changes in dissolved organic matter, photosynthetic production, and microbial community composition in Lake Bonney, Southern Victorialand, Antarctica. *In*: Llano, G. A. (ed.), Adaptations within Antarctic ecosystems. Smithsonian Inst., Washington, D.C., pp. 873—895.

Parker, P. L. & Leo, R. F. (1965) Fatty acids in blue-green algal mat communities. Science 148: 373—374.

Patrick, R. (1969) Some effects of temperature on freshwater algae. *In*: Krenkel, P. A. & Parker, F. L. (eds.), Biological aspects of thermal pollution. Vanderbilt Univ. Press, Tenn., pp. 161—185.

Peary, J. & Castenholz, R. W. (1964) Temperature strains of thermophilic blue-green algae. Nature 202: 720—721.

Poljakoff-Mayber, A. & Gale, J. (eds.) (1975) Plants in saline environments. Springer-Verlag, New York, 213 pp.

Ponnamperuma, C. (ed.) (1976) Chemical evolution of the giant planets. Academic Press, New York, 240 pp.

Por, F. D. (1968) Solar Lake on the shores of the Red Sea. Nature 218: 860—861.

Post, F. J. (1979) Oxygen-rich gas domes of microbial origin in the salt crust of the Great Salt Lake. Geomicrobiol. J. 2: 127—139.

Post, F. J. (1981) Microbiology of the Great Salt Lake north arm. Hydrobiologia 81: 59—69.

Rao, J. K. M. & Argos, P. (1981) Structural stability of halophilic proteins. Biochemistry 20: 6536—6543.

Rawson, D. S. & Moore, J. E. (1944) The saline lakes of Saskatchewan. Can. J. Res. D 22: 141—201.

Reimold, R. J. & Queen, W. H. (eds.) (1974) Ecology of halophytes. Academic Press, New York, 605 pp.

Remy, R., Bebee, G. & Moyse, A. (1976) Electrophoretic analysis of pigment protein complexes from *Porphyridium* (Rhodophyta) and *Spirulina* (Cyanophyta) thylakoids. Phycologia 15: 321—327.

Revsbech, N. P., Jørgensen, B. B., Blackburn, T. H. & Cohen, Y. (1983) Microelectrode studies of the photosynthesis and O_2, H_2S and pH profiles of a microbial mat. Limnol. Oceanogr. 28: 1062—1074.

Rhodes, K. S. (1981) Oxygen sensitivity of nitrogen fixation in the cyanobacterium *Anabaenopsis arnoldii*. Ph.D. thesis, Univ. of Michigan, Ann Arbor, 191 pp.

Riccardi, G., Sora, S. & Ciferri, O. (1981) Production of amino acids by analog-resistant mutants of the cyanobacterium *Spirulina platensis*. J. Bact. 147: 1002—1007.

Ridley, H. N. (1930) The dispersal of plants throughout the world. Reeve, Ashford, U.K., 744 pp.

Robinson, S. J., DeRoo, C. S. & Yocum, C. F. (1982) Photosynthetic electron transfer in preparations of the cyanobacterium *Spirulina platensis*. Plant Physiol. 70: 154—161.

Rushforth, S. R., Brotherson, J. D. Fungladda, N. & Evenson, W. E. (1981) The effects of dissolved heavy metals on attached diatoms in the Uintah Basin of Utah, U.S.A. Hydrobiologia 83: 313—323.

Satake, J. and Saijo, Y. (1974) Carbon dioxide content and metabolic activity of microorganisms in some acid lakes in Japan. Limnol. Oceanogr. 19: 331—338.

Schopf, J. W. (1983) Earth's earliest biosphere, its origin and evolution. Princeton Univ. Press, Princeton, N.J., 632 pp.

Schulthorpe, C. D. (1967) The biology of aquatic vascular plants. Edward Arnold Publ., London, 610 pp.

Sen, D. N. & Rajpurohit, K. S. (eds.) (1982) Contributions to the ecology of halophytes. Dr. W. Junk Publ., The Hague, 272 pp.

Sharma, R, S. & Singh, P. K. (1981) Growth of planktonic blue-green algae in mixed cultures. Microbios Lett. 16: 75—78.

Shelef, G. & Soeder, C. J. (eds.) (1980) Algae biomass, production and use. Elsevier/North Holland Biomedical Press, Amsterdam, 852 pp.

Shilo, M. (ed.) (1976) Strategies of microbial life in extreme environments. Dr. S. Bernhard, Dahlem Konferenzen, Berlin, 514 pp.

Smithsonian Science Information Exchange (1980) International directory of current biosaline research projects. National Technical Information Service, Springfield, Virginia.

Södergren, A. & Gelin, C. (1983) Effect of PCB's on the rate of carbon-14 uptake in phytoplankton isolates from oligotrophic and eutrophic lakes. Bull. Environm. Contam. Toxicol. 30: 191—198.

Soong, P. (1980) Production and development of *Chlorella* and *Spirulina* in Taiwan. *In*: Shelef, G. & Soeder, C. J. (eds.), Algae biomass. Elsevier/North Holland Biomedical Press, Amsterdam, pp. 97—113.

Squires, L. E., Rushforth, S. R. & Brotherson, J. D. (1979) Algal response to a thermal effluent: study of a power station on the Provo River, Utah, U.S.A. Hydrobiologia 63: 17—32.

Steinhorn, I. *et al.* (1979) The Dead Sea: Deepening of the mixolimnion signifies the overturn of the water column. Science 206: 55—57.

Stephens, D. W. & Gillespie, D. M. (1976) Phytoplankton production in the Great Salt Lake, Utah, and laboratory study of algae response to enrichment. Limnol. Oceanogr. 21: 74—87.

Stewart, G. R. & Lee, J. A. (1974) The role of proline accumulation in halophytes. Planta 120: 279—289.

Stewart, G. R., Larher, F. Ahmad, I. A. & Lee, J. A. (1979) Nitrogen metabolism and salt-tolerance in higher plant halophytes. In: Jefferies, R. L. & Davy A. J. (eds.), Ecological processes in coastal environments. Blackwell Scientific Publ., Oxford, pp. 211—227.

Stewart, W. D. P. (1970) Nitrogen fixation by blue-green algae in Yellowstone thermal areas. Phycologia 9: 261—268.

St. John, H. & Courtney, W. D. (1924) The flora of Epsom Lake. Amer. J. Bot. 11: 100—107.

Stockner, J. G. (1967) Observations of thermophilic algal communities on Mount Rainer and Yellowstone National Parks. Limnol. Oceanogr. 12: 13—17.

Stockner, J. G. (1968) Algal growth and primary productivity in a thermal stream. J. Fish. Res. Bd. Can. 25: 2037—2058.

Stokes, P. M., Hutchinson, T. C. & Krauter, K. (1973) Heavy-metal tolerance in algae isolated from contaminated lakes near Sudbury, Ontario. Can. J. Bot. 51: 2155—2168.

Szalay, A. A. & MacDonald, R. E. (1980) Genetic engineering of halotolerance in microorganisms: a summary. In: Rains, D. W., Valentine, R. C. & Hollaender, A. (eds.), Genetic engineering of osmoregulation. Plenum Press, New York, pp. 321—329.

Talling, J. F. & Talling, I. B. (1965) The chemical composition of African lake water. Int. Revue ges. Hydrobiol. 50: 421—463.

Talling, J. F., Wood, R. B., Prosser, M. V. & Baxter, R. M. (1973) The upper limit of photosynthetic productivity by phytoplankton: evidence from Ethiopian soda lakes. Freshwat. Biol. 3: 53—76.

Tansey, M. R. & Brock, T. D. (1972) The upper temperature limit for eukaryotic organisms. Proc. Natl. Acad. Sci. U.S.A. 69: 2426—2428.

Tansey, M. R. & Brock, T. D. (1978) Microbial life at high temperature: ecological aspects. In: Kushner, D. J. (ed.), Microbial life in extreme environments. Academic Press, London, pp. 159—216.

Taub, F. B. (1974) Closed ecological systems. Ann. Rev. Ecol. Syst. 5: 139—160.

Tel-Or, E., Boussiba, S. & Richmond, A. E. (1980) Products and chemicals from Spirulina platensis. In: Shelef, G. & Soeder, C. J. (eds.), Algae biomass. Elsevier/ North Holland Biomedical Press, Amsterdam, pp. 611—618.

Téodoresco, E. C. (1905) Organisation et développement du Dunaliella, nouveau genre de Volvocacée-Polyblepharidée. Beih. Bot. Zentralblatt 18: 215—232.

Thomas, W. H. (1972) Observations on snow algae in California. J. Phycol. 8: 1—9.

Tilzer, M. (1972) Dynamics and productivity of phytoplankton and pelagic bacteria in high mountain lakes. Arch. Hydrobiol. 40: 201—273.

Tindall, D. R., Yopp, J. H., Schmid, W. E. & Miller, D. M. (1977) Protein and amino acid composition of the obligate halophile Aphanothece halophytica (Cyanophyta). J. Phycol. 13: 127—133.

Titu, H., Popovici, G., Boldor, O., Spirescu, I. & Stanca, D. (1980) The ultrastructure of hormogonial cells in blue-green alga Spirulina platensis (Nordst.) Geitl. Rev. Roum. Biol.-Biol. Veget. 25: 143—150.

Tominaga, H. & Fukui, F. (1981) Saline lakes at Syowa Oasis, Antarctica. Hydrobiologia 82: 375—389.

Trent, J. D., Chastain, R. A. & Yayanos, A. A. (1984) Possible artefactual basis for apparent bacterial growth at 250 °C. Nature 307: 737—740.

Tuite, C. (1981) Standing crop densities and distribution of *Spirulina* and benthic diatoms in East African alkaline saline lakes. Freshwat. Biol. 11: 345—360.

Turco, R. P., Toon, O. B., Ackerman, T. P., Pollack, J. B. & Sagan, C. (1983) Nuclear winter: global consequences of multiple nuclear explosions. Science 222: 1283—1292.

Ueno, M. (1958) The disharmonious lakes of Japan. Verh. Intern. Verein. Limnol. 13: 217—226.

Ungar, I. A. (1974) Inland halophytes of the United States. *In*: Reimold, R. J. & Queen, W. H. (eds.), Ecology of halophytes. Academic Press, New York, pp. 235—305.

Uphof, J. C. Th. (1941) Halophytes. Bot. Rev. 7:1—58.

Vallentyne, J. R. (1963) Environmental biophysics and microbial ubiquity. Ann. N.Y. Acad. Sci. 108: 342—352.

Van Auken, O. W. & McNulty, I. B. (1973) The effect of environmental factors on the growth of a halophilic species of algae. Biol. Bull. 145: 210—222.

Vareschi, E. (1978) The ecology of Lake Nakuru (Kenya). I. Abundance and feeding of the lesser flamingo. Oecologia 32: 11—35.

Vareschi, E. (1979) The ecology of Lake Nakuru (Kenya). II. Biomass and spatial distribution of fish (*Tilapia grahami* Boulenger = *Sarotherodon alcalicum grahami* Boulenger). Oecologia 37: 321—335.

Vareschi, E. (1982) The ecology of Lake Nakuru (Kenya). III. Abiotic factors and primary production. Oecologia 55: 81—101.

Vareschi, E., Melack, J. M. & Kilham, P. (1981) Saline waters. *In*: Symoens, J. J., Burgis, M. J. & Gaudet, J. J. (eds.), The ecology and utilization of African inland waters. UNEP, Nairobi, pp. 93—102.

Vareschi, E. & Vareschi, A. (1984) The ecology of Lake Nakuru (Kenya). IV. Biomass and distribution of consumer organisms. Oecologia 61: 70—82.

Vareschi, E. & Jacobs, J. (1984) The ecology of Lake Nakuru (Kenya). V. Production and consumption of consumer organisms. Oecologia 61: 83—98.

Venkataraman, L. V., Nigam, B. P. & Ramanathan, P. K. (1980) Rural oriented freshwater cultivation and production of algae in India. *In*: Shelef, G. & Soeder, C. J. (eds.), Algae biomass. Elsevier/North Holland Biomedical Press, Amsterdam, pp. 81—95.

Verhoeven, J. T. A. (1979) The ecology of *Ruppia*-dominated communities in Western Europe. I. Distribution of *Ruppia* representatives in relation to their autecology. Aquat. Biol. 6: 197—268.

Verhoeven, J. T. A. (1980a) The ecology of *Ruppia*-dominated communities in Western Europe. II. Synecological classification. Structure and dynamics of the macroflora and macrofauna communities. Aquat. Biol. 8: 1—85.

Verhoeven, J. T. A. (1980b) The ecology of *Ruppia*-dominated communities in Western Europe. III. Aspects of production, consumption and decomposition. Aquat. Bot. 8: 209—253.

Viala, G. (1966) L'astaxanthine chez le *Chlamydomonas nivalis* Wille. Compt. Rend. hebd. Séances Acad. Sci. (Paris) 263: 1383—1386.

Vincenzini, M., Ferrari, F., Margheri, M. O. & Florenzano, G. (1980) Quinonoid and tocopherol levels in *Spirulina platensis*. Microbiologica 3: 131—136.

von Knorring, O. & duBois, G. G. B. (1961) Carbonatitic lava from Fort Portal area in western Uganda. Nature 192: 1064—1065.

Waisel, Y. (1972) Biology of halophytes. Academic Press, New York, 395 pp.

Walker, K. F. (1973) Studies on a saline lake ecosystem. Aust. J. mar. Freshwat. Res. 24: 21—71.

Walker, K. F. (1975) The seasonal phytoplankton cycles for two saline lakes in central Washington. Limnol. Oceanogr. 20: 40—53.

Walter, M. R. (1972) A hot spring analog for the depositional environment of

378

Precambrian iron formations of the Lake Superior region. Econ. Geol. 67: 965—980.

Walter, M. R. (ed.) (1976) Stomatolites. Developments in Sedimentology, 20. Elsevier, Amsterdam, 790 pp.

Walter, M. R., Bauld, J. & Brock, T. D. (1972) Siliceous algal and bacterial stromatolites in hot spring and geyser effluents of Yellowstone National Park. Science 178: 402—405.

Walter, M. R., Golubic, S. & Priess, W. V. (1973) Recent stromatolites from hydromagnesite and aragonite depositing lakes near the Coorong Lagoon, South Australia. J. Sed. Petrol. 43: 1021—1030.

Waring, G. A. (1965) Thermal springs of the United States and other countries of the world. A summary. U.S. Geol. Surv. Prof. Paper 492, 383 pp.

Weiss, R. L. (1983) Fine structure of the snow alga (*Chlamydomonas nivalis*) and associated bacteria. J. Phycol. 19: 200—204.

Weller, D., Doemel, W. & Brock, T. D. (1975) Requirements of low oxidation-reduction potential for photosynthesis in a blue-green alga (*Phormidium* sp.). Arch. Mikrobiol. 104: 7—13.

Wetzel, R. G. (1964) A comparative study of the primary productivity of higher aquatic plants, periphyton and phytoplankton in a large shallow lake. Int. Rev. ges. Hydrobiol. Hydrograph. 49: 1—61.

White, R. V. (1984) Hydrolytic stability of biomolecules at high temperature and its implications for life at 250 °C. Nature 310: 430—432.

Whitton, B. A., Gale, N. L. & Wixson, B. G. (1981) Chemistry and plant ecology of zinc-rich wastes dominated by blue-green algae. Hydrobiologia 83: 331—341.

Wiegert, R. G. & Fraleigh, P. C. (1972) Ecology of Yellowstone thermal effluent systems: net primary productivity and species diversity of a successional blue-green algal mat. Limnol. Oceanogr. 17: 215—228.

Williams, W. D. (1981) Inland salt lakes: an introduction. Hydrobiologia 81: 1—14.

Winter, K. (1979) Photosynthesis and water relationships of higher plants in a saline environment. *In*: Jefferies, R. L. & Davy, A. J. (eds.), Ecological processes in coastal environments. Blackwell Scientific Publ., Oxford, pp. 297—320.

Wirsen, C. O. & Jannasch, H. W. (1975) Activity of marine psychrophilic bacteria at elevated hydrostatic pressures and low temperatures. Mar. Biol. 31: 201—208.

Woolhouse, H. W. (1981) Toxicity and tolerance in the responses of plants to metals. *In*: Lange, O. L., Nobel, P. S., Osmond, C. B. & Zeigler, H. (eds.), Physiological plant ecology. III. Responses to chemical and biological environment. Springer-Verlag, Berlin, pp. 245—300.

Wright, S. W. & Burton, H. R. (1981) The biology of Antarctic saline lakes. Hydrobiologia 82: 319—338.

Yancey, P. H., Clark, M. E., Hand, S. C., Bowlus, R. D. & Somero, G. N. (1982) Living with water stress: evolution of osmolyte systems. Science 217: 1214—1221.

Yopp, J. H. Miller, D. M. & Tindall, D. R. (1978) Regulation of intracellular water potential in the halophilic blue-green alga *Aphanothece halophytica* (Chroococcocales). *In*: Caplan, S. R. & Ginzburg, M. (eds.), Energetics and structure of halophilic microorganisms. Elsevier/North-Holland Biomedical Press, pp. 619—624.

Zarrouk, C. (1966) Contribution a l'étude d'une cyanophycée: influence de divers facteurs physiques et chimiques sur la croissance et la photosynthèse de *Spirulina maxima* (Setch et Gardner) Geitler. D. S. (Appl.) thèse, Univ. Paris., 74 pp.

ZoBell, C. E. (1970) Pressure effects on morphology and life processes of bacteria. *In*: Zimmerman, A. M. (ed.), High pressure effects on cellular processes. Academic Press, New York, pp. 85—130.

INDEX

No attempt has been made to index every occurrence of all terms. Page numbers shown in boldface type indicate the first page of a chapter or of a section in a paper in which a topic is dealt with repeatedly.